Lecture Notes in Engineering

Edited by C. A. Brebbia and S. A. Orszag

IFIP

48

P. Thoft-Christensen (Editor)

Reliability and Optimization of Structural Systems '88

Proceedings of the 2nd IFIP WG7.5 Conference
London, UK, September 26–28, 1988

Springer-Verlag
Berlin Heidelberg New York
London Paris Tokyo Hong Kong

Editor
P. Thoft-Christensen
Institute of Building Technology
and Structural Engineering
The University of Aalborg
Sohngaardsholmsvej 57
DK-9000 Aalborg
Denmark

ISBN 3-540-51283-7 Springer-Verlag Berlin Heidelberg New York
ISBN 0-387-51283-7 Springer-Verlag New York Berlin Heidelberg

Printed in Germany

Offsetprinting: Mercedes-Druck, Berlin
Binding: B. Helm, Berlin
2161/3020-543210

PREFACE

The Proceedings contain 30 papers presented at the 2nd Working Conference on "Reliability and Optimization of Structural Systems", London, UK, September 26 - 28, 1988. The Working conference was organized by the IFIP Working Group 7.5. The Proceedings also include 3 papers which were submitted, but for various reasons not presented at the Working Conference. The Working Conference was attended by 47 participants from 15 countries.

The Conference was the second scientific meeting of a new IFIP Working Group 7.5 on "Reliability and Optimization of Structural Systems". The purpose of the Working Group 7.5 is

- to promote modern structural system optimization and reliability theory,
- to advance international cooperation in the field of structural system optimization and reliability theory,
- to stimulate research, development and application of structural system optimization and reliability theory,
- to further the dissemination and exchange of information on reliability and optimization of structural system optimization and reliability theory,
- to encourage education in structural system optimization and reliability theory.

At present the members of the Working Group are:

A. H. S. Ang, USA
G. Augusti, Italy
M. J. Baker, United Kingdom
P. Bjerager, Norway
A. Der Kiureghian, USA
O. Ditlevsen, Denmark
D. M. Frangopol, USA
S. Garribba, Italy
M. R. Gorman, USA
M. Grimmelt, Germany, F. R.
N. C. Lind, Canada
H. O. Madsen, Denmark
F. Moses, USA
Y. Murotsu, Japan
R. Rackwitz, Germany F. R.
P. Thoft-Christensen, Denmark (Chairman)

Members of the organizing committee are:

M. J. Baker, United Kingdom (Conference Organizer)
R. Rackwitz, Germany F. R.
P. Thoft-Christensen, Denmark (Conference Director)

The Conference was financially supported by IFIP, Imperial College of Science, Technology and Medicine, and The University of Aalborg.

On behalf of TC7 and WG 7.5 I would like to thank Michael J. Baker for his excellent organization of the Working Conference. I would also like to thank the organizing committee members for their valuable help in organizing the Working Conference, and all the authors for preparing papers for the Proceedings. Special thanks to Mrs. Kirsten Aakjær, University of Aalborg, for her efficient work as Conference secretary before, during and after the institute.

December 1988 P. Thoft-Christensen

CONTENTS

TIME VARIANT RELIABILITY ANALYSIS UTILIZING RESPONSE SURFACE APPROACH

C. G. Bucher, Y. M. Chen, G. I. Schuëller
Institute of Engineering Mechanics
University of Innsbruck, Austria

ABSTRACT

A new concept to determine response functions is introduced into time variant systems reliability analyses. The class of response functions considered consists of second order polynomials. The coefficients for these polynomials are determined from combinations of the basic variables close to the limit state surface. The method is shown to give highly accurate results while reducing computational efforts considerably in comparison to full analysis.

INTRODUCTION

Due to the increasing need for acceptably accurate reliability estimates in many fields of engineering, several methods have been developed and applied successfully. It appears that the class of (advanced) Monte Carlo simulation methods provides the best available combination of versatility, accuracy and computational efficiency. However, there are many engineering problems which are far too complex to be solved by direct simulation techniques (as far as computer time is concerned). Thus the need for suitable approximate methods becomes apparent.

In fact, it has been a long-standing practice in engineering to approximate the mechanical behavior of systems in a way to obtain simple solutions without significant loss of accuracy. A well known example is the discretization of continuous structures by the Finite Element Method. From this point of view it seems to be the obvious way to follow this practice in order to reduce computational efforts required for reliability analysis to an acceptable level. This approach, consequently, leads to the response surface method (RSM), i.e. the approximation of the mechanical behavior of a system by simple functions. These functions are obtained from sensitivity analyses of the system thus providing sufficient information on the system behavior.

Since the response surface (RS) is determined in terms of physical quantities, i.e. no transformation of random variables is involved, it is possible to accept or reject a particular RS on the basis of engineering judgement. Once this RS is defined suitably, any advanced accurate probabilistic method for reliability analysis can be applied easily.

The present paper deals with the application of the RSM to time variant reliability analysis. The numerical examples are chosen to show accuracy and numerical efficiency of the RSM. The numerical computations required for reliability analysis are carried out within the environment of the multi-purpose computer code ISPUD [Bourgund and Bucher, 1986].

RESPONSE SURFACE METHOD

In the following, **y** denotes the vector of basic variables and g(**y**) is the limit state function in which g(**y**) denotes failure. In the context of probabilistic analyses the surface g(**y**)=0 defines the integration limits. Hence any approximate limit state function g*(**y**) should be capable of reproducing the limit surface as closely as possible.
Since the sensitivity of g(**y**) with respect to the y_i's should be represented properly it is suggested to define the response surface in terms of reduced variables x_i:

$$x_i = \frac{(y_i - \mu_i)}{\sigma_i} \qquad (1)$$

where μ_i and σ_i denote mean values and standard deviations of the basic variables y_i, respectively [Bucher and Bourgund, 1987].
In order to meet the basic requirements of simplicity and generality a second order polynomial for g*(**x**) is suggested:

$$g^*(\mathbf{x}) = a_0 + \sum_{i=1}^{n} b_i x_i + \sum_{i=1}^{n}\sum_{j=1}^{n} c_{ij} x_i x_j \qquad (2)$$

in which x_i, x_j denote the reduced variables and a_0, b_i, c_{ij} are coefficients to be determined suitably.
As mentioned above, g*(**x**) should represent the actual limit state surface g(**x**) = 0 closely, hence the free coefficients are determined from interpolation points close to the limit surface. This is carried out conveniently in the following way. Let g*(**x**) be separated into two parts:

$$g^*(\mathbf{x}) = g_1^*(\mathbf{x}) + g_2^*(\mathbf{x}) \qquad (3)$$

where

$$g_1^*(\mathbf{x}) = a_0 + \sum_{i=1}^{n} b_i x_i + \sum_{i=1}^{n} c_{ii} x_i^2 \qquad (4)$$

$$g_2^*(\mathbf{x}) = \sum_{i=1}^{n-1} \sum_{j=i+1}^{n} 2\, c_{ij} x_i x_j \qquad (5)$$

In eqn.(5) symmetry $c_{ij} = c_{ji}$ is utilized, implying differentiability of $g^*(x)$. Hence the total number of coefficients contained in $g^*(x)$ is seen to be $(n+1)(n+2)/2$. Those coefficients can be determined sequentially, since they do not affect each other for $i \neq j$. [Ouypornprasert and Bucher, 1988].
In the first step, the coefficients of $g_1^*(x)$ are determined from separate interpolation on each axis. This provides information on the most important region (quadrant) for combinations of two variables. In the next step, the most important combinations of pairs x_i, x_j are selected and interpolation between the axes is carried out. This results in the coefficients c_{ij} for $g_2^*(x)$ thus defining the response surface completely.
It should be noted that for many applications it need not be necessary to determine all the c_{ij}'s since combinations of less important variables will hardly have significant effects on the results of a subsequent reliability analysis, however, the quality of the RS will, generally, be improved by mixed terms (cf. Fig.1)

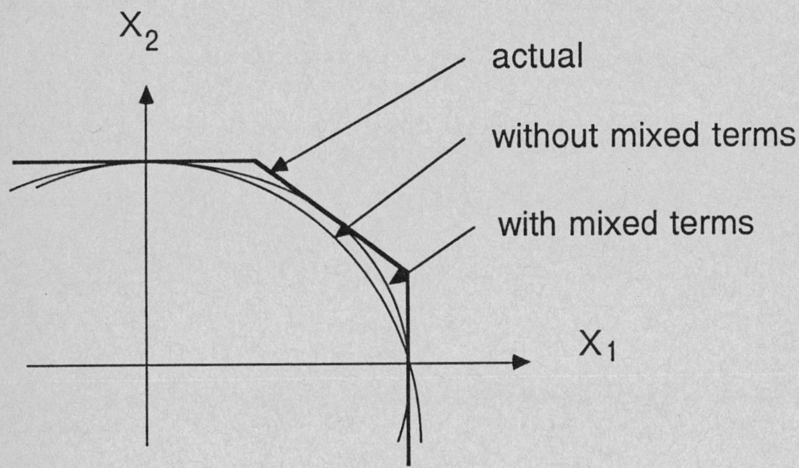

Fig.1: Schematic Sketch of Actual and Approximate Limit Surfaces

TIME VARIANT RELIABILITY ANALYSIS

To make the time variant reliability analysis of a structure under stochastic loadings tractable, the dynamic effect due to the load processes acting on the structure will not be considered, i.e. the structure will respond statically or quasi-statically. For simplicity, it is further assumed that the uncertainties considered in reliability analysis are only due to random time histories of the loadings and random structural resistances which are assumed to be time invariant variables and statistically independent from

the occurrence of load processes. Therefore, a limit state function for a considered structure is generally composed of a resistance vector $\underline{R}$ and load vector process $\underline{L}(t)$ and is defined by $G[\underline{R},\underline{L}(t)]$.
For practical purposes, in evaluating the time variant reliability, quite often the time variance is reduced to a time invariant problem [Shinozuka, 1981]. To accomplish this, it is necessary to evaluate the maximum effect of time varying loads which is expected to occur during a given life-time of the structure. Taking into account the life-time maximum effect of combined load processes, the failure probability of a structure within the life-time [0,T] can be stated as the probability that the limit state function yields values less than zero under the maximum load effect, i.e.

$$P_f(T) = \mathrm{Prob}\,(\,G[\underline{R}, \max_{[0,T]} S\{\underline{L}(t)\}] \le 0\,) = \mathrm{Prob}\,(\,G[\underline{R}, S_m(T)] \le 0\,) \tag{6}$$

where S{.} is the load effect of the loads $\underline{L}(t)$ and subscript m indicates the maximum in [0,T]. Alternatively, if the statistical properties of the maximum load effect can be determined, the failure probability in eqn.(6) can be also defined by

$$P_f(T) = \int_{\underline{r}} P_f^c(T\,|\,\underline{r})\; f_{\underline{R}}(\underline{r})\, d\,\underline{r} \tag{7}$$

where

$$P_f^c(T\,|\,\underline{r}) = \mathrm{Prob}\,(\,G[\underline{r}, S_m(T)] \le 0\,) \tag{8}$$

is the conditional failure probability for a given resistance vector $\underline{r}$. $f_{\underline{R}}$ is the joint density function of resistance variables. Due to the independence between resistances and load processes, the conditional failure probability is equivalent to the probability that the load effect process exceeds the given resistance level, i.e.

$$P_f^c(T\,|\,\underline{r}) = 1 - F_{S_m}(T\,,\underline{r}) \tag{9}$$

where F_{S_m} is the distribution function of the life-time maximum effect S_m of combined load processes. Due to the spatial as well as temporal stochastic nature of load processes, the exact estimation of F_{S_m} is an involved task. For case of multiple load processes, the life-time maximum load effect strongly depends on the particular models of load processes as well as the load combination rules. Based on the specific modeling of time varying loads and load combination analysis, the approximate solutions of the life-time maximum load effect can be found e.g. [Veneziano et al., 1977; Schwarz and Schuëller, 1979; Pearce and Wen, 1985]. However, questions with respect to the practical application and efficiency of these methods still remain open. For this reason, a convenient method presented in [Chen et al., 1988] is utilized here, in which the life-time maximum load effect is analyzed on the basis of existing linear load combination models for the specific sustained and transient load processes. The corresponding probability distribution F_{S_m} is estimated by Monte Carlo simulation techniques combined with statistical tests. For details of the procedure as well as the validity of the method it is referred to [Chen and Schuëller, et al., 1988].

Hence, by introducing the life-time maximum effect of combined load processes, the reliability analysis under time varying loads can be performed in the time invariant domain. For evaluating the failure probability in eqn.(6) or eqn.(7), the adaptive sampling method (ASM) [Bucher, 1988] is utilized. Using the importance sampling procedure [Schuëller et al., 1987], the expected value of the estimator of the failure probability, extended to the time variant problem, can be expressed as:

$$P_f(T) = \frac{1}{N} \sum_{i=1}^{N} P_f^c(T \mid \underline{x}_i) \frac{f_{\underline{X}}(\underline{x}_i)}{h_{\underline{Y}}(\underline{x}_i)} \tag{10}$$

in which P_f^c is defined by eqn.(8). $f_{\underline{X}}$ and $h_{\underline{Y}}$ are the actual joint probability density function and the weighting probability density function, respectively. $\underline{x}_i$ denotes the i-th simulation vector of time invariant variables.

NUMERICAL EXAMPLES

EXAMPLE 1

To show the efficiency of computational efforts by the response surface method (RSM), a simply supported compression member with initial imperfections, which has been analyzed in [Shiraki, 1977] as shown in Fig.2, is first considered.

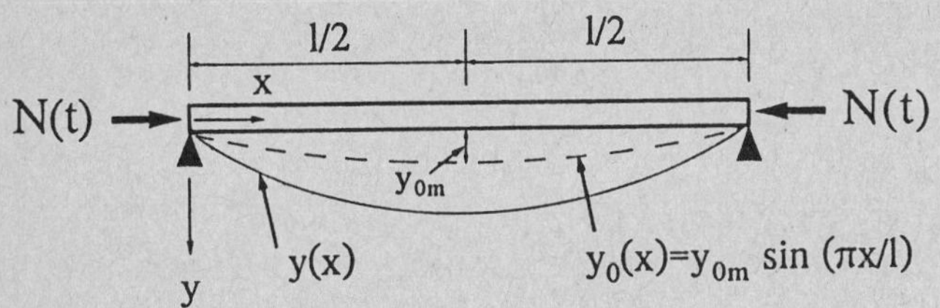

Fig. 2: Simple Compression Member with Initial Imperfections

First, let us consider the time invariant problem. The limit state for a compression member can be described by buckling failure i.e.

$$g(y) = R_{bc} - S_{bc} \tag{11}$$

where S_{bc} is the axial stress acting on the member. R_{bc} is the average buckling stress related to the yield stress R_y and the geometrical imperfection γ_0 and can be determined according to the theory of buckling of beam-columns [Gerstle, 1967] as:

$$|\gamma_0| = \sqrt{3}\,\left(\frac{R_y}{R_{bc}} - 1\right)\left(1 - \sqrt[3]{\frac{R_{bc}}{R_E}}\right)$$

$$= \sqrt{3}\,\left(\frac{R_y}{R_{bc}} - 1\right)\left(1 - \sqrt[3]{\frac{R_{bc}}{\pi^2 E_0/\lambda_0^2}}\right) \tag{12}$$

where R_E is Euler's elastic buckling stress, respectively. λ_0, E_0 are the slenderness ratio and Young's modulus, respectively. $y = (R_y, \lambda_0, \gamma_0, S_{bc})$ are assumed to be basic random variables which are independent. The member fails when $R_{bc} \le S_{bc}$. From eqn.(12) the variable R_{bc} cannot be expressed explicitly. Therefore, eqn.(11) is an non-explicit failure surface. Now the response surface method is applied to determine an approximate surface of eqn.(11). For the numerical calculations, the mean values and standard deviations are taken as

$$\mu = (\mu_{R_y}, \mu_{\lambda_0}, \mu_{\gamma_0}, \mu_{S_{bc}}) = (2.74\cdot10^5, 60.0, 0.0, \min(\mu_{R_y}, \mu_{R_E})/\Delta_c)$$

$$\sigma = (\sigma_{R_y}, \sigma_{\lambda_0}, \sigma_{\gamma_0}, \sigma_{S_{bc}}) = (2.74\cdot10^4, 6.0, 0.18, 0.2\,\mu_{S_{bc}}) \tag{13}$$

where Δ_c is the nominal central safety factor. The Young's modulus is equal to $1.96\cdot10^8$ KN/m^2. Based on the response surface, the failure probability P_f of eqn.(11) is evaluated by the adaptive sampling method for various central safety factors Δ_c. The relation between P_f and Δ_c is given in Fig.3 with T=0, where the results obtained by the response surface method RSM are compared with those (dashed line) by direct adaptive sampling method (ASM) on the original surface (eqn.(11)). The simulation number in both RSM and ASM is 1000. The CPU times required for both methods are compared in Table 1. It can be seen that the computational efforts by RSM are significantly lower. This example quite evidently demonstrates the accuracy and efficiency of the response surface method.

Table 1 - CPU Times (sec.) Required by RSM and ASM

Δ_c	2.0	2.5	3.0	3.5
(1)	(2)	(3)	(4)	(5)
RSM	33.9	32.5	32.0	31.7
ASM	115.9	150.9	147.0	147.9

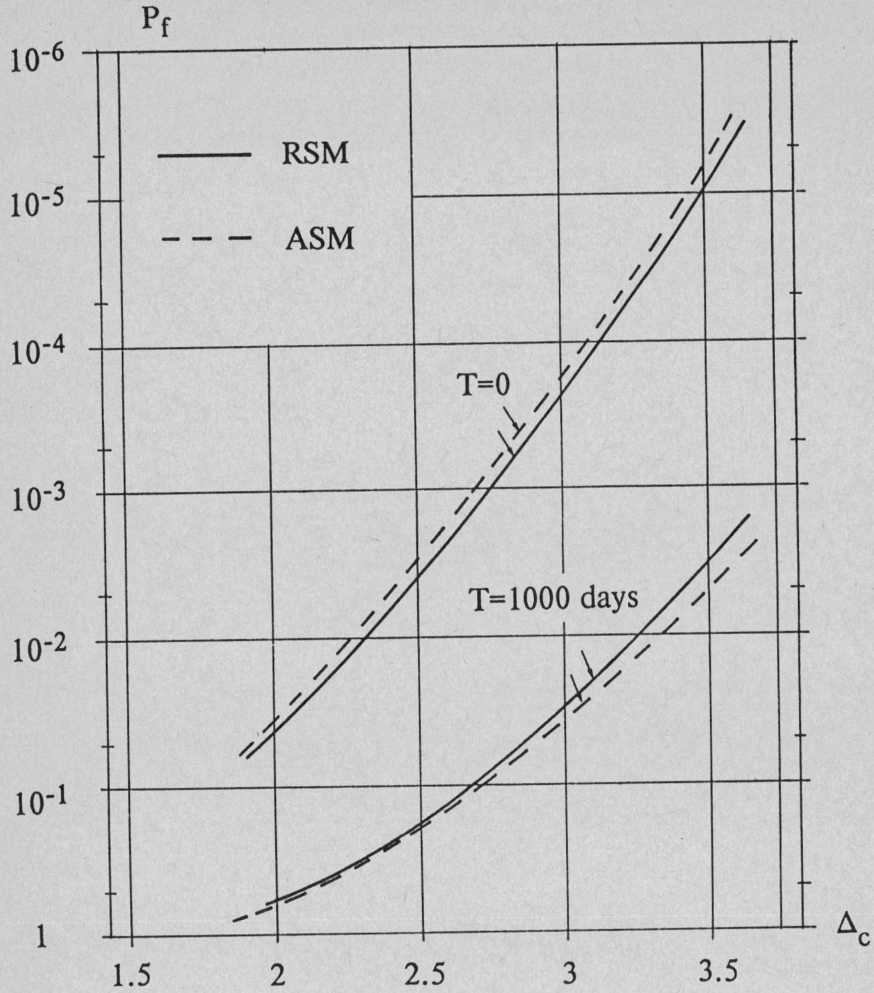

Fig. 3: Failure Probability Versus Nominal Central Safety Factor

Now, let us consider that the compression beam member in Fig. 2 is subjected to time variant axial stress $S_{bc}(t)$ modeled as a stationary normal process. In this case, instead of the variable S_{bc} (time invariant axial stress), the maximum of the process during the considered life time must be taken into account. The probability of the life-time maximum effect of the axial stress can be easily obtained as

$$F_{S_{bcm}}(x,T) = \exp\{-T\exp(-\frac{\sigma_{\dot{S}_{bc}}}{2\pi\,\sigma_{S_{bc}}}\exp[-\frac{(x-\mu_{S_{bc}})^2}{2\,\sigma^2_{S_{bc}}}])\} \tag{14}$$

where $\mu_{S_{bc}}$ and $\sigma_{S_{bc}}$ are the mean value and standard deviation of $S_{bc}(t)$; and $\sigma_{\dot{S}}$ is the standard deviation of the process $\dot{S}_{bc}(t)=dS_{bc}(t)/dt$. Using the same numerical data as given in eqn.(13) and taking the additional data for $\dot{S}_{bc}(t)$ i.e. zero mean and $\sqrt{2}$ times standard deviation of S_{bc}, the reliability analysis is carried out by utilizing both adaptive sampling method (ASM) and response surface method (RSM). The failure probabilities for a life time with 1000 days versus the central safety factor are also shown in Fig. 3 and compared with the time invariant results T=0. As expected, the CUP time needed by RSM is about 3 times lower as compared to ASM.

EXAMPLE 2

Next, a two-storey one-bay frame structure is analyzed as shown in Fig. 4. The structure is subjected to four transient loads modeled as rectangular Poisson pulse processes $L_i(t)$ [Wen, 1977], specified by mean rate υ_i, mean duration d_i and load intensity L_i. The numerical data and statistics (mean value μ_i, standard deviation σ_i) of load intensities and structural resistances are given in Table 2, where R_c and R_b are the resistances in column and beam members, respectively. For simplicity, all variables are asssumed to be independent and normally distributed. Under the consideration of ductile frame neglecting the P-Δ effects and carrying out a static plastic collapse analysis, the failure modes of the structure are defined as the formation of collapse mechanisms and can be described in terms of linear performance functions (limit state functions) of the applied loads and resistances. They can be expressed in the following form:

$$G[\underline{R}, \underline{L}(t)] = a_c R_c + a_b R_b - \sum_{j=1}^{4} b_j L_j(t) \tag{15}$$

By performing branching and bounding operations [Murotsu et al., 1981 and 1984; Grimmelt et al., 1983], eleven dominant failure modes are generated as shown in Fig.5. The structural coefficients a_i and b_i in eqn.(15) as well as the time invariant failure probability of individual failure modes are listed in Table 3. In the time invariant case where applied loads represent the intensities of load processes, the system failure probability is obtained as

$$P_f = \text{Prob}\,(\bigcup_{i=1} G_i(\underline{R}, \underline{L}) \le 0) = 0.142 \cdot 10^{-2} \tag{16}$$

with 2.1% statistical error of the estimation by the adaptive sampling simulations (N=10,000). The time variant failure probabilities of the structural system i.e.

$$P_f(T) = \text{Prob}\,(\bigcup_{i=1} G_i\,[\underline{R}, S_{m,i}(T)] \le 0) \tag{17}$$

are evaluated by both adaptive sampling method and response surface method. The results are shown in Table 4. It can be seen that the results obtained by the response surface compared to the accurate results by adaptive sampling seem to be quite reasonable, especially for long design life. The maximum error of the results by the response surface method in this case is less than 20% (see Table 4).

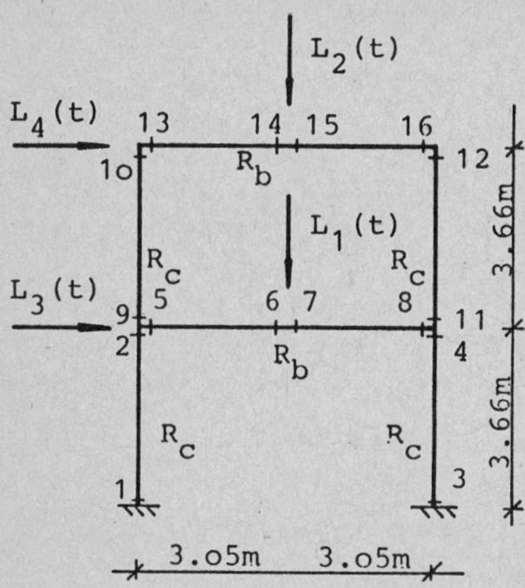

Fig. 4: Two-Storey One-Bay Frame

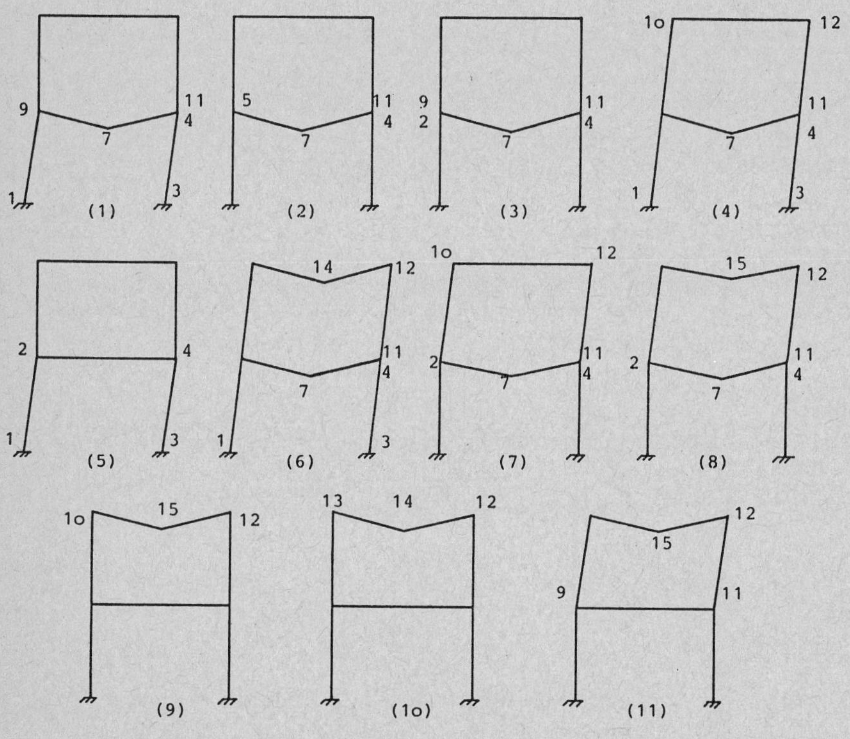

Fig. 5: Dominant Failure Modes of Two-Storey One-Bay Frame

Table 2 - Statistics of Load and Resistance Variables

Variables		μ	σ	υ (/year)	d(year)	Type
(1)		(2)	(3)	(4)	(5)	(6)
	L_1	180.0	27.0	10.0	0.08	Normal
	L_2	90.0	22.5	12.0	0.05	Normal
Load:	L_3	32.0	8.0	20.0	0.03	Normal
	L_4	16.0	4.0	20.0	0.01	Normal
Resistance:						
	R_c	110.0	16.5	--	--	Normal
	R_b	275.0	41.5	--	--	Normal

Table 3 - Coefficients and Probabilities of Eleven Failure Modes

$G_i[.]$	b_1	b_2	b_3	b_4	a_c	a_b	P_{fi}
(1)	(2)	(3)	(4)	(5)	(6)	(7)	(8)
1	3.05	0	3.658	3.658	6	2	$0.9790 \cdot 10^{-3}$
2	3.05	0	0	0	2	3	$0.5775 \cdot 10^{-3}$
3	3.05	0	0	0	4	2	$0.4961 \cdot 10^{-3}$
4	3.05	0	3.658	7.316	8	2	$0.1705 \cdot 10^{-3}$
5	0	0	3.658	3.658	4	0	$0.1650 \cdot 10^{-3}$
6	3.05	3.05	3.658	7.316	8	4	$0.6572 \cdot 10^{-4}$
7	3.05	0	0	3.658	6	2	$0.4430 \cdot 10^{-4}$
8	3.05	3.05	0	3.658	6	4	$0.3350 \cdot 10^{-4}$
9	0	3.05	0	0	2	2	$0.4890 \cdot 10^{-5}$
10	0	3.05	0	0	1	3	$0.1890 \cdot 10^{-5}$
11	0	3.05	0	3.658	4	2	$0.9610 \cdot 10^{-7}$

Table 4 - Comparison of Failure Probability functions $P_f(T)$

T (year)	5	15	25	35	45	50
(1)	(2)	(3)	(4)	(5)	(6)	(7)
Adaptive Sampling						
$P_f(T)^*$	0.021	0.035	0.042	0.048	0.058	0.061
Response Surface						
$P_f(T)^{**}$	0.025	0.038	0.045	0.052	0.057	0.060
$(P_f(T)^* - P_f(T)^{**}) / P_f(T)^*$	19%	8.6%	7.1%	8.3%	1.7%	1.6%

EXAMPLE 3

Finally, a nonlinear SDOF oscillator with random system parameters (Fig.6) subjected to a rectangular load pulse with random duration and amplitude (Fig.7) is analyzed.

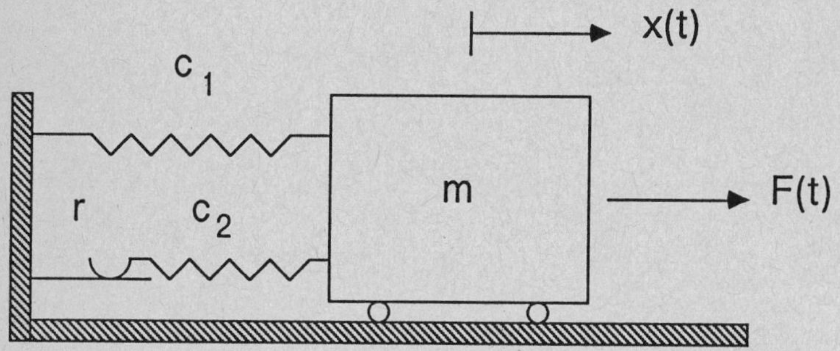

Fig. 6: Nonlinear SDOF Oscillator For Example 3

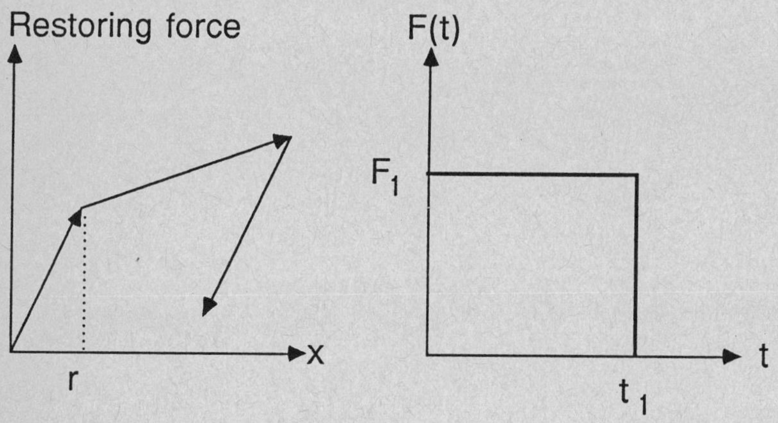

Fig.7: Nonlinear Restoring Force and Pulse Load For Example 3

Let the limit state function be defined by

$$g(y) = 3\,r - |x_{max}| \tag{18}$$

in which x_{max} is the maximum displacement response of the system. The statistical parameters of the basic variables are listed in Table 5.

Table 5 - Statistical Data for Example 3

Number	Variable	Mean Value	Standard Deviation
1	m	1.00	0.05
2	c_1	1.00	0.10
3	c_2	0.10	0.01
4	r	0.50	0.05
5	F_1	1.00	0.20
6	t_1	1.00	0.20

All variables are assumed to be independent and log-normally distributed.
The result from direct simulation based on adaptive sampling [Bucher, 1988] using 3×500 simulations (i.e. 1500 runs of nonlinear dynamic analysis) is $P_f = 3.81 \cdot 10^{-2}$ with a standard sampling error of 4.5%. The result from the response surface approach using three mixed terms only yields $P_f = 3.67 \cdot 10^{-2}$ with a standard error of 4.9% with the same number of simulations. These results correspond extremely well. It is noted that a comparative analysis without mixed terms yields $P_f = 3.04 \cdot 10^{-2}$ which is still accpetable, however, the inclusion of only three mixed terms improves the result considerably. The computer time is reduced by a factor of 35, as far as the effort for Monte Carlo simulation is concerned. Including the CPU time required for the determination of the response surface the reduction still amounts to a factor of 12.

CONCLUSIONS

The numerical examples given in the preceding section shown that the RSM in its presented formulation is a suitable and versatile tool for the probabilistic analysis of various structural problems. It is seen that the method can be applied successfully and with high accuracy to problems with multiple limit states (failure modes), non-explicit limit state functions and non-linear dynamic problems. The procedure is shown to be most efficient for non-linear dynamics in which case the computer time reduction amounts to more than one order of magnitude.
Obviously, since the interpolating procedure must cover combinations of different variables to ensure the quality of the RS, the number of basic random variables should not be too large. This fact limits the method to cases with less than, say, 20 random variables unless some mixed terms can be neglected. Without or with few mixed terms the method is suitable for up to 100 variables. Significant advantages over direct simulation techniques may be anticipated for cases with very complex systems and rather low number of basic variables.

ACKNOWLEDGEMENT

This research was supported by the Austrian Research Council (FWF) under Contract No. P5573 which is gratefully acknowledged by the authors. Special thanks are due to W. Ouypornprasert for helpful discussions.

REFERENCES

[1] Bourgund, U. and Bucher, C.G.: "Importance Sampling Procedure Using Design Points - ISPUD - A User's Manual", Rep.8-86, Inst. Eng. Mech, University of Innbruck, Austria, Nov. 1986.

[2] Bucher, C.G.: "Adaptive Sampling: An Iterative Fast Monte-Carlo Procedure", Structural Safety, Vol. 5, No.2, June, 1988, pp. 119-126.

[3] Bucher, C.G. and Bourgund, U.: "Efficient Use of Response Surface Methods", Rep.9-87, Inst. Eng. Mech., University of Innsbruck, Austria, 1987.

[4] Chen, Y.M., Schuëller, G. I. and Bourgund, U.: "Reliability of Large Structural Systems under Time Variant Loads", ASCE EMD/GTD/STD Joint Specialty Conference on Probabilistic Methods, Virginia, May 25-27, 1988, pp.420-423.

[5] Chen, Y.M., Schuëller, G. I. and Bourgund, U.: "Reliability of Large Structural Systems under Time Variant Loads", Submitted for publication, 1988.

[6] Gerstle, K.H.: "Basic Structural Design", McGraw-Hill, New York, 1967, pp.305-308.

[7] Grimmelt, M., Schuëller, G.I., Murotsu, Y.: "On the Evaluation of Collapse Probabilities", Proc., 4th ASCE Spec. Conf. on Rec. Adv. in Eng. Mech., W.Lafayette, Vol. II, May, 1983, pp.859-862.

[8] Murotsu, Y.: "Reliability Assessment of Redundant Structures", Proc. ICOSSAR' 81, Eds. T. Moan, M. Shinozuka, Elsevier, 1981,

[9] Murotsu, Y., Okada, H., Taguchi, K., Grimmelt, M., and Yonezawa, M.: "Automatic Generation of Stochastically Dominant Failure Modes of Frame Structures", Structural Safety, 2, 1984, pp.17-25.

[10] Ouypornprasert, W. and Bucher, C.G.: "An Efficient Scheme to Determine Response Functions For Reliability Analyses", Int. Working Report No.30, Inst. Eng. Mech., University of Innsbruck, Austria, Sept. 1988.

[11] Pearce, H.T., Wen, Y.K.: "On Linearization Points for Nonlinear Combination of Stochastic Load Processes", Structural Safety, 2, 1985, 169-176.

[12] Schuëller, G. I.; Bucher, C. G.; Bourgund, U.; Ouypornprasert, W.: "On Efficient Computational Schemes to Calculate Structural Failure Probabilities", in Lecture Notes in Eng., Vol.31, Stochastic Structural Mechanics, Eds. Y.K.Lin, G.I.Schuëller, Springer-Verlag, Berlin, New York, 1987, pp.388-408.

[13] Schwarz, R.F. and Schuëller, G.I.: "Reliability of Structures Under Combined Loading", Proc. ASCE Specialty Conf. on Prob. Mechanics and Struct. Rel., Tucson, AZ, Jan., 1979, pp.112-120.

[14] Shinozuka, M.: "Stochastic Characterization of Loads and Load Combinations", Proc. ICOSSAR' 81, Eds. T. Moan, M. Shinozuka, Elsevier, 1981, pp.57-75

[15] Shiraki, W. and Takaoka, N.: "Reliability Analysis of Compression Members with Rectangular Cross-Section", Procedings of the 27th Japan National Congress for Applied Mechanics, 1977, pp.249-258.

[16] Veneziano, M., Grigoriu, M. and Cornell, C.A.: "Vector-Process Models for System Reliability", J. of Eng. Mech. Div., ASCE, Vol. 103, No. EM3, June, 1977, pp.441-460.

[17] Wen, Y.K.: "Statistical Combination of Extreme Loads", J. of Str. Div., ASCE, Vol. 103, No. ST5, May, 1977, pp.1079-1093.

ITERATIVE ELASTIC AND PLASTIC OPTIMAL DESIGN OF STEEL FRAMES BY A MIXED METHOD INCLUDING DOUBLE SECANT METHOD FOR NONLINEAR MATHEMATICAL PROGRAMMING

Chang-Ping Chiou, Tsung-Wu Lin, Jenn-Rong Wu
Department of Civil Engineering, National Taiwan University
Taipei, Taiwan 10764, R.O.C.

Introduction

Recent applications of optimization methods in building frame design practice include optimality criteria method [1,2,3] and the conventional LP method. To use the recurrence relationship in the optimality criteria method, the Lagrange multipliers associated with each active constraint, and the gradients of the weight function and the active constraints must first be determined. But there derivations are lengthy and tedius. Therefore, we try to use the 'double secant method for nonlinear mathematical programming 'developed by Tsung-Wu Lin [4] for elastic optimal design. The major advantage of the latest method is that it provides double secants to make the search of correct direction quickly with smaller maximum error at each iterate. Nevertheless, there are still too many iterations required to reach an optimum design if the double secant method alone is used. It can become more effective if we determine first the initial design variables, or the initial member sizes, by applying the stress-ratio method and the shifting method alternatively. The amount of calculation can also be reduced by applying the reanalysis technique or Gauss-Seidal iteration method in the process of nonlinear mathematical programming. Numerical examples have shown that after the initial design variables were determined after two or more iterations, only several iterations were needed for the double secant method.

This paper also tried a simplified approach for a plastic optimal design of steel frame. This approach makes use the result obtained from the elastic optimal design as the preliminary design variables and then apply the 'plastic hinge method 'as in the conventional elasto-plastic analysis. A 'load-ratio'method is used to obtain the plastic optimal design. Some of the code requirements are used as constraints.

Elastic Optimal Design

Optimization problem. —

1. Stress constraints:

$$\sigma_j^L \leq \sigma_{j,p,l_n} \leq \sigma_j^U \tag{1}$$

$$j = 1, \ldots, m; \qquad p = 1, \ldots, n_p; \qquad l_n = 1, \ldots, n_l$$

where m is the number of member ; n_p is number of different points of the member j ; n_l denote loading conditions.

2. Displacement constraints:

$$u_{i,l_n} \leq u_i^U, \qquad i = 1, \ldots, n \tag{2}$$

where n is the number of displacement contraints.

3. Size constraints:

$$x_j^L \leq x_j \leq x_j^U, \qquad j = 1, \ldots, m \tag{3}$$

4. Move limits:

$$\delta x_i^L \leq \delta x_i \leq \delta x_i^U, \qquad i = 1, \ldots, n_x \tag{4}$$

5. Objective function:

$$W = \sum_{j=1}^{m} \rho_j l_j A_j \tag{5}$$

in which W is the total weight of steel frame, ρ_j = specific weight, l_j = the length, A_j = the cross-sectional area of element j.

6. Design variables : The area A is defined as the independent design variable. The moment of inertia I and the section modulus S were related to A by the following function [5]:

$$\left.\begin{array}{l} I = (A/0.465)^2 \quad in^4 \\ s = (60.6I + 84100)^{1/2} - 290 \quad in^3 \end{array}\right\} \quad 0 < A \leq 44.12 \quad in^2 \tag{6}$$

$$\left.\begin{array}{l} I = 256A - 2300 \quad in^4 \\ s = (I + 500)/18.87 \quad in^3 \end{array}\right\} \quad 44.12 < A \leq 88.28 \quad in^2 \tag{7}$$

Determination of Initial Design Variables.—

Before the application of double secant method, it is beeter to use a stress-ratio method coupled with displacement shift to bring the initial design close to the optimum, in the first two or three cycles [5]. The procedure will be as follows:

1. assign any values for the design variables (say, 1 in^2 for all A_i).

2. perform structural analysis to find member stresses σ_i.

3. find the size variables for the next cycle by

$$x_i = x_i * \sigma_i / \sigma_i^a \tag{8}$$

in which σ_i^a = the allowable stress for member i.

4. re-analyze the structure and consider the displacement constraints by using the shifting factor $\Omega = Max(u_i/u_i^U)$, then the new size variable will be calculated from

$$x_i = x_i * \sigma_i / \sigma_i^a * \Omega \tag{9}$$

the shifting factor will transfer the design point to the boundary of the feasible region.

5. repeat the procedure 4 once or twice to obtain the initial design variables for the double secant method.

Reanalysis Technique.—[6]

The reanalysis technique is intended to deal with situations in which the element stiffnesses are modified on account of changes in cross-sectional properties. If the changes to the original stiffness matrix $[K^0]$ are denoted as $[\delta K^0]$, the new stiffness equations can be written in the form:

$$\{P\} = [[K^0] + [\delta K^0]]\{\Delta\} \tag{10}$$

where $\{\Delta\}$ denotes the displacement vector or degrees of freedom. Solution of $\{\Delta\}$ will be

$$\{\Delta\} = [[I] + [K^0]^{-1}[\delta K^0]]^{-1}\{\Delta^0\} \tag{11}$$

where $\{\Delta^0\} = [K^0]^{-1}\{P\}$.

If all degrees of freedom associated with stiffness changes are designated as $\{\Delta_1\}$ and the remainder $\{\Delta_2\}$, equation (11) can be expressed by [5]

$$\{\Delta\} = \begin{Bmatrix} \Delta_1 \\ \Delta_2 \end{Bmatrix} = \begin{Bmatrix} q^{-1}\Delta_1^0 \\ \Delta_2^0 - D_{21}^0 \delta K_{11}^0 q^{-1} \Delta_1^0 \end{Bmatrix} \tag{12}$$

where D_{21}^0 is contained in $[K^0]^{-1} = \begin{bmatrix} D_{11}^0 & D_{12}^0 \\ D_{21}^0 & D_{22}^0 \end{bmatrix}$ and $[q] = [I + D_{11}^0 \delta K_{11}^0]$

Because the only matrix inverse in equation (12) is that of q, the computation cost of a complete reanalysis is greatly reduced.

Plastic Optimal Design

When an optimum plastic design of plane rigid frame is required, 'plastic hinge method'is used for structural analysis and a simplified 'load rato method'is used to modify the design variables at each load step when one or several plastic hinges are formed in the nonlinear elasto-plastic analysis. We implement the above methods to avoid tedious task in finding all possible collapse mechanisms required if 'linear programming method'for the conventional minimum weight design is adopted.

Load Ratio Method.—

When the plastic moment capacities, with fixed relative ratios, of all members in a plane frame are given, then the limit load which may force the frame to become a collapse mechanism or to attain the specfied drift limit can be obtained. If the obtained limit load is greater than the prescribed factored load specified in a structural code such as part 2 of AISC specifications, it is reasonable to reduce the given plastic moment capacities of all mambers by a ratio to minimize the weight. This is the so-called 'load ratio'implemented in this paper. As soon as the load ratio reaches 1 or the specified drift limit is attained in the step-by-step elasto-plastic analysis, then we can say that an optimum plastic design is obtained. One thing which is necessary to support this approach is that it is required to use the final result of optimum elastic design, because the relative sizes are almost "optimized"at the very beginning. This enables us to adopt a simplified approach instead of solving an enormous LP problem.

Illustrative examples

A fifteen-story steel frame with or without bracings (Figure 1 and 2) were taken for example. The structure of A36 steel with a specific weight $\rho = 0.28356$

lb/cu in. and a modulus of elasticity $E = 30,000$ *ksi*, has a fixed base and carries a gravity load of 2.4*kips/ft* on each floor beam and 1.2*kips/ft* on the roof beam, as adopted from Tabak and Wright's example [5]. The numerical values of the horizontal wind load distribution from roof down to the first floor are 8.93, 17.86, 17.86, 17.86, 17.86, 16.37, 14.88, 14.88, 14.88, 14.88, 13.39, 11.91, 10.42, 8.93, 8.93 in *kips* acting at each floor level. The load acts in three combinations: (1) 75% of gravity load plus wind from the left ; (2) Full gravity load ; and (3) 75% of gravity load plus wind from the right . The allowable stresses are +22 *ksi* and −22 *ksi* for all members and the horizontal displacement at roof level is limited to 3.6 inches. The maximum vertical deflection for beams is limited to 1/360 of the beam span. While optimal plastic design is required, the displacement limits are multiplied by a factor 1.3 or 1.7 for different load combinations.

For the frames with bracings of Fig. 1 , the final total weight of 100.39 *kips* (50.20 *tons*) for elastic design and that of 93.36 *kips* (46.68 *tons*) for plastic design with the application of codes in part 2 of AISC specification were obtained as shown in Fig. 3. The final member sizes are shown in Table 1. If part 2 of AISC specifications is not applied , then the final weight of 80.93 *kips* was obtained for the plastic design. The iteration history is shown in Fig. 3. For the frames without bracings of Fig. 2, the final total weight of 147.58 *kips* for the plastic design was obtained, without the application of codes.

Conclusions

The proposed mixed method for the optimal structural design has the benifit of reducing the design cycles as revealed in the illustrative examples. But it should be kept in mind , the stress constraints of ±22 *ksi* have no thing to do with any code requirements. Therefore, the elastic design is at present still not applicable to structural design office and is needed for further improvements. But it is useful to adopt the result of elastic design as the preliminary design of the plastic design.

References

1. Uri Krisch, "Optimum Structural Design", McGraw-Hill Book Company, 1981.
2. Carloe I. Perguera, "Integrated Analysis and Design of Steel Frames with Interactive Computer Graphics", Department of Structural Engineering Report, Number 84-4, Cornell University, march 1984.
3. Richard H.Gallagher , "Computation and its Role in the Unification of Design "Lecture Note from "Unification of Finite Element Software Systems", H. Kardestuncar, Editor, University of Connecticut, May 1985.
4. Tsung-Wu Lin , "Double Secant Method For Nonlinear Mathematical Programming ", paper presented in the 2nd IFIP Working Conference on Reliability and Optimization of Structural Systems, London, U.K.,Sep. 26-28,1988.
5. Enrique I. Tabak and Peter M. Wright, "Optimality Criteria Method For Building Frames "J. of Structural Division , ASCE ,Vol. 107, No. ST.7, July 1981, pp.1327~ 1342.
6. William Mcguire, P.H. Gallagher, Matrix Structural analysis, Ch.10, John Wiley Book Co., N.Y., 1979.

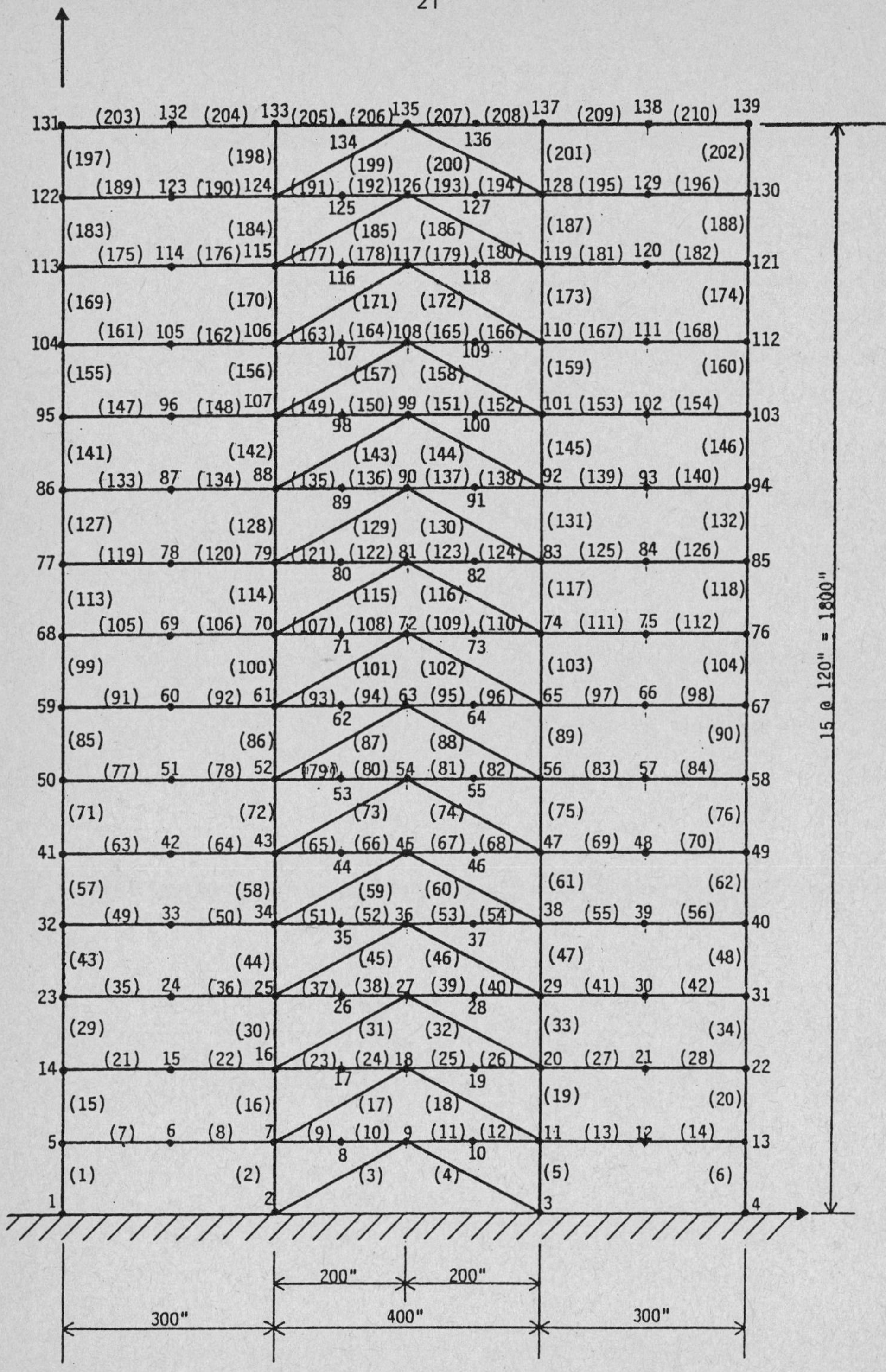

Fig. 1 15-Story Steel Frame with Bracings

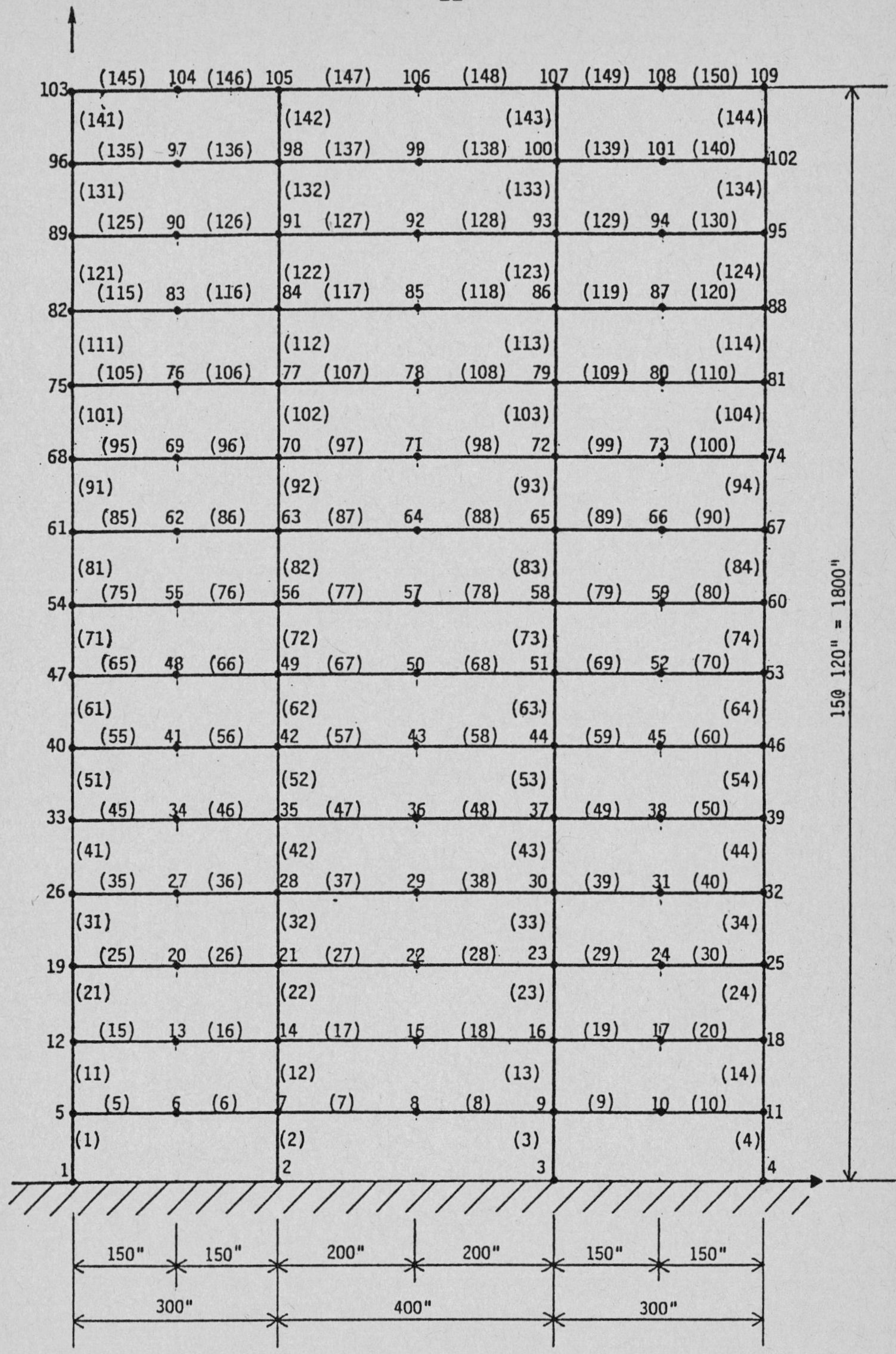

Fig. 2 15-Story Steel Frame without Bracings

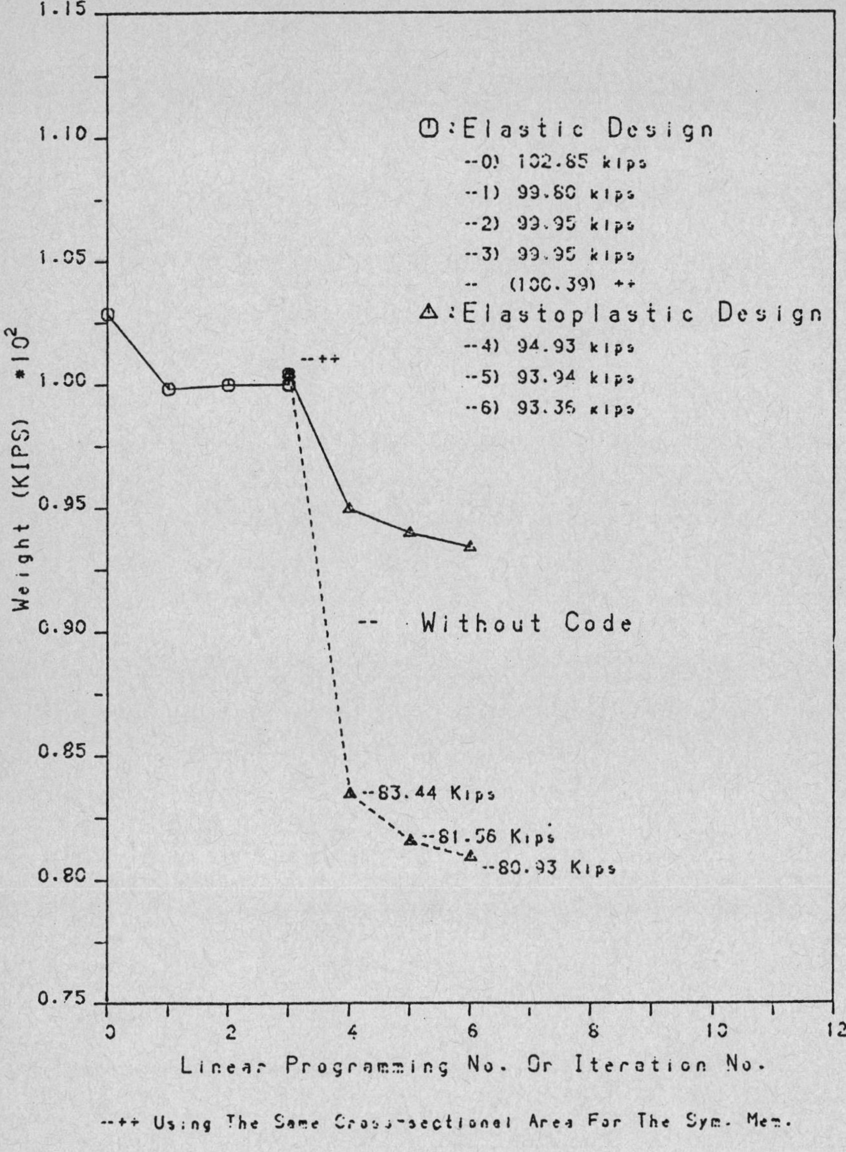

Fig. 3 The Iteration History of Steel Frames with Bracings

Table 1 The Final Member Sizes of Steel Frame With Bracings

member	elastic design			elasto-plastic design	
	A (in2)	I (in4)	stress ratio	A (in2)	I (in4)
1	24.92	2872.77	1.00	31.16	4491.31
2	66.21	14650.83	1.00	54.71	11706.53
3	3.19	46.93	0.69	2.32	24.94
7	12.28	697.09	1.00	8.95	370.54
9	12.28	697.09	0.99	8.95	370.54
15	21.91	2220.60	1.00	29.35	3985.27
16	41.37	7914.76	1.00	51.13	10789.79
17	5.26	128.06	1.00	3.84	68.05
21	13.41	831.44	0.99	9.77	441.90
23	10.55	514.80	1.00	7.69	273.64
29	20.64	1969.91	1.00	27.93	3606.97
30	38.39	6816.99	1.00	51.13	10789.79
31	5.15	122.80	1.00	3.76	65.28
35	13.72	870.93	1.00	10.01	462.94
37	10.74	533.17	1.00	7.83	283.40
43	19.55	1767.73	1.00	24.84	2854.09
44	35.12	5702.85	1.00	50.05	10512.80
45	4.91	111.31	1.00	3.58	59.17
49	13.79	879.06	1.00	10.05	467.21
51	10.53	513.25	1.00	7.68	272.78
57	18.53	1588.81	1.00	21.67	2172.56
58	32.16	4782.83	1.00	45.92	9455.01
59	4.61	98.18	1.00	3.36	52.18
63	13.75	874.40	1.00	10.02	464.80
65	10.32	492.15	1.00	7.52	261.60
71	17.52	1420.10	1.00	18.67	1611.38
72	29.23	3951.40	1.00	42.22	8242.30
73	4.30	85.45	1.00	3.13	45.42
77	13.61	856.91	1.00	9.92	455.48
79	10.11	472.73	1.00	7.37	251.27
85	16.53	1263.32	1.00	16.80	1305.31
86	26.31	3201.76	1.00	38.82	6970.27
87	3.98	73.18	1.00	2.90	38.89
91	13.49	842.20	1.00	9.84	447.62
93	9.92	454.78	1.00	7.23	241.75
99	15.56	1120.09	1.00	14.92	1029.37
100	23.38	2528.90	1.00	34.49	5502.13
101	3.66	61.90	1.00	2.67	32.90
105	13.46	837.42	1.00	9.81	445.07
107	9.73	437.41	1.00	7.09	232.48

Table 1 (continue)

member	elastic design			elasto-plastic design	
	A (in2)	I (in4)	stress ratio	A (in2)	I (in4)
113	14.62	988.92	1.00	13.02	783.88
114	20.47	1937.49	1.00	32.14	4776.15
115	3.34	51.54	1.00	2.43	27.40
119	13.36	825.91	1.00	9.74	439.02
121	9.66	431.68	1.00	7.04	229.47
127	13.70	867.99	1.00	11.10	570.03
128	17.54	1422.59	1.00	29.12	3920.91
129	3.01	41.98	1.00	2.20	22.30
133	13.21	807.23	1.00	9.63	429.07
135	9.86	449.22	1.00	7.18	238.75
141	12.84	762.64	1.00	9.36	405.35
142	14.58	983.19	1.00	23.57	2568.64
143	2.66	32.82	1.00	1.94	17.44
147	13.01	782.40	1.00	9.48	415.90
149	10.10	471.90	1.00	7.36	250.80
155	11.94	659.29	1.00	8.70	350.45
156	11.52	613.37	1.00	18.05	1505.94
157	2.30	24.55	1.00	1.68	13.05
161	12.75	752.22	1.00	9.30	399.83
163	10.41	501.00	1.00	7.59	266.29
169	11.35	595.97	1.00	8.28	316.76
170	8.71	350.75	1.00	12.56	729.35
171	1.93	17.14	1.00	1.40	9.12
175	12.73	749.19	1.00	9.28	398.20
177	10.79	538.84	0.99	7.87	286.37
183	9.98	460.38	1.00	7.27	244.70
184	5.69	149.74	1.00	6.99	226.29
185	1.91	16.80	1.00	1.39	8.92
189	13.47	838.65	0.99	9.82	445.80
191	10.73	532.86	0.99	7.83	283.25
197	13.00	781.79	1.00	9.48	415.55
198	2.70	33.71	1.00	2.05	19.45
199	1.80	15.04	1.00	1.32	8.00
203	12.77	754.49	1.00	9.31	401.03
205	10.90	549.55	1.00	7.95	292.08

elastic design	elasto-plastic design
total weight : 100.39 kips	total weight : 93.36 kips
displ. constraint :	displ. constraint :
in in in	in in in
2.25 -0.30 -2.25	4.68 -0.56 -4.68

RELIABILITY APPROACH TO THE DESIGN, CONSTRUCTION AND USE OF NATURAL DRAUGHT COOLING TOWERS

Tadeusz Chmielewski, Marian Golczyk
Technical University of Opole
ul. Katowicka 48, PL-45-951 Opole, Poland

1. Introduction

Structures or structural elements should be designed and constructed such they are suited to their intended use. But principles for the design, construction, use and maintenace are changing evolutionary. It is due to fundamental research and the progress in technology.

Contemporary stage tends to treat civil engineering structures like others objects, especially electronic and mechanical systems, for which exist and is applied theory of reliability, [1,2].

This paper consider the possibility of introduction the reliability approach to the design, construction and use of natural draught cooling towers. The following definition of the reliability has been assumed. The reliability of cooling tower is its ability to fullfil its design purpose for some specified time under the environmental conditions encountered.

2. Cooling towers construction development in Poland

Most natural draught cooling towers in Poland as in other countries are designed as reinforced concrete thin shell supported along the circumference by a system of columns, in general having a height less than 10% of the total height of the tower. The V - shaped columns are resting on single footings or on a ring foundation, generally being independed of the cooling systems. The fill inside the tower or cooling elements are supported by a system of beams with adequate support posts. Among all these structural units the shell part requires most of the attention in design and construction, [3,4].

Cooling towers in general and shell structures in particular belong to the largest structures in civil engineering. The development in Poland between the years 1955 and 1985 is shown in Fig. 1 marking the highest cooling tower of each year with the corresponding plant power unit and capacity. The limit 100m in the height was reached in 1960.

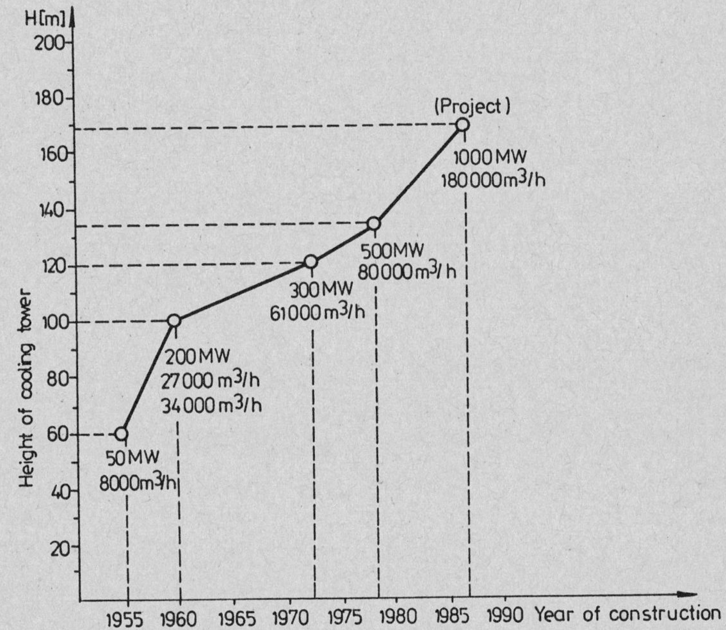

Fig.1. Development in the size of the Cooling Towers and Block Capacites.

With increasing capacity of the power plants it was necessary to increase both the height and the diameter. For example in France cooling towers have been built such that the height is bigger than 180m with the diameter over 150m. This development during a relatively short time could not be realized without fundamental research of these remarkable structures and the application the principles of reliability seems now unavoidable.

When traveling throughout Poland, whenever we came across a cooling projects, monumental reinforced concrete hyperboloidal cooling towers, we can always be sure that they all are the result of the minds and hands of the staff of the Cooling Towers Construction Enterprise "Chłodnie Kominowe" in Gliwice. These achievments have been done to its close cooperation with some universities and research centres in Poland, [6].

3. Application of reliability principles for the design, construction and use of cooling towers

The problem which should be considered during processes of the design, construction and use of cooling towers necessary for reliability analysis are shown in Fig. 2. This is an algorithm which show the set and run of information on: basic variables, design process, construction, use, maintenace and repair. Such approach unifies the consideration of use, maintenace and repair in design theory to assure structural quality. One should remember that good structural design produces reliable and durable structure.

3.1 Basic variables

Practical experiences and tests in situ have shown that basic variables have random character and should be prescribed as random variables or random fields. From the design point of view it is important to know dispersions of these variables. During the design process the decision should be taken if these dispersions are negligible or not to the mean values of basic variables.

3.1.1 Geometry of cooling towers

The shape of the meridian curve of the cooling tower shell generally is a hyperbola described by

$$R = r_o + a\sqrt{1 + \frac{z^2}{b^2}}$$

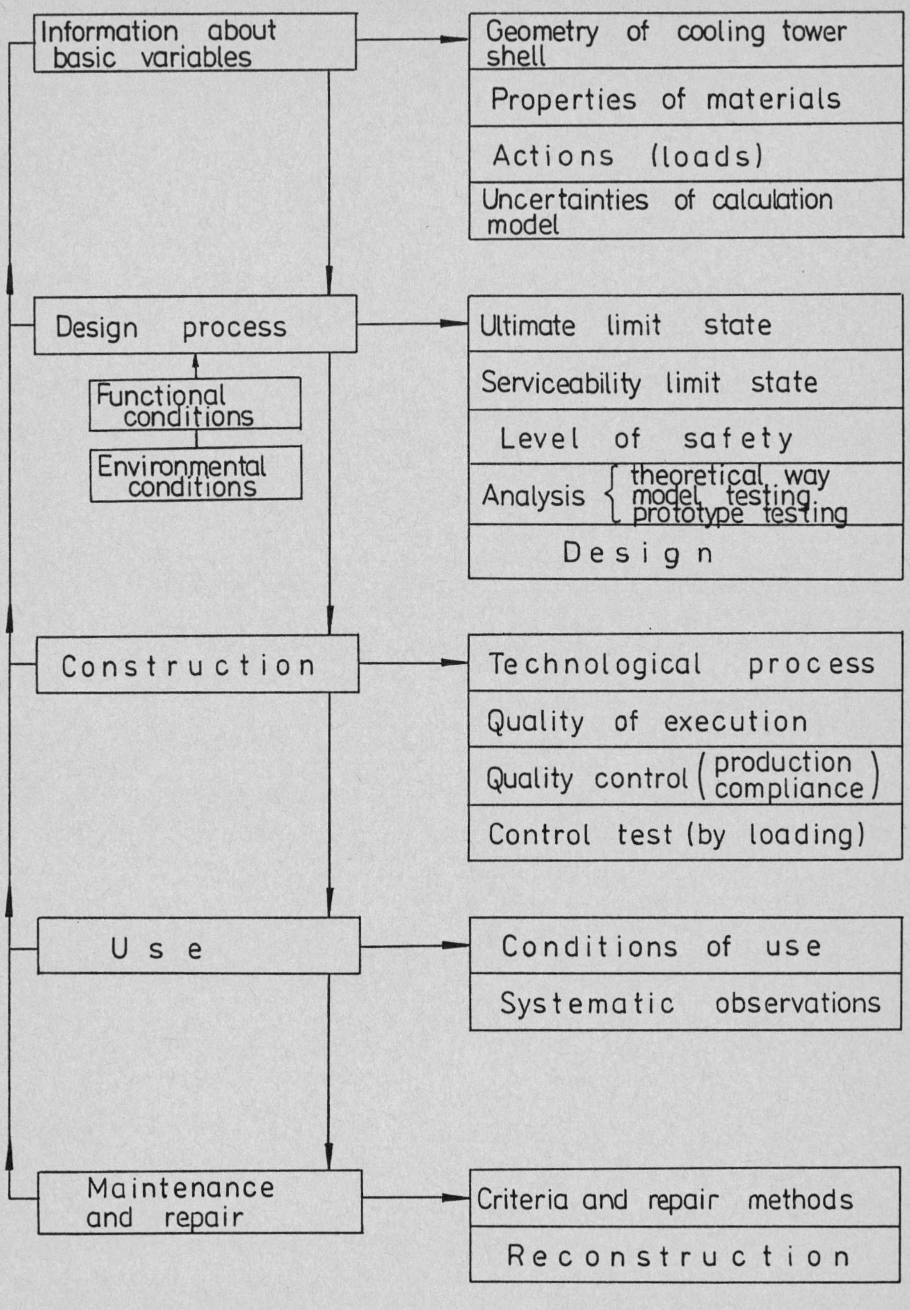

Fig. 2. Elements of reliability analysis

in which R, z and r_o are shown in Fig. 3; r_o being the distance between the axis of rotation and the axis of the hyperbola, [5].

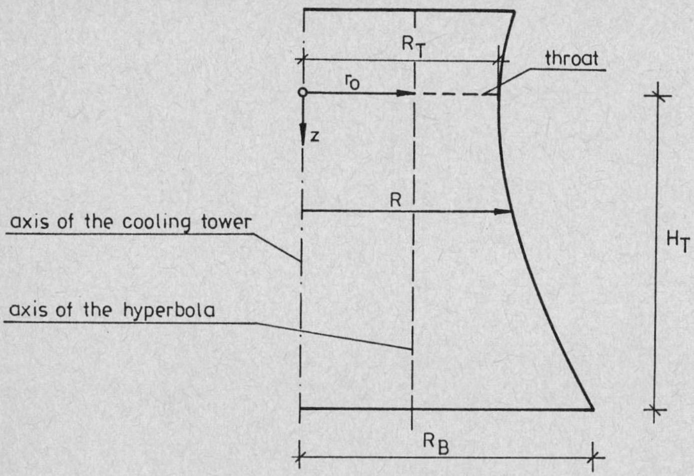

Fig. 3. Geometry of Cooling Tower Shell

For $r_o = 0$, parameter a is the radius of the cooling tower at the throat. The parameter describing the curvature is b, given by

$$b = \frac{H_T}{\sqrt{\left(\frac{R_B - r_o}{R_T - r_o}\right)^2 - 1}}$$

in which R_B and R_T are the radii of the cooling tower at the base and the throat, respectively, and H_T is the vertical distance between the base and the throat.

For given thermodynamic data the values of H_T, R_B and R_T are related with each other and offer only a restricted possibility for variation. On the other hand the offset distance r_o influences not only the stress distribution in the shell, but also the buckling load and the natural frequency.

3.1.2 Properties of materials

For the construction of cooling towers high class materials are used. Concrete should have enough resistance against cracking and freezing. Concrete class B25 and steel class A-III /high strength steel/, 34GS are used in Poland.

3.1.3 Actions on cooling towers

An action is considered to be one single action if it can be assumed to be stochastically independed, in time and space, of any other action acting on the cooling tower.

In reality, actions which are introduced simultaneously are stochastically dependent to a certain extent. To simplify calculations, those which, where they are presented, depend closely on each other and attain their upper values at the same time are considered together as a single action. Actions where the dependence is small /low correlation/ can be considered as independent.

To facilitate the calculation of the action effects, it is convenient to classify all actions acting on cooling towers into three subsets, according to their variations in time, i. e. all actions are divided into: permanent, variable and accidental actions.

The permanent actions include:

a/ the weight of cooling tower,

b/ the actions resulting from support settlements,

c/ the actions resulting from shrinkage of concrete,

d/ the effect of imperfections in the cooling tower shell /which should be as small as possible/.

The variable actions include:

a/ wind loads,

b/ temperature changes /axi-symetrical temperature change = 35°C and temperature change in the wall due to sun-heat/,

c/ the weight of a scaffolding, pipes and pomps for concrete.

The accidental actions include:

a/ earthquakes /for some regions may be considered as variable action/,

b/ The actions resulting from minimg subsidense /when a region is exposed to them/.

3.2 Computation methods

3.2.1 In the past /three and two decades ago/ the membrane solution of the continually supported hyperboloidal shell under axisymmetric dead load and non-axisymmetric wind action treated statically with the application of dynamic coefficient constituted the standard method of analysis. The disturbances of the stiffened upper and lower boundaries were accounted for by a simplified bending theory for edge disturbancies. The region of the shell immediately above the supporting columns, termed the lower lintel, was designed according to deep beam theory.

Due to rapid development of the computers and computing methods there is today no problem in the analysis of the shell part having the geometric form of a hyperboloid of revolution according to the general bending theory for thin elastic shells. The numerical evaluation of the bending theory is based on finite element or finite difference methods. For design purposes it is appropriate to assume the reinforced concrete to be isotropic, homogeneous and without cracks. A refinement of the analysis is achieved if the supporting columns too are introduced into analysis.

In this place it must be pointed out that any of the analysis methods described may be accurate enough, provided that care is given to a sound structural modelling of the cooling tower.

3.2.2 Wind forces

The problem arising with increasing dimensions of the cooling towers is not to improve the method of analysis but rather the assessment of the forces induced through wind. Besides the dead load, wind is the main loading of a cooling tower and the design of the reinforcement is based on the difference of two stresses of comparable magnitude, namely tension due to wind and compression due to dead load. On the other hand, the buckling of the cooling tower shell is substantially influenced by the compressive wind stresses added to the compression due to dead load. For these reasons to seize the wind effects realistically is of vital importance for the analysis of the cooling tower. This is why a considerable part of the research on cooling tower concentrated during the last fifteen years on this item. Measurements both on aeroelastic models tested in wind tunnel and on real reinforced concrete cooling towers have yielded detailed information on the magnitude and distribution of the wind forces along circumference and the height of the structure. For a group of cooling towers wind tunnel tests are considered inevitable in order to obtain the wind effects on each cooling tower of the group. However, according to measurements on aeroelastic cooling tower models in a boundary layer wind tunnel, cooling towers erected in a distance more than 1,5 times the base diameter or twice of the throat diameter between the cooling tower axes can be treated as singles, [5].

3.3 Construction technology of cooling towers

The particular role played by the technical progress was appreciated by the Cooling Tower Construction Enterprise throughout the last 43 years of its activity, [6].

Intensive research and develpoment studies resulted in the seventies in devising its original shell construction method that is in use at present. Is should be emphasized that it was not only the shell construction technology that was modernized but all other cooling tower elements as well.

In the new shell construction technology the formwork is independent from scaffolding structure. The formwork panels are made of waterproof laminated plywood strengthened on the borders with steel sections. The cages of the adjustable scaffolding are fixed to the shell only at inside. The internal cage ends at its top with a latticed support to which the external cage is freely suspended. Each two telescopic stagings are suspended to the external and internal scaffoldings. The whole of the cage scaffolding is made of an aluminium-hydronalium alloy. The cages are shifted in the meridional direction with the use of hydraulic hoists one by one or by groups.

The vertical and horizontal transportation of materials is performed with the use of a crane situated centrally in the cooling tower.

The transport of labour force is performed by a special cargo-passenger lift traveling inside along the meridian of the cooling tower shell. A special structure change cage has been provided for bringing workers into the internal working stagings.

According to the opinion of specialists, the new cooling tower shell construction technology applied in Poland is the best method used so far.

This construction method minimize the imperfections in the cooling tower shell. In Fig. 4 the imperfections of the shell surface for one

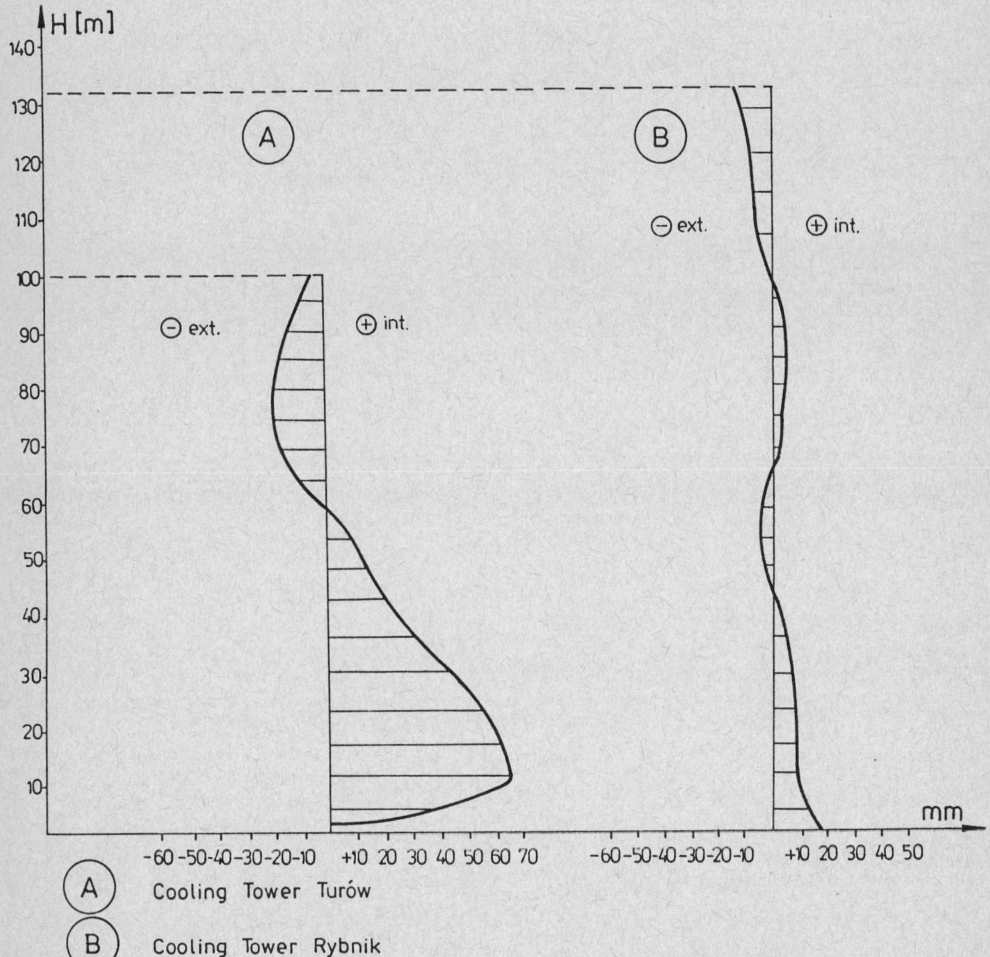

Fig. 4. Imperfections Along the Meridian Curve

meridian of two cooling tower shells executed with the application of two types of scaffoldings are shown schematically. The maximum deviation of the wall thickness has been found to be about 10%. It is obvious that such low values can be tolerated and need not to be considered in design.

3.4 Maintenance and repair

Cooling towers should be maintained in such a way that they can be used during their intended life for the purpose for which they were designed. When a cooling tower has been damaged or deteriorated to such an extent that its use would entail an unacceptable risk or cost, then repairs should be undertaken.

Damage or deterioration may be discovered through inspcctions which should be undertaken at certain time intervals, for example 3-4 years during first 20 years of exploatation and then after 2 years. For such systematic observations the user should be responsible.

3.5 Quality control

All steps in planing, design, construction and use of a cooling tower should be controlled to an extent which depends on the possible consequences of error and costs of control.

The purpose of quality control is to ensure an acceptable quality of design, materials, work on the construction site, and to detect gross errors.

It is assumed that participants in the building process of the cooling tower intend to carry out their respective tasks with care and skill. This can only be excersised through individual selfchecking.

In the Enterprise "Chłodnie Kominowe" the design work is checked by an internal control. On the construction site the shape, shape corrections, wall thickness are controlled by a special geodesy enterprise and quality of concrete is also under the internal control. After hardening of the concrete, the final shape is surveyed once more and real wall thickness is measured.

4. Assessment of reliability of cooling towers

4.1 Today's situation

When a cooling tower designer selects the materials for the reinforced concrete shell, it is fairly easy for him to satisfy the required function of the component at "time zero", i. e. at the time of delivery of the finished structure. He base his choice of materials and his design on reasonably well-defined properties that are listed in product brochures from the producer or utilize general data listed in handbooks, etc; such properties include the 28-day strength of concrete. The considerably more complex properties that describe the "ageing" of the material in situ, i. e. the time-dependent interaction between two adjecent materials, or the interaction between the materials and the surrounding enviroment, or the effect of possible changes in the shell or in the environment, are of no importance for the function at time zero. Those properties are also very seldom accounted for by the producer or in handbooks. In many cases, the producer and the designer do not even know which material properties are of importance for the long-term function.

4.2 Reliability steering system of cooling tower

Achievment of certain level of reliability of a cooling tower may be done at the stage of good design and construction. Optimal choice of the meridian curve, application materials with good properties, best descriptions of actions and good estimation of the response of cooling tower due to them, technological development in the erection and quality control in all important stages leads to achieve higher level of the reliability. Therefore the authors propose the following realibility steering systems of cooling towers, which is shown in Fig. 5.

In Fig. 6 it is shown the run of information among: designer /D/, constructor /C/ and user /U/.

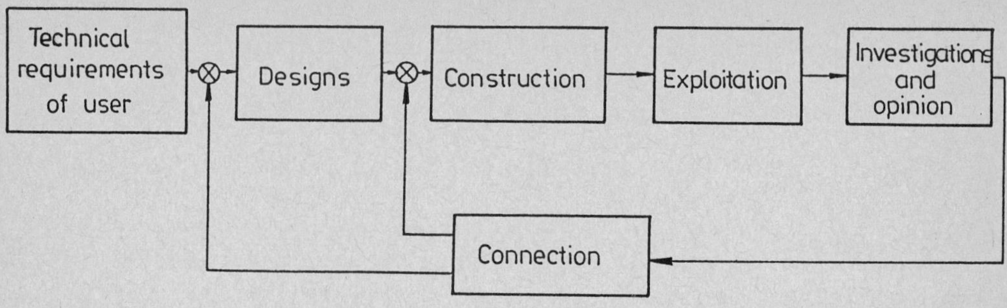

Fig. 5. Reliability Steering System

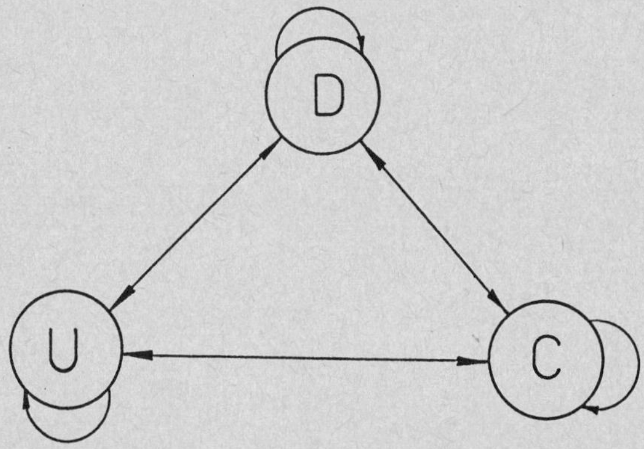

Fig.6. Run of Information Among: Designer, Constructor and User

One of the way to increase the reliability of cooling towers are inspections in situ: routine, periodic and special /according to needs/. These data should be the base for designer and constructor to improve their work.

4.3 Lifetime distribution and reliability

Planning, design and construction of cooling towers systems must be accomplished without the benefit of complete information, consequently, the assurance of performance can seldom be perfect. Moreover, many decisions that are required during the process of planning and design are invariably made under conditions of uncertainty. Therefore, there is invariably some chance of nonperformance or failure and its associated adverse consequences, hence risk often unavoidable. Under such conditions, it is not feasible /practically or economically/ to assure absolute safety or performance of cooling towers /like many others engineering systems/.

Reliability problem of cooling tower involve a determination and assurance of adequacy of a system over its useful life. The problem, from authors point of view, is time - depend because the capacity of each cooling towers deteriorates with time and usage.

Let t denote the time elapsed since a cooling tower was put in service, i. e. its age. Let T consider as the random variable which prescribe a lifetime of the cooling towers. For the set of all cooling towers built up in Poland a plot can be made of the proportion of the structures that are in a failed state at age t, i. e. if T is the random variable, its probability can be described by its cumulative distribution function, [7]

$$F(t) = P(T \leqslant t) \qquad \text{for all } t.$$

From structural point of view F(t) is called the lifetime distribution function for the set of cooling towers. The derivative $f(t)=dF(t)/dt$ is called the failure rate function.

All cooling towers built in Poland consist a population. They are available for inspections. That is why the functions F(t) and f(t) can be estimated statistically from observation of the attributes of cooling towers and their enviroment.

Let t_d be a time of particular interest, e. g. a projected life or design life. Then $P_F = F(t_d)$ is the failure probability and its complement, $P_R = 1 - P_F$, is the reliability of the population.

If function F(t) is known one can calculate other rates of the reliability of cooling towers, for example the mean lifetime T_o may be calculated according the formula

$$T_o = \int_0^\infty t\, f(t)\, dt.$$

All cooling towers built in Poland from 1955 to 1985 are under inspections. The histogram of work time of cooling towers without failure or with failures which needed repair for 46 structures /for which the authors had data/ is shown in Fig. 7.

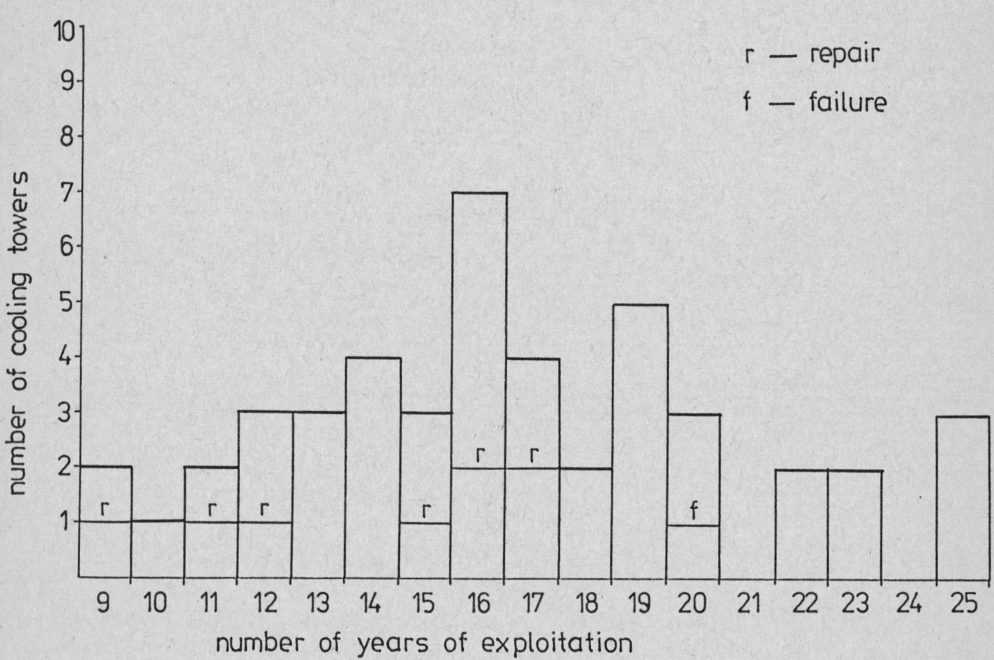

Fig. 7. Histogram of number of years of cooling towers exploatation

REFERENCES

1. Thoft-Christensen P., Baker M. J., Structural Reliability Theory and Its Applications, Springer-Verlag, Berlin, Heidelberg, New York, 1982.
2. Ziemba S. and others, Physical Aspects of Durability and Reliability of Technological Objects, PWN, Warszawa-Poznań, 1976 /in Polish/.
3. Ledwoń J., Golczyk M., Cooling Towers and Fan Cooling Projects, Arkady, Warszawa 1967 /in Polish/.
4. Proceedings of the 2nd International Symposium: NATURAL DRAUGHT COOLINGS TOWERS, Ruhr-Universität Bochum, Springer-Verlag, September 5-7, 1984.
5. Zerna W., Impulses of the Research on the Development of Large Cooling Towers, Proceedings of the 2nd International Symposium: NATURAL DRAUGHT COOLING TOWERS, Ruhr-Universität Bochum, September 5-7, 1984 pp. 1-18.

6. Cooling Towers Construction Enterprise "CHŁODNIE KOMINOWE", 1945-1985.

7. Chmielewski T., Probabilistic Methods in Structural Dynamics, Technical University of Opole, 1982 /in Polish/.

PERCEPTION OF RISKS AND REACTIONS TO ACCIDENTS

F. Vasco Costa
Rua Joaquim Ant. Aguiar 27,9
Lisbon 1000, Portugal

Summary

Whatever the function to be fulfilled by a structural system, the system has to be designed taking into due consideration how human beings are going to use it, how they will perceive the risks involved in its use and how they will behave in case of an accident. The participation of a social psychologist or of a behaviourist in future meetings of the IFIP Working Group 7.5 is therefore advocated.

Introduction

The design of any structural system is usually approached as an economical enterprise, having as scope the maximization of benefits or the minimization of the initial cost plus the expectation of future expenses with its operation.

As all structural systems, without exception, are intended to be used by human beings, they need to be designed taking into due consideration their ability to make an adequate use of the systems.

No doubt that the behaviour of human beings, in particular when they misuse a system or after an accident, will appear to the designer as irrational, and therefore not deserving to be considered. But, in order to be able to design structures that will be properly used, experts in structural systems need to take into account how people behave.

Human factors and ergonomics

The taking into consideration of human factors in the design of structural systems can be denoted in several ways, namely: human engineering, engineering psychology and ergonomics. This last term is being applied, in particular, to the study of the efficiency of workers.

Makers of luxury products, like perfumes and whiskies, just like politicians before electoral periods, never fail to consult experts on human behaviour, in order to be successful. Designers of structural systems should follow their example in order to be able, by taking into account how human beings will use the systems, perceive risks and behave in case of accident, to design structural systems that will last longer and render people happier.

Hazards and Risks

Technicians always take care to attribute precise meanings to the terms they use. In a field as elusive as the perception of risks and the behaviour of human beings, particular care is required with the meanings that can be attributed to some words.

Danger is a general term considered as meaning liability to injury, possibly implying losses of life. Hazard suggests forseable but uncontrolable danger, mainly due to natural causes, as storms, floods and earthquakes. Peril conveys the impression of a great and imminent danger, as one can feel while making a sharp turn at a too high velocity. Risk implies the voluntary acceptance of danger, having in view gaining some kind of benefit.

Psychologists usually only speak of hazards, including in them the risks voluntary incurred, as when one, because being in a hurry, crosses a street outside the crossing lanes.

Designers of structural systems usually only speak of risks, including in them all the hazards resulting from natural causes, like floods and earthquakes.

As this conference is being attended mainly by designers of structural systems, which use always implies the voluntary incurrence in dangers of diverse nature, in the following paragraphs the only term used will be "risks".

Perception and evaluation of risks

When confronted with situations involving risks, human beings show a great variety of attitudes. Such variety is determined not only by the lack of information about the possible consequences of the distinct attitudes, but depends as well on how different persons care about such consequences.

There are persons who always avoid incurring in risks, independently of the importance of the benefits that they can bring, whereas other persons appear to be unconcerned or even to enjoy incurring in risks. We all know people who avoid being promoted just to escape increased responsabilities, while others spend huge amounts of money just for the fun of driving fast cars or of going ski-jumping.

Designers of structural systems are trained to judge situations involving risks as objectively as possible. By taking into account how human beings behave the designers will be able to render structures even better for long years of use.

Human errors

Human errors are the remote cause of most accidents. Opinions on the percentage of accidents that can be attributed to human errors vary from author to author, depending on what they can consider as human error. In some cases values as high as 80% have been advanced. Notwithstanding, most engineering structures are designed and built under the implicit assumption that accidents will only occur in case of an extremely large load with a very low probability of occurence being applied on an extremely weak element of a structure, also with a very low probability of occurence.

Structural systems need be designed taking into consideration all possible kinds of human errors during the design, the erection, the use and even after possible accidents.

Professional Ethics

Technicians who design structural systems have obligations towards employers, clients, users of the systems and also to professional institutions. Such obligations should bring them to try as objectively as possible, but not ignoring subjective factors, to design structural systems that will render, if properly used, the best possible service permitted by the state of the art.

When technicians propose new solutions to old problems, or when they design new types of structures, they incur in new risks, not only in risks inherent to the lack of past experience on their use and to unforeseen modes of failure, but to the risks that can result from an improper use of the new types of structures.

To what extent shall the technicians play it safe in order not to become the scapegoats in case of accidents having causes beyond their control? If technicians are not encouraged to try new designs the living standards of society will cease to improve.

Final Considerations and Recommendations

In case of structures being designed taking into account how people will use the system, how they will perceive risks and how they will behave in case of accidents, the following benefits can be expected:

- periods of training to use the systems will become shorter;
- frequency of accidents due to misuse will decrease;
- consequences of accidents will not be so serious;
- systems will be kept longer in service;
- new models and alterations in old models will be introduced with greater confidence.

The assistance of a behaviourist, or social psychologist, during the design of structural systems is to be regarded as an essencial requirement to achieve the above benefits.

Suggested reading

Canter (1981) Fires and Human Behaviour, Wiley, NY.

Fischhoff, Svenson & Slovic (1987) "Active response to environmental hazards: perceptions and decision making", in D. Stokols and I. Alfman, Handbook of Environmental Psychology, Wiley, NY.

Fishbein & Ajzen (1975) - Belief, attitude, intention and Behavior, Addison-Wesley, MA.

Ives & Furuseth (1988) - "Immediate response to headwater flooding in Charlotte, NC", Environment and Behavior, 4:512-525.

Murrell (1969) - Ergonomics: man in his working environment, Chapman and Hall, London.

Sanders & McCormick (1987) - Human factors in Engineering Design, McCgraw Hill, NY.

A NEW MONTE CARLO TECHNIQUE IN STRUCTURAL RELIABILITY WITH A PLASTIC FRAME EXAMPLE

Attila Csenki
Centre for Software Reliability, The City University
Northampton Square, London EC1V 0HB, Great Britain

ABSTRACT

A new Monte Carlo method is presented for obtaining accurate estimates of component failure probabilities based on small sample sizes. It is an importance sapling technique which is based on the exclusion from sampling of a maximum hypersphere contained in the safe region. After finding the maximum hypersphere in the standard normal space of basic variables a two-stage sampling procedure involving conditional distributions is employed. The proposed method is illustrated with an example from plastic theory.

Key Words: Monte Carlo method; importance sampling; failure probability; non-central chi-square distribution; plastic analysis;

1. INTRODUCTION

Given a limit state function $g(Z_1,\ldots,Z_n)$ in terms of the standard normal basic variable vector $\underline{Z}=(Z_1,\ldots,Z_n)^T$ the expression for the failure probability P_f associated with g can be written as (see, e.g., [1])

$$P_f = P(\ g(\underline{Z}) \leq 0\ ). \tag{1}$$

An approximate value of the failure probability in eqn. (1) is based on the reliability index β which is defined as the distance of the failure surface

$$\{\ \underline{z} \in \underline{R}^n : g(\underline{z}) = 0\ \}$$

from the origin. β can be calculated by an iterative method [1] and P_f is then approximated as

$$P_f \simeq \Phi(-\beta), \tag{2}$$

where Φ denotes the standard normal distribution function. The approximation given by eqn. (2) can be improved by subsequently employing a Monte Carlo method whereby the β-sphere with centre at the origin, which of course is contained in the safe region, is excluded from sampling [2].

The method considered here is a generalization of the idea of enhancing the effectiveness of the Monte Carlo procedure by fitting a "maximum" region into the safe region and excluding it from sampling. In effect we are trading in an increased optimization effort for some reduction of the sample size which is needed to achive a prescribed accuracy of our estimate. The class of procedures considered here can in outline be described as follows. Take a (measurable) region B which contains the failure region

$$\{ \underline{z} \in \underline{R}^n : g(\underline{z}) \leq 0 \} \subset B.$$

Then eqn. (1) can be rewritten as

$$P_f = P(g(\underline{Z}) \leq 0, \underline{Z} \in B) = P(g(\underline{Z}) \leq 0 \mid \underline{Z} \in B) P(\underline{Z} \in B). \tag{3}$$

The first term on the r.h.s. of eqn. (3) denotes the conditional failure probability given that $\underline{Z} \in B$. It will be estimated by employing a suitable Monte Carlo procedure while the second term on the r.h.s. of eqn. (3) is to be calculated numerically. This method coincides with the crude Monte Carlo method [3] if B is the whole n-dimensional space. If B is the complement of the β-sphere the method is as given in [2]. In general,

$$\hat{p}(N) = \frac{1}{N} \sum_{i=1}^{N} I_{\{ g(\underline{z}_i) \leq 0 \}} P(\underline{Z} \in B) \tag{4}$$

is an unbiased estimator of P_f if $\underline{z}_1,\ldots,\underline{z}_N$ are N independent observations from the n-dimensional parent distribution Q defined by

$$Q(A) = P(\underline{Z} \in A \mid \underline{Z} \in B), \quad A \subset \underline{R}^n. \tag{5}$$

($I_{\{\ldots\}}$ denotes the indicator function of the set $\{\ldots\}$). The variance of the estimator $\hat{p}(N)$ is given by

$$\begin{aligned} \mathrm{Var}_Q(\hat{p}(N)) &= [P(\underline{Z} \in B)]^2 N^{-2} N \, \mathrm{Var}_Q(I_{\{ g(\underline{Z}) \leq 0 \}}) \\ &= N^{-1} [P(\underline{Z} \in B)]^2 P(g(\underline{Z}) \leq 0 \mid \underline{Z} \in B) [1 - P(g(\underline{Z}) \leq 0 \mid \underline{Z} \in B)] \\ &= N^{-1} P_f [P(\underline{Z} \in B) - P_f]. \end{aligned}$$

Hence, the coefficient of variation of the estimator $\hat{p}(N)$ is given by

$$\mathrm{COV}_Q(\hat{p}(N)) = N^{-1/2} [P(\underline{Z} \in B) - P_f]^{1/2} P_f^{-1/2}. \tag{6}$$

Eqn. (6) shows that in order to achieve a small coefficient of variation of $\hat{p}(N)$ the region B has to be selected such that it approximates the failure region from above as closely as possible. Furthermore, in order to calculate the estimator $\hat{p}(N)$ from eqn. (4) we have to restrict ourselves to domains B the probability content of which can be evaluated numerically. Finally, a method has to be available for obtaining independent samples according to the distribution Q.

This paper presents the theory for regions B which are complements of hyperspheres with the centre not necessarily at the co-ordinate origin. It will be seen that this simple idea leads to a procedure which is mathematically involved and by no means trivial. It would be interesting to see further extensions of this method by trying to fit more sophisticated regions into the safe region thereby enabling us to further reduce the sample size by learning more about the failure surface prior to sampling. A natural candidate is the class of hyperellipsoids for which the optimization and sampling theories should be much more complicated than the present ones.

The first step in our method entails seeking a hypersphere $S=B^c$ in the safe region with maximum probability content (or near so). Subsequent sampling from Q as given by eqn. (5) yields a value for $\hat{p}(N)$. It appears that such an optimization effort can result in a further order of magnitude reduction of the number N of limit state function evaluations when compared with [2] while maintaining the same coefficient of variation.

It should be noted that the method presented here is an importance sampling technique [3] whereby the crude Monte Carlo estimator, i.e., the indicator function of the failure region is replaced by the ratio of that function and I_B which then after normalization is a new estimator mimicing the random behaviour of the old one. (Since B contains the failure region the above ratio prior to normalization coincides with the old estimator). The optimization procedure is intended to minimize the random variation of the new estimator, while the sampling procedure allows independent samples from Q to be generated.

The paper is concluded with an example in which results by the present method are compared with those obtained when Harbitz's method [2] is employed.

Recent papers discussing Monte Carlo methods in Structural Reliability rarely go beyond the recognition that importance sampling might be an effective approach [4,5,6]; Schuëller and Stix [7] mention importance sampling without giving definitive advice on the choice of the new sampling distribution. Harbitz's approach [8] requires an analytic reformulation of the problem which makes it unsuitable for automatic implementation on a computer. Bucher [9] suggests a multivariate normal distribution as the new sampling distribution whose first and second moments should be chosen to coincide with estimated values of the conditional expectations $E(\underline{Z} \mid g(\underline{Z}) \leq 0)$ and $E(\underline{Z}\,\underline{Z}^T \mid g(\underline{Z}) \leq 0)$ respectively. This gives rise to an adaptive procedure in which these estimated values are updated as the sample size increases.

2. THE OPTIMIZATION PROCEDURE

An algorithm for finding a hypersphere contained in the safe region with (near) maximum probability content can be formulated by means of one of the well-known constrained optimization methods, e.g., the Method of Steepest Descent [10]. The problem then reduces to formulating subroutines which evaluate the objective function and the constraint(s). The situation is illustrated in Fig. 1 for n=2.

2.1 The objective function f

For a given point $\underline{a}=(a_1,\ldots,a_n)^T$ in the safe region $R(\underline{a})$ denotes the distance of $\underline{a}$ from the failure surface and $S(\underline{a};r)$ denotes the hypersphere with centre $\underline{a}$ and radius r

$$R(\underline{a}) = \min \{ \; ||\underline{z} - \underline{a}|| \; : \; g(\underline{z}) = 0 \; \}, \tag{7}$$

$$S(\underline{a};r) = \{ \; \underline{z} \; \varepsilon \; \underline{R}^n \; : \; ||\underline{z} - \underline{a}|| \leq r \; \}. \tag{8}$$

The objective function f evaluated at $\underline{a}$ is then simply given by

$$f(\underline{a}) = P(\; \underline{Z} \; \varepsilon \; S(\underline{a};R(\underline{a})) \;). \tag{9}$$

$R(\underline{a})$ is easily obtained by applying a usual level 2 reliability algorithm [1] to the "new" limit state function h defined by

$$h(\underline{w}) = g(\underline{w} + \underline{a})$$

and by writing the point $\underline{z}^*$ at which the minimum is attained in eqn. (7) as

$$\underline{z}^* = \underline{w}^* + \underline{a} \; ,$$

where $\underline{w}^*$ is the design point associated with the limit state function h.

The standard normal probability content of $S(\underline{a};r)$ is equal to the distribution function evaluated at r^2 of a non-central chi-square distribution with n degrees of freedom and non-centrality parameter $||\underline{a}||^2$. Even though methods are available to compute the distribution function of a quadratic form of a standard normal random vector [11,12,13] there seems to be no algorithm which is elementary (i.e., does not involve inversion of characteristic functions) and does not require truncation of an infinite sum of functions. A simple formula for the non-central chi-square distribution function is available only for odd degrees of freedom [14]. We provide a straightforward method of evaluation of the non-central chi-square distribution function in Appendix 1 for any degrees of freedom. The method is based on the introduction of polar co-ordinates and subsequent reduction of the dimension by means of a recurrence relation. The resulting one-dimensional integral is in terms of the standard normal distribution function Φ for which any of the usual high precision numerical approximations [15,16] may be taken to perform a numerical integration.

2.2 The constraints

The objective function $f(\underline{a})$ defined by eqn. (9) should ideally be maximized over the entire safe region, i.e., the inequality constraint

$$g(\underline{a}) > 0 \tag{10}$$

should be applied. In most cases, however, no maximum will exist over the whole safe region and therefore eqn. (10) has to be replaced by a constraint confining $\underline{a}$ to a compact subset of the safe region over which a maximum exists if f is continuous. The overall optimization procedure may be carried out as a sequence of optimization steps with the current region being defined as the optimum sphere from the previous step. The region for the first step may be defined as the β-sphere. The optimization procedure can be terminated if in any one step a local maximum is found, i.e., the point obtained is inside the region being considered, or after a predetermined number of steps.

Using the above approach involves inequality constraints the methods for which are computationally time-consuming and not always certain to converge. Therefore, it is preferable to use range constraints. Suitable range constraints are defined by any hypercube contained in the previous optimum sphere $S(\underline{a};r)$. Such a hypercube is given by, e.g.,

$$\{ \underline{z} \in \underline{R}^n : ||\underline{z} - \underline{a}||_\infty \le 0.5r\, n^{-1/2} \}.$$

If available, library optimization routines such as the NAG [17] may be used.

3. THE SAMPLING PROCEDURE

Having obtained a (near) optimum hypersphere $S(\underline{a};R)$ in the safe region it remains to estimate the first term on the r.h.s. of eqn. (3) by a Monte Carlo method. Since S will in general not be centered at the origin the conditional random radius and direction are not independent and thus the sampling procedure employed by Harbitz [2] is not applicable.

In this Section a sampling procedure is described which can be classified as a composition method sampling procedure [3]. It delivers first a sample $t^{1/2}$ of the random radius ρ and subsequently yields a sample $(\theta_1,...,\theta_{n-1})$ of the angles $(\Theta_1,...,\Theta_{n-1})$ representing the random direction. The latter sampling is conditioned on $\rho^2=t$ and both sampling steps are conditioned on the event that the optimum sphere is excluded.

3.1 Change of co-ordinates prior to sampling

New co-ordinates will be introduced for which the centre of S lies on the positive section of the n-th co-ordinate axis. (See Fig. 1).

The objective of the sampling procedure is to estimate the first term on the r.h.s. of eqn. (3) which is

$$P(\ g(\underline{Z}) \le 0 \ |\ ||\underline{Z} - \underline{a}|| > R\ ). \tag{11}$$

In Appendix 2 an orthogonal matrix $\underline{M}$ is constructed such that

$$\underline{M}^T \underline{a} = (0,\ldots,0,||\underline{a}||)^T, \tag{12}$$

and hence the probability in eqn. (11) can be reformulated to give

$$\begin{aligned} &P(\ g(\underline{M}\ \underline{M}^T\ \underline{Z}) \leq 0 \mid ||\underline{M}^T\ (\underline{Z} - \underline{a})|| > R\ ) \\ &= P(\ g_1(\underline{V}) \leq 0 \mid ||\underline{V} - (0,\ldots,0,||\underline{a}||)^T|| > R\ ), \end{aligned} \tag{13}$$

where

$$g_1(\underline{v}) = g(\underline{M}\ \underline{v})$$

is the g-function written in terms of the $\underline{v}$-co-ordinates with $\underline{V}=\underline{M}^T\underline{Z}$ being a standard normal n-dimensional random vector. Sampling will take place in the $\underline{v}$-space conditioned on the event that the (transformed) maximum hypersphere is excluded from sampling.

3.2 Sampling in the new space of variables

The components of $\underline{V}$ can be represented in polar co-ordinates as

$$\left.\begin{aligned} V_1 &= \rho \prod_{i=1}^{n-1} \cos \Theta_i, \\ V_j &= \rho\ \sin \Theta_{n-j+1} \prod_{i=1}^{n-j} \cos \Theta_i,\quad j = 2,\ldots,n\ , \end{aligned}\right\} \tag{14}$$

with independent random variables $\rho,\Theta_1,\ldots,\Theta_{n-1}$ such that ρ^2 has a chi-square distribution with n degrees of freedom and the angles Θ_i take values in $[-\pi/2,\ \pi/2]$ and $[0,\ 2\pi]$ for $i<n-1$ and $i=n-1$ respectively [18]. Samples of $\underline{V}$ will be obtained from eqn. (14) by first sampling ρ conditioned on the event as given in eqn. (13) and then sampling the random angles $\Theta_1,\ldots,\Theta_{n-1}$.

3.2.1 Sampling the random radius ρ

Denoting the conditioning event in eqn. (13) by E the (conditional) density of $T=\rho^2$ is obtained from

$$\begin{aligned} &P(\ t < T < t + \delta t \mid E\ ) = P(\{\ t < T < t + \delta t\} \cap E\ )/P(E) \\ &= P(\ t < T < t + \delta t,\ T - 2\ ||\underline{a}||\ V_n + ||\underline{a}||^2 > R^2\ )/P(E) \\ &= P(\ t < T < t + \delta t,\ T - 2\ T^{1/2}||\underline{a}||\ \sin \Theta_1 + ||\underline{a}||^2 > R^2\ )/P(E) \\ &= \chi_n^{\ 2}(t)\ L_n(t;y,R)\ \delta t/P(E) + o(\delta t),\quad (\delta t \downarrow 0), \end{aligned} \tag{15}$$

where

$$L_n(t;y,R) = P(\ t - 2\ t^{1/2}\ y \sin \Theta_1 + y^2 > R^2\ ),\quad y = ||\underline{a}|| \geq 0,$$

and χ_n^2 stands for the density of the central chi-square distribution with n degrees of freedom. Notice that the independence of T and Θ_1 has been used to arrive at eqn. (15). Observing that the density of Θ_1 is proportional to $\cos^{n-2}\theta$ ([18]) we get

$$L_n(t;y,R) = \begin{cases} 0 & \text{if } 0 \le t \le t_1 \text{ and } R \ge y, \\ 1 & \text{if } 0 \le t \le t_1 \text{ and } R < y, \\ 1 & \text{if } t > t_2, \\ 0.5 + \theta_0/\pi & \text{if } n = 2,\ t_1 < t \le t_2, \\ c_n \int_{-\pi/2}^{\theta_0} \cos^{n-2}\theta \, d\theta & \text{if } n \ge 3,\ t_1 < t \le t_2, \end{cases} \tag{16}$$

with

$$t_{1(2)} = (y \underset{(+)}{-} R)^2,$$

$$\theta_0 = \arcsin(\max\{-1, \min\{1, \zeta\}\}), \tag{17}$$

$$\zeta = (t + y^2 - R^2)\, y^{-1}\, t^{-1/2}/2, \tag{18}$$

and a normalizing constant c_n whose inverse is given by

$$c_n^{-1} = \int_{-\pi/2}^{\pi/2} \cos^{n-2}\theta \, d\theta = \begin{cases} \dfrac{1\ 3\ 5\ldots(n-3)}{2\ 4\ 6\ldots(n-2)}\,\pi & \text{if } n > 3 \text{ is even,} \\ \dfrac{2\ 4\ 6\ldots(n-3)}{1\ 3\ 5\ldots(n-2)}\,2 & \text{if } n \ge 3 \text{ is odd.} \end{cases} \tag{19}$$

Eqn. (19) and the integral in eqn. (16) may be obtained from the following recurrence relation (n = 2,3,...)

$$\int_{-\pi/2}^{\theta_0} \cos^n\theta \, d\theta = \frac{1}{n}\left\{ \sin\theta_0 \cos^{n-1}\theta_0 + (n-1)\int_{-\pi/2}^{\theta_0} \cos^{n-2}\theta \, d\theta \right\}.$$

Three cases will be distinguished when sampling from the distribution of T given E whose distribution function will be denoted by G.

<u>Case 1:</u> y=0. The maximum hypersphere is the β-sphere and G is given by

$$G(t) = \begin{cases} 0 & \text{if } 0 \le t \le R^2, \\ [H_n(t) - H_n(R^2)]/[1 - H_n(R^2)] & \text{if } t > R^2, \end{cases}$$

where H_n stands for the central chi-square distribution function with n degrees of freedom. Samples of T can be obtained by the inverse method by observing that

$$G^{-1}(u) = H_n^{-1}(u + (1-u) H_n(R^2)), \quad 0 < u < 1.$$

Case 2: $0<y<R$. The co-ordinate origin lies in the maximum hypersphere but does not coincide with its centre. Using eqns. (15) and (16) it can be seen that

$$P(T > t_2 \mid E) = [1 - H_n(t_2)]/P(E) = \gamma, \tag{20}$$

say, and

$$P(t_1 < T \leq t_2 \mid E) = 1 - \gamma.$$

The distribution function G can be represented as a convex combination of distribution functions G_1 and G_2 as follows

$$G(t) = \gamma G_1(t) + (1 - \gamma) G_2(t), \tag{21}$$

with

$$G_1(t) = \int_{-\infty}^{t} I_{(t_2,\infty)}(s)\, \chi_n^2(s)/[\gamma P(E)]\, ds,$$

$$G_2(t) = \int_{-\infty}^{t} I_{(t_1,t_2]}(s)\, \chi_n^2(s)\, L_n(s;y,R)/[(1 - \gamma) P(E)]\, ds. \tag{22}$$

Since γ is known from eqn. (20), samples according to G are obtained from independent samples ξ_1 and ξ_2 drawn from G_1 and G_2 respectively when they are selected independently with probability γ and $(1-\gamma)$ respectively, i.e., if η is the realization of an independent $B(1,\gamma)$-variable then the distribution function of $\eta \xi_1 + (1-\eta) \xi_2$ is G. ξ_1 can be generated by the inverse method by observing that

$$G_1(t) = [(H_n(t) - H_n(t_2))/(1 - H_n(t_2))]\, I_{(t_2,\infty)}(t)$$

and hence

$$G_1^{-1}(u) = H_n^{-1}(u + (1-u) H_n(t_2)).$$

The sample ξ_2 is obtained by the rejection technique [3] as follows. Independent samples u_1 and u_2 are drawn from uniform distributions on $[t_1,t_2]$ and $[0,m]$ respectively, where m is defined by

$$m = \max \{ \chi_n^2(t)\, L_n(t;y,R) : t_1 \leq t \leq t_2 \}.$$

If

$$u_2 < \chi_n^2(u_1)\, L_n(u_1;y,R),$$

then $\xi_2 = u_1$ is accepted as a sample from G_2 but sampling of u_1 and u_2 is repeated otherwise. The acceptance ratio C of this sampling procedure is given by

$$C = \int_{t_1}^{t_2} \chi_n^2(t)\, L_n(t;y,R)\, dt/[m\,(t_2 - t_1)]$$

$$= (1 - \gamma)\, P(E)/(4yRm) = (P(E) + H_n(t_2) - 1)/(4yRm).$$

The expected number K of independent rectangular samples necessary in order to generate one sample of T is

$$K = 1 + \gamma + 2(1 - \gamma)/C = 1 + (1 + 8yRm - H_n(t_2))/P(E).$$

Case 3: $y \geq R$. The maximum hypersphere does not contain the co-ordinate origin. This unusual case will occur very seldom and it signifies that a level 2 reliability method is not an appropriate one to be used. Nevertheless, the present Monte Carlo method might be useful to improve upon the level 2 result considerably. The sampling procedure is similar to the previous case and the formulae have to be modified as follows. Eqn. (21) is valid with

$$\gamma = P(\,T \leq t_1 \text{ or } T > t_2 \mid E\,) = (1 + H_n(t_1) - H_n(t_2))/P(E),$$

$$G_1(t) = \int_{-\infty}^{t} I_{[0,t_1] \cup (t_2,\infty)}(s)\, \chi_n^2(s)/[\gamma P(E)]\, ds$$

and G_2 is again defined by eqn. (22). ξ_1 will be generated by the inverse method by observing that

$$G_1(t) = \begin{cases} \dfrac{H_n(t)}{1 + H_n(t_1) - H_n(t_2)} & \text{if } t \leq t_1, \\[2ex] \dfrac{H_n(t_1)}{1 + H_n(t_1) - H_n(t_2)} & \text{if } t_1 < t \leq t_2, \\[2ex] \dfrac{H_n(t) + H_n(t_1) - H_n(t_2)}{1 + H_n(t_1) - H_n(t_2)} & \text{if } t > t_2, \end{cases}$$

and hence

$$G_1^{-1}(u) = \begin{cases} H_n^{-1}(\, u\,[1 + H_n(t_1) - H_n(t_2)]\,) & \text{if } 0 < u < u_0, \\ H_n^{-1}(\, u + (1+u)\,[H_n(t_2) - H_n(t_1)]\,) & \text{if } u_0 \leq u < 1, \end{cases}$$

where

$$u_0 = H_n(t_1)/[1 + H_n(t_1) - H_n(t_2)].$$

Sampling the distribution G_2 is identical to the previous case except that the acceptance ratio C now becomes

$$C = [P(E) + H_n(t_2) - H_n(t_1) - 1]/(4yRm).$$

The expected number of independent rectangular samples needed to obtain one sample of T is now

$$K = 1 + [1 + H_n(t_1) + 8yRm - H_n(t_2)]/P(E).$$

C and K are measures of the effectiveness of the above sampling procedure. Because of the rather complicated structure of the density of G_2 it cannot be expected to find a transformation of the random radius in a closed form which would be analogous to the one that significantly increases the number of successful samples when the β-sphere is excluded from sampling [2]. Nevertheless, the limit state function need not be calculated at the sampling stage, and therefore in cases where it is time-consuming to evaluate g the total computing time will still be dominated by the time spent in the limit state calculations even though a certain proportion of the random numbers drawn to generate samples of T will have to be discarded. The present procedure has the advantage of a reduced average number of limit state function evaluations as implied by eqn. (6).

3.2.2 Sampling the angles $\Theta_1,\ldots,\Theta_{n-1}$

Let

$$P_{T=t}(\ \underline{\Theta}\ \varepsilon\ .\ |\ E\ ) \tag{23}$$

denote the conditional distribution of $\underline{\Theta}$ given T=t conditioned on the event E. For a given measurable subset A of the (n-1)-dimensional space the above conditional probability is thus obtained as the Radon-Nikodym derivative of a measure ν with respect to a probability measure μ [19], i.e.,

$$\nu(D) = \int_D P_{T=t}(\ \underline{\Theta}\ \varepsilon\ A\ |\ E\ )\ \mu(dt), \tag{24}$$

where ν and μ are defined for Borel subsets of the real line by

$$\nu(D) = P(\ T\ \varepsilon\ D,\ \underline{\Theta}\ \varepsilon\ A\ |\ E\ ), \tag{25}$$

$$\mu(D) = P(\ T\ \varepsilon\ D\ |\ E\ ).$$

Assuming A as an (n-1)-fold Cartesian product, $A=A_1\times\ldots\times A_{n-1}$, and using the conditional density of T given E (eqn. (15)) and the (unconditional) independence of

$T, \Theta_1, \ldots, \Theta_{n-1}$, eqn. (25) may be evaluated as follows

$$\nu(D) = P(\{T \in D\} \cap \bigcap_{i=1}^{n-1} \{\underline{\Theta}_i \in A_i\} \cap \{T - 2T^{1/2}y \sin \Theta_1 + y^2 > R^2\})/P(E)$$

$$= \int_D P(\Theta_1 \in A_1,\ t - 2t^{1/2}y \sin \Theta_1 + y^2 > R^2)P(T \in dt) \frac{\prod_{i=2}^{n-1} P(\Theta_i \in A_i)}{P(E)}$$

$$= \int_D \frac{P(\Theta_1 \in A_1,\ t - 2t^{1/2}y \sin \Theta_1 + y^2 > R^2)}{L_n(t;y,R)} \frac{L_n(t;y,R)\ \chi_n^2(t)}{P(E)} dt \prod_{i=2}^{n-1} P(\Theta_i \in A_i)$$

$$= \int_D \frac{P(\Theta_1 \in A_1,\ t - 2t^{1/2}y \sin \Theta_1 + y^2 > R^2)}{L_n(t;y,R)} \prod_{i=2}^{n-1} P(\Theta_i \in A_i)\ P(T \in dt \mid E). \tag{26}$$

Comparing eqns. (24) and (26) shows that the conditional probability in eqn. (23) is a product measure whose marginal distributions coincide with the respective (unconditional) marginal distributions of $\underline{\Theta}$ except for the first one which has the distribution function

$$F_n(s;t,y,R) = P(\Theta_1 \le s \mid t - 2t^{1/2}y \sin \Theta_1 + y^2 > R^2). \tag{27}$$

Several cases have to be distinguished when sampling from $F_n(\cdot;t,y,R)$. The <u>case $n=2$</u> will be examined first. If $y=0$ then, of course, $R \geq y$ and hence for L_2 to be non-zero $t > t_1$ $(=t_2=R^2)$ may be assumed (see eqn. (16)). Thus, the conditioning event in eqn. (27) is the full event and therefore F_2 is the unconditional distribution function of Θ_1 which is uniformly distributed on $[0,2\pi]$. If $y>0$ then the conditioning event in eqn. (27) is equal to $\{\sin \Theta_1 < \zeta\}$ with ζ defined by eqn. (18). $\zeta \le -1$ cannot occur since it would imply $t \le t_1$ and $R>y$ and hence from eqn. (16) $L_2=0$. The conditional distribution of Θ_1 is therefore uniform on a set A_0 with

$$A_0 = \begin{cases} [0,\ 2\pi] & \text{if } \zeta > 1, \\ [0,\ \arcsin \zeta] \cup [\pi - \arcsin \zeta,\ 2\pi] & \text{if } 0 < \zeta \le 1, \\ [\pi - \arcsin \zeta,\ 2\pi + \arcsin \zeta] & \text{if } -1 < \zeta \le 0. \end{cases}$$

<u>If $n \geq 3$</u> then the unconditional distribution of Θ_1 is concentrated on $[-\pi/2,\pi/2]$ with density $c_n \cos^{n-2}\theta$ (see eqn. (19)). $y=0$ again implies that F_n is equal to the unconditional distribution function of Θ_1. Equally, $y>0$ and $\zeta>1$ implies the same, while $\zeta \le -1$ can be excluded if $y>0$ for reasons already explained above. Hence, for $y>0$ and s between $-\pi/2$ and θ_0 (defined by eqn. (17)) we have

$$F_n(s;t,y,R) = \frac{P(\Theta_1 \le s)}{P(\Theta_1 \le \Theta_0)} = \int_{-\pi/2}^{s} \cos^{n-2}\theta\, d\theta \Big/ \int_{-\pi/2}^{\theta_0} \cos^{n-2}\theta\, d\theta.$$

Samples from F_n may easily be obtained by the inverse method by applying the Newton-Raphson scheme which will converge for the starting point $s_1=\min\{0,\theta_0\}$ since F_n is convex for $s\le s_1$ and concave for $s\ge s_1$.

Finally, samples of Θ_i for $n\ge 3$ $(i=2,\ldots,n-1)$ are obtained as follows. Θ_{n-1} is uniformly distributed on $[0,2\pi]$. Θ_i $(2\le i\le n-2)$ has a density which is proportional to $\cos^{n-1-i}\theta$ on $[-\pi/2,\pi/2]$. The inverse method is again applicable with the origin as the starting point for the Newton-Raphson iteration.

4. AN EXAMPLE: FAILURE PROBABILITY OF A PLANE FRAME

A general FORTRAN 77 program has been written and applied to accurately calculate the failure probability of a plastic frame when a certain failure mechanism is assumed. This example can be found in [20]; the frame is shown on Fig. 2 and the distribution assumptions are summarized in Table 1. All basic variables are assumed to be independent. Using the principle of virtual displacements it is seen that on the failure surface the plasic moments $M_1,\ldots,M_4$, and the forces H and V satisfy the equation

$$H\,h + V\,h = M_1 + 2\,M_2 + 2\,M_3 + M_4.$$

The structure is therefore safe if

$$g(H,V,M_1,\ldots,M_4) = M_1 + 2\,(M_2 + M_3) + M_4 - h\,(H + V) > 0.$$

Tvedt [21] analysed this structure and estimated by numerical integration the probability of failure to be between 1.215×10^{-2} and 1.233×10^{-2}. The level 2 estimate is 9.433×10^{-3}.

Results obtained by the present method are shown in Tables 2 and 3. Table 2 shows the results of several optimization cycles. Starting from the β-sphere with probability content 0.520, the first optimization step gives a hypersphere whose probability content is 0.806 whereas two more optimization steps resulted in a hypersphere with probability content 0.947. Radius and distance from the co-ordinate origin are increasing accordingly. Accepting 1.22×10^{-2} to be the "exact" estimate for the failure probability we realize that for, say, N=100 limit state function evaluations the respective relative errors of the estimates are 57%, 20%, and 16% if the β-sphere or the maximum hyperspheres are excluded from sampling. The corresponding figures for N=10,000 limit state function evaluations are 7.2%, 2.7%, and 2.2%. This shows that engaging in several optimization cycles beyond the initial one may not significantly increase the accuracy of the estimate for any given number of limit state function evaluations. It also demonstrates that performing one optimization step improves the estimate considerably for a fixed number of limit state function evaluations N.

Values of C and K indicate that many more rectangular samples have to be discarded in the course of the sampling procedure for the random radius ρ as the (probability content of the) maximum hypersphere increases. C an K can be evaluated prior to the actual sampling procedure in order to decide which hypersphere to choose for the sampling procedure for minimum CPU time.

The program for the above example uses the NAG program library [17] and is capable of analysing problems with a wide range of assumptions. It is available from the author upon request.

5. CLOSING REMARKS

A Monte Carlo method has been presented which is based on the exclusion from sampling of a maximum hypersphere in the standard normal space of basic variables. The number of limit state function evaluations needed to achieve a given accuracy of the estimate of the failure probability by this method is orders of magnitude smaller than that required when the β-sphere is excluded from sampling. A plastic frame example has been given demonstrating the merits of the technique. Futher possibilities to extend the present theory have been briefly discussed.

REFERENCES

[1] Thoft-Christensen, P. and Baker, M. J.: Structural Reliability Theory and Its Applications. Springer-Verlag, Berlin, Heidelberg, New York (1982).

[2] Harbitz, A.: An Efficient Sampling Method for Probability of Failure Calculation. Structural Safety 3, 109-115 (1986).

[3] Hammersley, J. M. and Handscomb, D. C.: Monte Carlo Methods. Chapman and Hall, London (1979).

[4] Ouypornprasert, W. and Schuëller, G. I.: Discussion to: Harbitz, A., An Efficient Sampling Method for Probability of Failure Calculation. Structural Safety 4, 311-312 (1987).

[5] Rackwitz, R.: Discussion to: Harbitz, A., An Efficient Sampling Method for Probability of Failure Calculation. Structural Safety 4, 313-314 (1987).

[6] Melchers, R. E.: Monte Carlo Methods for Structural Reliability Analyses. Proceedings of The Tenth Australasian Conference on the Mechanics of Structures and Materials, The University of Adelaide, 20-22 August 1986, 7-12, Adelaide (1986).

[7] Schuëller, G. I. and Stix, R.: A Critical Appraisal of Methods to Determine Failure Probabilities. Structural Safety 4, 293-309 (1987).

[8] Harbitz, A.: Efficient and Accurate Probability of Failure Calculation by Use of the Importance Sampling Technique. Proceedings of the Fourth International Conference on Applications of Statistics and Probability in Soil and Structural Engineering, Universita di Firenze (Italy), 13-17 June 1983, 825-836, Pitagora Editrice, Bologna (1983).

[9] Bucher, C. G.: Adaptive Sampling - An Iterative Fast Monte Carlo Procedure. Structural Safety 5, 119-126 (1988).

[10] Siddal, J. N.: Optimal Engineering Design. Marcel Dekker, New York, Basel (1982).

[11] Rice, S. O.: Distribution of Quadratic Forms in Normal Random Variables - Evaluation by Numerical Integration. SIAM J. Sci. Stat. Comput. 1, 438-448 (1981).

[12] Fiessler, B., Neumann, H.-J. and Rackwitz, R.: Quadratic Limit States in Structural Reliability. J. Eng. Mech. Div., ASCE, Proc. Paper 14739, 105(EM4), 661-676 (1979).

[13] Narula, S. C. and Desu, M. M.: Computation of Probability and Non-centrality Parameter of a Non-central Chi-squared Distribution. Algorithm AS170, Appl. Statist. 30, 349-352 (1981).

[14] Han, C.-P.: Some Relationships between Noncentral Chi-Squared and Normal Distributions. Biometrika 62, 213-214 (1975).

[15] Rosenblueth, E.: On Computing Normal Reliabilities. Structural Safety 2, 165-167 (1985); Errata in Structural Safety 3, 67 (1985).

[16] Hart, J. F., Cheney, E. W., Lawson, C. L., Maehly, H. J., Mesztenyi, C. K., Rice, J. R., Thacher, H. G. and Witzgall, C.: Computer Approximations. Robert E. Krieger, Huntington and New York (1978).

[17] Numerical Algorithms Group (NAG): FORTRAN Library Manual, Mark 11. Oxford (1984).

[18] Kendall, M. G.: A Course in the Geometry of n Dimensions, Griffin's Statistical Monographs & Courses. Charles Griffin, London (1961).

[19] Rényi, A.: Probability Theory. North-Holland, Amsterdam, London (1970).

[20] Madsen, H. O., Krenk, S. and Lind, N. C.: Methods of Structural Safety. Prentice-Hall, Englewood Cliffs, New Jersey (1986).

[21] Tvedt, L.: Two Second-Order Approximations to the Failure Probability. Technical Report, A/S Veritas Research, Hovik, Norway (1986).

[22] Abramowitz, M. and Stegun, I. A. (Eds.): Handbook of Mathematical Functions. Dover Publishers, New York (1965).

[23] Leon, S.J.: Linear Algebra with Applications. Macmillan, New York (1980).

APPENDICES

Appendix 1: The standard normal probability content of an offset hypersphere in $n \geq 2$ dimensions

The standard normal probability content of a hypershere $S(\underline{a};R)$ defined by eqn. (8) can be written as

$$P(\ \underline{Z}\ \varepsilon\ S(\underline{a};R)\ ) = P(\ \underline{Z}\ \varepsilon\ S(\underline{c};R)\ ), \tag{28}$$

where $\underline{c}$ denotes the n-dimensional vector whose only non-zero entry is the last component which is equal to the Euclidean norm of $\underline{a}$. The explicit form of a suitable orthogonal transformation justifying eqn. (28) is immaterial for our present purposes but it is nevertheless given in Appendix 2 because it will be used in the sampling procedure. Eqn. (28) can simply be viewed as a consequence of the invariance of the standard normal density under orthogonal transformations. The above probability can therefore be written as

$$P(\underline{Z} \in S(\underline{c};R)) = \int\limits_{||\underline{u}|| \le R} \cdots \int (2\pi)^{-n/2} \exp(-||\underline{u}+\underline{c}||^2/2)\, d(u_1,\ldots,u_n)$$

$$= \int\limits_{||\underline{u}|| \le R} \cdots \int (2\pi)^{-n/2} \exp\{-(||\underline{u}||^2+2u_n||\underline{a}||+||\underline{a}||^2)/2\}\, d(u_1,\ldots,u_n)$$

$$= \int_0^R \underbrace{\int_{-\pi/2}^{\pi/2} \cdots \int_{-\pi/2}^{\pi/2}}_{((n-2)\text{ times})} \int_0^{2\pi} (2\pi)^{-n/2} \exp\{-(r^2+2r||\underline{a}||\sin\theta_1 + ||\underline{a}||^2)/2\}\, r^{n-1} \times$$

$$\times \prod_{i=1}^{n-2} \cos^{n-1-i}\theta_i \, d\theta_{n-1} \ldots d\theta_1\, dr, \tag{29}$$

where the last step is by transition to polar co-ordinates [18]. Notice that the domains of integration for θ_1 in eqn. (29) are from 0 to 2π and from $-\pi/2$ to $\pi/2$ for $n=2$ and $n\geq 3$ respectively. Eqn. (29) thus can be rewritten as

$$\varepsilon_n \alpha_n \int_{x_0}^{x_1} J_n(\theta) \cos^{n-2}\theta \, d\theta, \tag{30}$$

where

$$x_0 = \begin{cases} 0 & \text{if } n = 2 \\ -\pi/2 & \text{if } n \geq 3, \end{cases}$$

$$x_1 = \begin{cases} 2\pi & \text{if } n = 2 \\ \pi/2 & \text{if } n \geq 3, \end{cases}$$

$$\varepsilon_n = \begin{cases} 1 & \text{if } n = 2 \\ 2\pi & \text{if } n \geq 3, \end{cases}$$

$$\alpha_n = (2\pi)^{-n/2} \prod_{i=2}^{n-2} \int_{-\pi/2}^{\pi/2} \cos^{n-1-i}\theta \, d\theta, \tag{31}$$

$$J_n(\theta) = \int_0^R r^{n-1} \exp\{-(r^2+2r||\underline{a}|| \sin\theta + ||\underline{a}||^2)/2\}\, dr.$$

(For $n = 2,3$ the product in eqn. (31) is understood to be unity). It can be shown that

$$\varepsilon_n \alpha_n = \begin{cases} (2\pi)^{-1} & \text{if } n = 2, \\ 2^{(n/2)-1}\ \pi^{-1}\ (n/2 - 1)!/(n - 2)! & \text{if } n \geq 4 \text{ is even}, \\ 2^{1-n/2}\ \pi^{-1/2}/((n-3)/2)! & \text{if } n \text{ is odd}. \end{cases}$$

Finally, integration by parts shows that the following recurrence relation holds true for $J_n(\theta)$

$$\left.\begin{aligned} J_2(\theta) &= a_0 + b_0 + c_0\ J_1(\theta), \\ J_n(\theta) &= (n-2)\ J_{n-2}(\theta) + c_0\ J_{n-1}(\theta) + b_0\ R^{n-2}, \quad n \geq 3, \end{aligned}\right\} \quad (32)$$

with

$$a_0 = \exp(\ -||\underline{a}||^2/2\ ),$$

$$b_0 = -\ \exp\left\{\ -(R^2 + 2R\ ||\underline{a}||\ \sin\theta + ||\underline{a}||^2)/2\ \right\},$$

$$c_0 = -\ ||\underline{a}||\ \sin\theta.$$

In order to evaluate $J_n(\theta)$ from eqn. (32) only $J_1(\theta)$ must be known which can be written as

$$J_1(\theta) = \sqrt{(2\pi)}\exp\left\{-(||\underline{a}||^2\cos^2\theta)/2\right\}[\Phi(R+||\underline{a}||\sin\theta) - \Phi(||\underline{a}||\sin\theta)]. \quad (33)$$

Taking in eqn. (33) a high-precision approximation for the standard normal distribution function Φ [15,16] enables us to obtain J_1 and hence J_n by means of the recurrence relation eqn. (32) without numerical integration. The calculation of the n-dimensional integral in eqn. (28) is therefore reduced to a simple one-dimensional integration. It should be noted that a simpler approximation of Φ which is usually sufficiently accurate for standard reliability calculations [22] proved inadequate for the purposes of this probability evaluation.

Appendix 2: Construction of the matrix $\underline{M}$

An orthogonal matrix $\underline{M}$ satisfying eqn. (12) is required. If $\underline{a}$ is non-zero (which may be assumed since otherwise the identity matrix is suitable) then the Gram-Schmidt orthogonalisation process [23] can be applied as follows. The basis to be orthogonalised is obtained from the standard basis of the n-dimensional space by replacing its i-th element by $\underline{a}$ with $i \in \{1,\dots,n\}$ chosen such that $a_i \neq 0$. Denoting the elements of this basis by $\underline{p}_1,\dots,\underline{p}_{n-1},\underline{a}$ the column vectors of $\underline{M}$ are defined by the following recurrence relation

$$\underline{m}_n = ||\underline{a}||^{-1}\ \underline{a},$$

$$\underline{m}_j = ||\underline{u}_j||^{-1}\ \underline{u}_j, \quad j = n-1,\ n-2, \ldots, 1,$$

where

$$\underline{u}_j = \underline{p}_j - \sum_{k=j+1}^{n} (\underline{p}_j^T\ \underline{m}_k)\ \underline{m}_k.$$

FIGURES

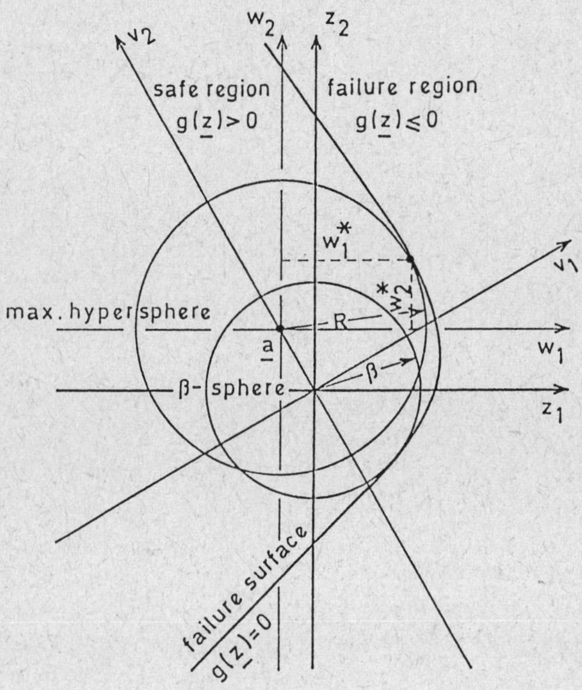

Fig. 1: The maximum hypersphere.

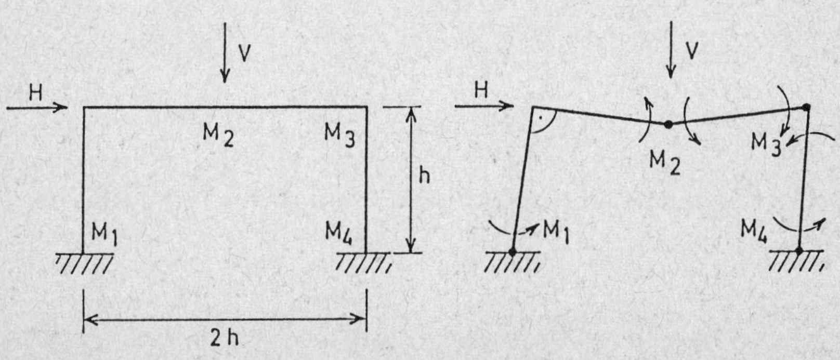

Fig. 2: Plane frame and its assumed plastic collapse mechanism.

Variables	Symbol	Basic variable	Distribution	Unit	Mean	St. Dev.
Load variables	H	X1	lognormal	kN	50.0	15.0
	V	X2	lognormal	kN	40.0	12.0
Resistance variables	M1,...,M4	X3,...,X6	lognormal	kNm	120.0	12.0
Other variable (Frame height)	h	–	deterministic	m	5.0	0.0

Table 1: Data for plastic frame example

	β-sphere	Maximum hypersphere 1	Maximum hypersphere 2
Centre	(0,...,0)	(-0.4793, -0.4793, 0.1653, 0.3284, 0.3284, 0.1653)	(-1.9170, -1.3871, 0.2305, 0.4552, 0.4552, 0.2305)
Radius R	2.3482	3.1135	4.8107
Probability	0.52022	0.80566	0.94747
y	0.00000	0.85429	2.4738
Y	–	0.78216 E-1	0.21685 E-7
C	–	0.58736	0.16693
K	–	4.2170	12.981
m	–	0.28667 E-1	0.66112 E-2

Table 2: Results of the optimization cycles

Number of g-evaluations N	Number of samples in failure region			100 × $\hat{p}(N)$			Sample values of K	
	β-sphere	Max.Sph.1	Max.Sph.2	β-sphere	Max.Sph.1	Max.Sph.2	Max.Sph.1	Max.Sph.2
10	1	1	3	4.7978	1.9434	1.5760	3.500	18.00
50	3	4	14	2.8787	1.5547	1.4709	4.220	11.48
100	4	5	27	1.9191	0.9717	1.4184	4.050	13.04
500	12	34	107	1.1515	1.3215	1.1242	4.180	13.42
1,000	18	68	223	0.8636	1.3215	1.1715	4.271	13.13
5,000	122	319	1,136	1.1707	1.2399	1.1936	4.220	12.81
10,000	236	645	2,272	1.1323	1.2535	1.1936	4.197	12.95
50,000	1,213	3,218	11,636	1.1639	1.2508	1.2226	4.223	12.92
100,000	2,498	6,318	23,043	1.1985	1.2279	1.2105	4.222	12.94

Table 3: Results of the sampling procedures; (Note: Level 2 failure probability is 9.433×10^{-3}.)

ON SPATIAL CORRELATION OF NODAL WAVE FORCES IN SYSTEM RELIABILITY ANALYSIS OF OFFSHORE STRUCTURES

Rabi S. De*, Ashish Karamchandani**, Peter Bjerager*** and C. Allin Cornell**

*Department of Naval Architecture & Offshore Engineering
University of California, Berkeley, California 94720, USA

**Department of Civil Engineering
Stanford University, Stanford, California 94305, USA

***A.S. Veritas, Oslo, Norway
(Formerly, Department of Civil Engineering
Stanford University, Stanford, California 94305, USA)

ABSTRACT

The current practice of wave load modeling for failure path based system reliability analysis consists of using a deterministic wave load pattern (corresponding to the position of a regular design wave with respect to the structure when the base shear is maximum) scaled by the total lateral load (base shear) which is considered random. However, local variations duc to uneven marine growth and uncertainties in modeling wave forces result in less than perfect correlation among the nodal wave forces.

In the present work, an analytical form is assumed for the spatial distribution of correlation between nodal wave forces. The parameters of the model include a base (non-ergodic level) correlation and rates of decay in each of the three dimensions in space. A framework is developed for the probabilistic description of nodal wave forces, consistent with the prescribed correlation structure and the available information on the base shear distribution. The structure analyzed is an eight-leg steel offshore jacket in 140 feet of water modeled as a space truss. The postfailure behavior of truss members was modeled by semibrittle elements. The system failure (according to a specified criterion) is considered for a number of different member-failure sequences, known as failure paths.

A parametric study is carried out to investigate the effects of variation of the correlation model parameters on both the member level and system level failure probabilities including changes in potential failure sequences.

Finally we calibrate our correlation model based on the published wave loading results from the Ocean Test Structure.

INTRODUCTION AND BACKGROUND

The deterministic approach of calculating wave loadings on fixed offshore jackets in shallow to moderate water depths involves the following steps:

i) selecting a design wave height and period (usually with 100 year mean return period) through extreme value analysis;
ii) selecting an appropriate regular wave theory (e.g., Airy wave, Stokes Vth order, etc.) to calculate fluid kinematics for the design wave;
iii) calculating the wave forces on the structure from the calculated fluid kinematics using the Morison equation with appropriate values of drag and inertia coefficients, C_D and C_M.

A number of wave load conditions are generated corresponding to the various positions of the design wave relative to the structure, and the structural members are designed to withstand them.

The wave forces acting on an offshore structure can truly be represented only by a random field varying both in time and space. In addition to the inherent randomness of the waves, considerable uncertainties are introduced in the reliability problem due to modeling errors. When modeled by a regular wave theory uncertainty in fluid kinematics may be due to the irregularity, directional spread, degree of nonlinearity of the incident wave pattern, and because of the shielding effects due to the presence of the structural members. A second set of uncertainties are involved in using a standard set of drag and inertia coefficients in force prediction due to changes in local flow parameters and uneven marine growth.

Due to limitations in our capability of analyzing nonlinear structural behavior and subsequent probability computations, a simplified probabilistic description of the wave loading is typically used in current system reliability analysis of offshore structures [1, 2, 3, 4 and 9]. The entire wave load pattern is considered deterministic while its magnitude is considered random. In the simplest case, the system reliability is evaluated under the static monotonic application of the load pattern scaled by a randomly varying magnitude. Multiple system reliability analyses can be carried out for a number of different deterministic load patterns corresponding to various positions of the design wave relative to the structure to take into account the fact that maximum load effects on different members are reached at different positions of the wave [3]. Such analysis considers the load pattern to be functionally independent of wave height. Alternatively, if it is assumed that for most of the critical members the maximum load effects are reached when the base shear (i.e., the total lateral load) is maximum, the system reliability can be evaluated using the load pattern corresponding to the position of the design wave producing maximum base shear. The variation of this wave load pattern with wave height can then be taken into account by evaluating the system reliability for different wave load patterns corresponding to different wave heights, and combining them according to the extreme wave height distribution [2]. While the practice of modeling the entire wave loading by a single random variable produces much desired computational ease, the wave loads acting on different members are far from perfectly correlated due to local randomness and modeling uncertainties (irregular surface wave elevation, local variations in fluid kinematics, drag and inertia coefficients, uneven marine growth, etc.) even though they may arise from the same wave characterized by a given height and period. In response to this Bjerager et al. [4] have modeled the drag and inertia coefficients and the excess thickness due to marine growth as spatial Gaussian white noise processes over the structure. In this work we have assumed a more general spatial correlation model among the nodal wave forces and calibrate the model from available information on spatial correlation among parameters influencing the wave kinematics and subsequent wave force prediction.

WAVE FORCE MODEL

In order to reduce computational effort in this sensitivity study we neglected the variation of mean wave load pattern with wave height. Following the usual practice, [1, 2, 9] we assumed the base shear (total lateral wave load), f, on the jacket structure to have an empirical relationship with the wave height, h, of the form:

$$f = c h^{\alpha} \qquad (1)$$

where, the constants c and α were obtained from the wave load data. For the particular jacket structure under study [Fig. 1, details in Reference 2] the south waves were found to be most critical. Again, as usual, we assumed a probabilistic model of the base shear, F, as follows:

$$F = \Gamma_a \, c \, (H)^{\alpha} \qquad (2)$$

where, F = base shear corresponding to the maximum life-time (20 years) wave from the south.

c, α = constants estimated from the wave load data using Equation 1.

Γ_a= analysis uncertainty factor associated with estimating the force given a wave height (e.g., Morison's equation, wave kinematics, etc.), assumed to have a unit mean and coefficient of variation, $\delta_{\Gamma_a} = 0.25$. The basis for this value of δ_{Γ_a} is the results from the Ocean Test Structure, and is reported in References 10 & 11.

H = height of life-time maximum wave from the south, assumed to have moments determined from extreme value analysis [2], e.g., c.o.v., $\delta_H = 0.20$ [2]

Γ_a and H are assumed to be independent random variables, and the moments of the base shear, F, can be conveniently evaluated if Γ_a and H are assumed lognormal. In any case, a first order approximation for second moment is: $\delta_B^2 = \delta_{\Gamma_a}^2 + \alpha^2 \delta_H^2$, where δ_B, δ_{Γ_a} and δ_H are coefficients of variation of base shear F, uncertainty factor Γ_a and wave height H respectively.

Following our earlier investigation [2] of the same jacket structure, we adopted the wave load pattern corresponding to the wave height (75 ft.) at which the system failure is most likely to occur as the mean nodal wave force pattern. This pattern includes the estimate of wave slamming forces on deck.

The spatial correlation between nodal wave forces was modeled as follows:

$$\rho_{ij} = \rho_c + (1 - \rho_c)\left[\exp\left\{-\left(\frac{x_i - x_j}{r_x}\right)^2 - \left(\frac{y_i - y_j}{r_y}\right)^2 - \left(\frac{z_i - z_j}{r_z}\right)^2\right\}\right] \tag{3}$$

where, ρ_{ij} = correlation between y-direction lateral nodal wave forces at node #i (x_i, y_i, z_i) and node #j (x_j, y_j, z_j)

ρ_c = base level (non ergodic) correlation arising from the fact that all nodal wave forces arise from the same wave characterized by its height and period.

r_x, r_y, r_z = correlation lengths in x, y and z directions respectively.

Lacking definitive spatial correlation data, the model presented here is purely hypothetical in nature, and a different model could have been used for this parametric study. An alternate model could be of the simple exponential decay type instead of the squared exponential form, as above.

It was further assumed that each of these y-direction wave nodal forces has the same coefficient of variation, δ. We also assumed that the wave nodal forces in x, y and z directions at each node, i, are fully correlated, and hence were characterized through a single random variable equal to F_i, the y-direction wave nodal force acting at node, i.

Our approach in the parametric study was to vary the correlation structure while adjusting the c.o.v. of the nodal wave forces in order to keep the c.o.v. of the base shear a constant. The base shear, F, in the y-direction is the sum of the nodal wave forces in the same direction :

$$F = \sum_i F_i$$

then,

$$E(F) = \sum_i E(F_i)$$

and,

$$\mathrm{Var}(F) = \sum_i \mathrm{Var}(F_i) + \sum_{i \neq j}\sum \rho_{ij} \sqrt{\mathrm{Var}(F_i)}\,\sqrt{\mathrm{Var}(F_j)}$$

To maintain the same c.o.v. of the base shear, $\delta_B = \frac{\sqrt{Var(F)}}{E(F)}$, the c.o.v of individual nodal wave forces in y-direction, $\delta = \frac{\sqrt{Var(F_i)}}{E(F_i)}$, can be calculated to be:

$$\delta = \delta_B \frac{\sum_i E(F_i)}{\sqrt{\sum_{i \neq j}\sum \rho_{ij} E(F_i) E(F_j) + \sum_i \{E(F_i)\}^2}} \qquad (4)$$

In the present study the individual nodal forces are considered jointly normal with mean vector **M** and covariance matrix $\mathbf{C} = [\rho_{ij}\, \sigma_i\, \sigma_j\,]$. The adopted deterministic wave force pattern to be used as mean is scaled so that the sum of the lateral wave forces equals to the mean base shear, F. The covariance matrix, C, is calculated using Equations 3 and 4. The transformation from correlated normal space X to the standard normal uncorrelated space U is given by:

$$\mathbf{U} = \mathbf{L}^{-1}\mathbf{D}^{-1}(\mathbf{X}-\mathbf{M}) \qquad (5)$$

or,

$$\mathbf{X} = \mathbf{DLU} + \mathbf{M} \qquad (6)$$

in which $\mathbf{D} = \text{diag}\,[\sigma_i = \delta\, E(F_i)]$ = the diagonal matrix of the standard deviations, and $\mathbf{L}$ = a lower triangular matrix obtained from Cholesky decomposition of the correlation matrix $\mathbf{R} = [\rho_{ij}] = \mathbf{L}\,\mathbf{L}^{-1}$.

If, however, a non-normal distribution function is used to characterize the nodal wave forces in the original X space a two step transformation is recommended [5]. The nodal force can first be mapped into correlated Gaussian space Z, using marginal univariate transformation of the form: $Z_i = \Phi^{-1}[\, F_{X_i}(X_i)]$; i=1,2,3,…. The correlation structure calculated in X space (Equation 3) can also be related to the correlation in the Gaussian space using a power series involving Hermite polynomials [6]. This series can often be truncated with good accuracy after cubic terms, leading to closed form estimation of the underlying Gaussian correlation. In particular, if the distribution functions are lognormal, an exact closed form solution to map correlation is available. Finally, the transformation from Z to U space is similar to the step described by Equation 5.

STRUCTURAL MODEL

The structure analyzed is an eight-leg (four bents by two rows) steel offshore jacket in 140 feet of water (Fig. 1). It was designed for a 63 foot wave with a period of 11.4 seconds, which has a mean return period of 100 years in the Gulf of Mexico. The design is idealized in the sense that it was designed for the inplace condition only. Forces due to loadout, transportation, and installation were not considered in the design. The importance of this should be kept in mind while generalizing specific numerical results of this study.

The jacket was modelled as a space truss. The member capacities were assumed to be normally distributed, with a coefficient of variation of 10% for the tension capacity, and 13% for the compression capacity. The postfailure behavior of truss members was modelled by semibrittle elements with ductile tensile failure, and an instantaneous drop to 40% of the original capacity when failing in compression.

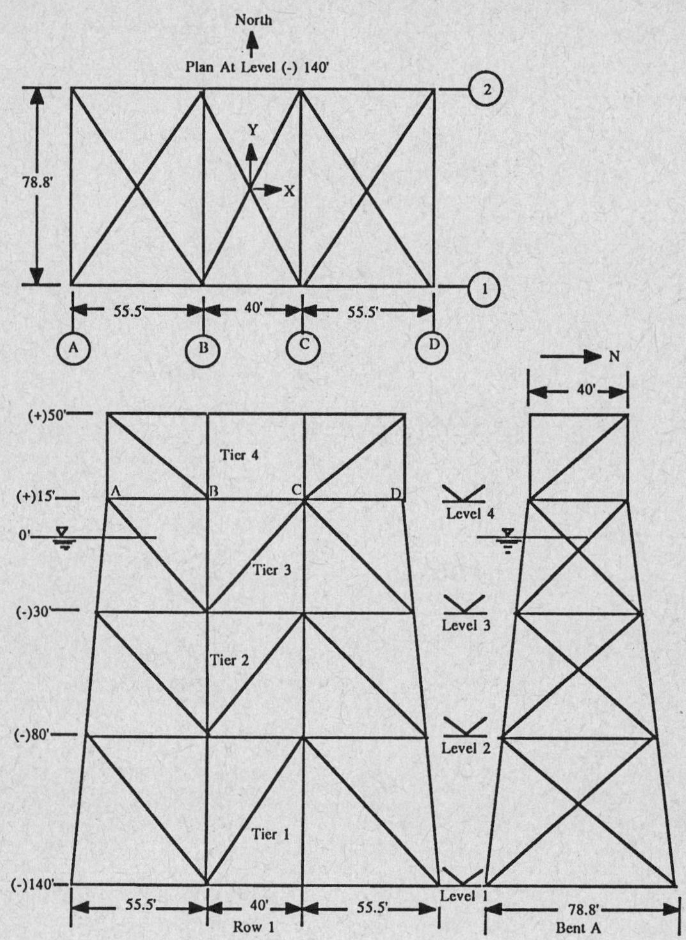

Fig. 1 Offshore Jacket Structure

SYSTEM RELIABILITY FRAMEWORK

System reliability analysis consists of analyzing failure events that are expressed as combination of single component failure events. Failure is defined for each component (truss member in our system) by a performance function describing a limit state. The limit state used for the truss members of the example jacket is the condition that the axial member force is equal to its elastic capacity. When a member fails it offers zero stiffness but continues to carry a constant load equal to its post-failure capacity. After a component fails the structure is reanalyzed to ascertain the forces in the remaining members to establish new component failure events. The criterion of system failure, as used in this study, is a major loss (50% reduction, based on earlier investigation of the same structure, Reference 2) of global stiffness. The stiffness is measured by the increase in deflection at the center of the deck due to an unit increase in load.

A system may fail as a result of a number of different member-failure sequences, known as a 'failure paths'. Each failure path event can be seen as an intersection of individual component failure events leading to the system failure. In this study, the probability of system failure was approximated as the union of six carefully culled, critical failure path events.

PARAMETRIC STUDY

In order to insure that the correlation matrix generated by Equation 3 remains positive definite under all parametric variations, we rewrote the equation as follows:

$$\rho_{ij} = \rho_c + (0.999 - \rho_c)\left[\exp\left\{-\left(\frac{x_i - x_j}{r_x}\right)^2 - \left(\frac{y_i - y_j}{r_y}\right)^2 - \left(\frac{z_i - z_j}{r_z}\right)^2\right\}\right] \qquad (5)$$

Table 1 shows samples of correlation values generated by Equation 5 across different horizontal levels (due to variation of z-coordinate only), and across different vertical bents (due to variation of x-coordinate only). Even though changes in correlation level in y-direction was modeled in Equation 5, we did not expect the changes to be significant, because the jacket dimension is relatively small in y-direction, the forces considered in this study are from a regular wave travelling in y-direction and also because the wave length is large compared to the jacket dimension in y-direction.

r_x (ft.)	r_z (ft.)	$\|x_i-x_j\|$ (ft.)	$\|z_i-z_j\|$ (ft.)	ρ_{ij}		
				ρ_c=0.6	ρ_c=0.4	ρ_c=0.2
-	1000	0	190	0.986	0.979	0.971
-	500	0	190	0.946	0.919	0.919
-	100	0	190	0.611	0.416	0.221
-	50	0	190	0.6	0.4	0.2
-	1	0	190	0.6	0.4	0.4
-	100	0	130	0.674	0.551	0.348
-	100	0	80	0.811	0.716	0.621
-	100	0	35	0.954	0.931	0.908
-	1	0	35	0.6	0.4	0.2
100	-	40	0	0.941	0.911	0.882
100	-	80	0	0.811	0.761	0.621
100	-	120	0	0.694	0.542	0.390
60	-	40	0	0.856	0.785	0.712
60	-	80	0	0.668	0.501	0.335
60	-	120	0	0.607	0.411	0.215

Table 1. Sample Correlations Generated by Equation 5.

For the parametric study we determined the moments of the base shear following an earlier investigation of the same jacket structure [2]. We assumed the constants $\alpha = 1.45$ and $c = 9.23$ (evaluated from an analysis of wave loads near the wave height 75 ft., at which the system failure is most likely to occur, Reference 2) in Equation 2. The mean value of life time maximum wave from the south is 47.1 ft. with coefficient of variation = 0.20. The lifetime maximum base shear mean is 2379 kips with coefficient of variation = 0.40.

The results of the system reliability analyses are presented in terms of the following probability of failures. For each member, the probability that it fails in the intact structure was calculated, and the member with the largest probability is called the most-likely-to-fail (MLM) member. We also computed the probability that any-first (member) failure (AFF) occurs. This is the probability of occurrence of the event corresponding to the union of all (in practice, about six members contributes to the probability of the union) first member failures. Also included are probabilities of occurrence of the most likely to fail sequence (MLS). In most of the cases studied, the same member and the same sequence remained most likely to fail, except in cases involving low levels of correlation among nodal forces. An asterisk (*) sign against the failure probabilities denotes that the most likely to fail member or sequence is different from that in the case of perfectly correlated nodal forces. The system factor is defined as the ratio of probability of system failure and probability of most likely to fail member (MLM). It is an overall measure of system effects including probabilistic and deterministic redundancies of the system. Table 2 summarizes results from system reliability analyses for parametric variations in ρ_c, r_x, r_y and r_z in Equation 5.

Case #	ρ_c	r_x (ft.)	r_y (ft.)	r_z (ft.)	Nodal Force c.o.v.(δ)	P(MLM)	P(AFF)	P(MLS)	P(Sys. Failure)	System Factor
A.1	**1.00**	**1**	**1**	**1**	**0.40**	**9.84e-4**	**1.89e-3**	**2.91e-5**	**5.01e-5**	**0.05**
A.2	0.90	1	1	1	0.42	9.98e-4	1.94e-3	3.10e-5	5.44e-5	0.054
A.3	0.60	1	1	1	0.51	1.07e-3	2.20e-3	4.16e-5	7.88e-5	0.07
A.4	0.40	1	1	1	0.62	1.20e-3	2.63e-3	6.08e-5	1.28e-4	0.11
A.5	0.20	1	1	1	0.85	1.55e-3	4.22e-3	1.71e-4*	3.87e-4	0.25
A.6	0.10	1	1	1	1.13	2.70e-3*	8.31e-3	7.57e-4*	1.36e-3	0.50
A.7	0.00	1	1	1	2.34	4.19e-2*	8.35e-2	2.39e-2*	3.80e-2	0.91
B.I.1	0.60	1000	1000	100	0.43	1.31e-3	2.53e-3	5.33e-3	9.62e-5	0.07
B.I.2	0.40	1000	1000	100	0.44	1.52e-3	2.99e-3	7.41e-5	1.37e-4	0.09
B.I.3	0.02	1000	1000	100	0.46	1.80e-3	3.57e-3	1.04e-4	2.00e-4	0.11
B.II.1	0.90	1000	1000	1	0.416	1.05e-3	2.08e-3	3.50e-5	6.65e-5	0.06
B.II.2	0.60	1000	1000	1	0.48	1.36e-3	2.73e-3	8.65e-5*	2.02e-4	0.15
B.II.3	0.40	1000	1000	1	0.54	1.78e-3	4.81e-3	2.80e-4*	5.38e-4	0.30
B.II.4	0.20	1000	1000	1	0.63	2.65e-3	9.06e-3	1.07e-3*	1.90e-3	0.72
C.I.1	0.60	60	1000	1000	0.44	1.09e-3	2.40e-3	5.51e-5	9.39e-5	0.09
C.I.2	0.40	60	1000	1000	0.47	1.16e-3	2.85e-3	8.00e-5	1.36e-4	0.12
C.I.3	0.20	60	1000	1000	0.505	1.26e-3	3.54e-3	1.22e-4	2.14e-4	0.17
C.II.1	0.79	1	1000	1000	0.44	1.03e-3	2.12e-3	4.09e-5	6.92e-5	0.06
C.II.2	0.60	1	1000	1000	0.49	1.10e-3	2.47e-3	6.18e-5	1.05e-4	0.10
C.II.3	0.40	1	1000	1000	0.56	1.23e-3	3.22e-3	1.13e-3	1.84e-4	0.15
C.II.4	0.20	1	1000	1000	0.69	1.51e-3	5.25e-3	2.83e-4	4.60e-4	0.30
D.1	0.40	1000	20	1000	0.495	1.18e-3	2.26e-3	4.43e-5	7.47e-5	0.06
E.1	0.4	60	1000	100	0.505	1.57e-3	3.87e-3	1.41e-4	2.55e-4	0.16
E.2	0.4	100	1000	100	0.474	1.58e-3	3.57e-3	1.17e-4	2.13e-4	0.14
F.1	0.4	100	100	100	0.48	1.59e-3	3.56e-3	1.17e-4	2.13e-4	0.14
F.2	0.2	100	100	1000	0.525	2.02e-3	4.90e-3	2.13e-4	4.01e-4	0.02

* most likely to fail member or sequence is different from the base case

Table 2. Summary of System Reliability Analyses with Parametric Variation in Spatial Correlation between Nodal Wave Forces

Case A.1 represents the base case where all the nodal wave forces are perfectly correlated, and consequently individual nodal wave forces have the same c.o.v. as that of the base shear. This is the case against which we compare the effects of modeling the correlation. Case A.7 represents the case of the other extreme when the nodal forces are independent, and is only of theoretical interest. Other cases in this set, Cases A.2 through A.6, describe results if the nodal wave nodal forces are equicorrelated ($\rho_{ij} = \rho_c$, for $i \neq j$). This set is of interest because it can easily be incorporated into existing system reliability software with minor modification. Within this set, we observe several trends which are also present within other sets. In general, a reduction in correlation among nodal force results in higher nodal force coefficient of variation, because the base shear coefficient of variation (Equation 4) was kept constant. With the decrease in correlation level (and increase in nodal force c.o.v.) both the member level and system level failure probabilities increase, but not to the same extent, as can be seen by the increase in the system factor. A careful inspection of the sensitivity vectors revealed that for the assumed normal distribution of random variables, the system failure probabilities are more sensitive to the variability in load vectors than are first member failures. For example, in Case A.1 the effective square sensitivity, α_L^2, of the nodal wave forces is 0.52 for the most likely to fail member, whereas for the system failure the load square sensitivity is 0.87. Hence a decrease in correlation level affects the system failure probabilities more even though more load averaging is involved in the failure of multiple components.

Cases B.I.1 through B.II.4 represent correlation variation of nodal forces only with vertical (z-direction) separation of the nodes. Cases B.I.1 through B.I.3 represent the case when the variation is gradual, whereas Cases B.II.1 through B.II.4 represent cases with sharp drop in correlation level (see Table 1) over distance.

Cases C.I.1 through C.II.4 represent correlation variation in x-direction (horizontal direction orthogonal to the direction of wave travel, and the broad side of the structure, Fig. 1) only. The variation is gradual (see Table 1) in Cases C.I.1 through C.I.3, and sharp in Cases C.II.1 through C.II.4.

Only y-direction variation of correlation has been allowed in Case D.1. Comparison with the perfectly correlated case (Case A.1) confirms our earlier intuition that such variations do not affect the results.

In Cases E.1 and E.2, correlation levels vary with both x and z-direction separation distance of the nodes for two different correlation lengths, r_x = 60 and 100. In Cases F.1 and F.2 correlation between nodal forces were allowed to vary with all three (x, y and z) directional separation between nodes. Comparison of Cases F.1 and E.2 again shows that the effect of variation of r_y is insignificant.

Taken together the above results show that the failure probabilities cannot be ranked according to the coefficient of variation of the nodal forces, and the correlation pattern influences the failure probabilities and system factor. The most important conclusion is that for Gaussian variables, the correlation level has to drop appreciably (below ρ_c = 0.6) before it starts affecting significantly the failure probabilities and system effects.

Failure of Sequences

Case #	ρ_c	Sequence 1 (Tier 3 & 4)		Sequence 2 (Tier 4)		P(System Failure)
		P(1st. mem.)	P(seq.)	P(1st. mem.)	P(seq.)	
A.1	1.0	5.01e-4	2.91e-5	1.31e-4	1.16e-5	5.01e-5
A.3	0.6	5.98e-4	4.16e-5	2.08e-4	2.16e-5	7.88e-5
A.5	0.2	12.2e-4	14.3e-5	8.86e-4	17.1e-5	38.7e-5

Table 3. Effect of Correlation of Nodal Forces on Sequence Failure Probability

In order to demonstrate the effect of correlation on the probability of failure of sequences we present in Table 3 the results from equicorrelated Cases A.1, A.3 & A.5. Sequence 1 comprises of members that are primarily in Tier 3 (one member is in Tier 4), whereas Sequence 2 comprises entirely of members in Tier 4, the topmost tier of the jacket. Results presented in Table 3 indicate that the changes in failure probability are larger for Sequence 2, both for the sequence and for the first member in the sequence. The reason for this is as follows: as we decrease the correlation between the nodal forces (keeping the base shear c.o.v. constant), the c.o.v. of the individual nodal forces increases. Correspondingly, the c.o.v. of the internal member forces also increases, but the increase is less because each internal member force is influenced by multiple nodal forces. The larger the number of nodal forces contributing to the internal force in a member, the greater is the averaging effect and the smaller is the increase in the internal force c.o.v. In our structure, the internal force in a member is influenced by all the nodal forces acting on the vertical level of the member as well as on the levels above. Therefore, there are more nodal forces influencing the internal force in a member in Tier 3 than in a member in Tier 4. This leads to a smaller increase in the c.o.v. of the internal force in a member in Tier 3 than in a member in Tier 4. This explains why the increase in the failure probability for Sequence 1 is smaller than that for Sequence 2.

In the perfectly correlated case (Case A.1, $\rho_c = 1.0$), sequence 1 has a higher failure probability than Sequence 2. However, as we decrease the correlation, the increase in failure probability of Sequence 2 is larger than that of Sequence 1. Finally, for Case A.5 ($\rho_c = 0.2$), the failure probability of Sequence 2 is larger than that of Sequence 1. At the system level, the sequences which contribute most significantly to the system failure probability include members from both Tier 3 and Tier 4. Hence, as expected, we find that the percentage increase in the system failure probability is larger than that for a sequence involving members primarily in Tier 3 (e.g., Sequence 1) and smaller than that for a sequence involving members in Tier 4 (e.g., Sequence 2).

As pointed out earlier, the change in failure probability is significant only after a large drop in the correlation (e.g., below $\rho_c = 0.6$). If the sequences involving members in the the topmost tier had dominated the system failure event, the change in the failure probability would have been more sensitive to the correlation level, and a significant increase in failure probability would have been caused by a smaller drop in correlation. Also, in our truss model, the internal member forces are influenced by quite a few nodal forces, and hence, the effect of an increase in the nodal force c.o.v. (due to a decrease in correlation) is diminished because of averaging. In contrast, if we had a frame model, in which members could fail in bending, and we considered the effect of local loads (e.g., vertical loads on members in the splash zone), then there would have been less averaging for some members. If these local force dominated member failures contribute significantly to the system failure event, the system failure probability would be more sensitive to the correlation level.

CALIBRATION

The open question now is what correlation pattern is realistic. In order to investigate if the additional computational effort to model the correlation is justified, we have followed an approximate procedure to calibrate our correlation model from the published results from the Ocean Test Structure [7, 8]. This highly instrumented 20x40x120 ft. platform was installed in 66 ft. water depth in the Gulf of Mexico in November 1976 to evaluate wave force calculation procedure.

The correlation scheme used in our parametric study modeled correlation between the nodal forces because of computation expediency. It is more natural to quantify the spatial correlation between y-direction forces per unit length, S, at any two points on the structural members. Accordingly, we assume a very simplified probabilistic model for the wave force (ignoring inertia force) per unit length at a point of the form:

$$S = c_1 \bar{C}_D f'(z) H^2 \varepsilon_i \phi_i \qquad (6)$$

where, c_1 = a deterministic constant determined from density of water, time period of the wave, etc.

$\bar{C}_D$ = mean coefficient of drag force used in wave force calculation

ε_i = uncertainty in local value of drag coefficient $C_{D_i} = \dfrac{C_{D_i}}{\bar{C}_D}$

ϕ_i = uncertainty in fluid kinematics (velocity square term) $= \dfrac{v^2_{i\,(\text{actual})}}{v^2_{i\,(\text{calculated})}}$

$f'(z)$ = function of depth for the decay of fluid velocity square with depth

H = random wave height.

H, ε_i, and ϕ_i are independent random variables. We assumed that the error terms, ε and ϕ are unbiased; their coefficients of variation were estimated to be: $\delta_\varepsilon = 0.25$ [7] and $\delta_\phi = 0.30$ (for calculations based on Stokes Vth order theory) [8] from the OTS results. We further assumed the spatial correlation of ε and ϕ of the form:

$$\rho_{\varepsilon_i \varepsilon_j} = \exp\left\{ - \left(\frac{x_i - x_j}{r_{x\varepsilon}}\right)^2 - \left(\frac{z_i - z_j}{r_{z\varepsilon}}\right)^2 \right\} \tag{7}$$

$$\rho_{\phi_i \phi_j} = \exp\left\{ - \left(\frac{x_i - x_j}{r_{x\phi}}\right)^2 - \left(\frac{z_i - z_j}{r_{z\phi}}\right)^2 \right\} \tag{8}$$

The above models were calibrated analyzing the limited OTS results reported in References 7 & 8. The x and z direction correlation lengths (r_x and r_z) were assumed to be equal in the absence of any information to the contrary. We estimated the correlation lengths to be: $r_{x\varepsilon} = r_{z\varepsilon} = 48$ ft., and $r_{x\phi} = r_{z\phi} = 50$ ft.

In order to relate the above model for the correlation between local wave forces, S, to the nodal wave forces, F_i, we proceeded as follows. We used the model in Equation 6 to represent the local force per unit length at each point of two adjacent members AB and BC at level 4 (=15 ft. elevation) of the jacket structure under study (Fig. 1). For this larger jacket structure, under study, we used for the mean, c.o.v. and correlation lengths of the error terms ε and ϕ the same numerical values that were estimated for the OTS structure. We obtained the moments of and the correlation between the four member end reactions at nodes A, Band C by integrating the product of the load influence function and the spatial random wave force process across the lengths of the members AB and BC (Fig. 1). Because more than one member frames into one node, one more level of averaging was needed to get the correlation between the total nodal wave forces at nodes A, B, C (and extrapolated to D). We did this approximately by assuming an intermediate model for correlation between the four member end reactions at A B and C (due to members AB and BC) of the form:

$$\rho_{ij} = \rho'_c + (1 - \rho'_c) \left[\exp\left\{ - \left(\frac{x_i - x_j}{r'_x}\right)^2 - \left(\frac{z_i - z_j}{r'_z}\right)^2 \right\} \right] \tag{9}$$

A non-ergodic term ρ'_c is needed because unlike ε and ϕ, the member end reactions arise out of a common random variable, wave height H. We calibrated the above model, i.e., estimated $\rho'_c = 0.51$, and $r'_x = 50$ ft. from the correlations between the four member end reactions calculated before. For the purpose of calibration we assumed that the member end reactions are located at the member third points following the shape of the influence function. The c.o.v. of each of the equal member end reactions were calculated to be equal to 0.574.

Next, we extended the model to include forces from all the members framing into a node. Using the calibrated model in Equation 9, and considering the relative mean magnitude (but with equal c.o.v. = 0.574) of the member end reactions due to different members framing in at nodes A, B, C and D (Fig. 1), we calculated the correlation between total nodal forces at the above nodes, and also the coefficient of variations of the total nodal forces. We finally obtained the following crude estimates:

$\rho_{F_AF_B} = 0.80$, $\rho_{F_AF_C} = 0.58$, $\rho_{F_AF_D} = 0.54$,

and also the c.o.v.s: $\delta_{F_A} = 0.558$, $\delta_{F_B} = 0.558$, $\delta_{F_C} = 0.558$, and $\delta_{F_D} = 0.558$.

For this relatively high level of correlation the c.o.v. of the total nodal forces, at nodes A, B, C and D, were in fact numerically equal up to three significant figures. We assumed the c.o.v. of the total nodal forces at each node of the jacket structure to be equal to 0.56.

Finally we used the above correlations between the total nodal forces to calibrate the model in Equation 3, and obtained: $\rho_c = 0.53$ and $r_x = 55$ ft. We further assumed $r_z = r_x$, and that the variation in y direction is not significant, i.e., $r_y = 1000$ (large). Now, for the calibrated correlation structure and the calculated value of nodal force c.o.v., we calculated the base shear c.o.v., δ_B, to be 0.45.

If in the probabilistic modeling of base shear (by Equation 2) we had used $\alpha = 2$, consistent with the local wave force per unit length modeling (by Equation 6) in our calibration study, we would have obtained for $\delta_{\Gamma_a} = 0.25$ and $\delta_H = 0.20$:

$$\delta_B = \sqrt{\delta^2_{\Gamma_a} + \alpha^2\, \delta^2_H} = 0.47$$

Note that the calculated value of δ_B is in fact smaller than 0.47. This is the result of the greater spatial averaging that takes place over this larger structure compared to the OTS structure which was the basis of using $\delta_{\Gamma_a} = 0.25$. The additional averaging reduces the uncertainty on base shear given the wave height. In fact, the implied value of δ_{Γ_a} for this structure is equal to $\sqrt{0.45^2 - 2^2 \mathrm{x} 0.2^2} = 0.21$ rather than 0.25.

We finally reevaluated the system reliability. The mean nodal wave force pattern and the mean base shear were kept the same as before. The probability of system failure is 6.99 x10^{-4} with the correlation and the nodal force c.o.v. specified by the calibrated model, and is shown as Case G.2 in Table 4. For comparison, we show as Case G.1 the perfectly correlated case in which the base shear, and hence the nodal force, c.o.v. is same as the calibrated case, i.e., = 0.45. Finally the case G.3 corresponds to the perfectly correlated case with base shear c.o.v. = 0.47 (for $\delta_{\Gamma_a} = 0.25$ with $\alpha = 2$).

Case #	ρ_c	r_x (ft.)	r_y (ft.)	r_z (ft.)	Nodal Force c.o.v.(δ)	P(MLM)	P(AFF)	P(MLS)	P(Sys. Failure)	System Factor
G.1	1.00	1	1	1	0.45	1.85e-3	3.62e-3	1.09e-4	1.91e-4	0.10
G.2	**0.53**	**55**	**1000**	**55**	**0.56**	**2.79e-3**	**6.61e-3**	**3.52e-4**	**6.99e-4**	**0.25**
G.3	1.00	1	1	1	0.47	2.33e-3	4.57e-3	1.71e-4	2.99e-4	0.13

Table 4. Summary of System Reliability Analyses with Calibrated Correlation Model

Effect of Modeling by Lognormal Distribution

Lognormal distributions are skewed to the right and have fatter tails compared to normal distributions. Noting that the c.o.v. of the load is higher than that of the capacity, the most obvious effect of modeling load and capacity (resistance) variables by lognormal distributions instead of normal distributions (for the same mean and c.o.v. of the variables) would be an increase of probability of failure both at the member and the system level. Another effect of using lognormal distribution is that the member and system failure probabilities would be more sensitive to the load variability because the effective sensitivity square on the nodal wave loads, α_L^2, is higher than that in the Gaussian case for the same load and resistance c.o.v.[1] In our structure, if we fix the resistances at their mean values then the load at which the system fails is not very much higher than the load at first member failure. Hence, in the lognormal case where load effects dominates (i.e., higher sensitivity) the failure probability of the most likely to fail member (MLM) is expected to be of the same order as the failure probability of the system. This would result in higher system factors for the lognormal case.

---

[1]Consider a simple load and resistance problem in which the mean load and resistance are μ_L and μ_R; the load and resistance c.o.v.s are δ_L and δ_R, and the failure criterion is L > R. If the random variables are normally distributed, then the square of the sensitivity to the load, α_L^2, is $(\mu_L\delta_L)^2/[(\mu_L\delta_L)^2+(\mu_R\delta_R)^2]$ and the square of the sensitivity to the resistance, α_R^2, is $(\mu_R\delta_R)^2/[(\mu_L\delta_L)^2+(\mu_R\delta_R)^2]$. In our case, for the critical members in Tiers 3 & 4, δ_L is larger than δ_R, but μ_R is larger than μ_L. The products $\mu_R\delta_R$ and $\mu_L\delta_L$ are about the same, and hence, $\alpha_R^2 \approx \alpha_L^2$. For the sequence and the system failure events, we have multiple member failures all of which involves the same load, and hence the sensitivity to the load is higher in these cases. If we have lognormal random variables, then $\sigma_{\log L} = \delta_L$ and $\sigma_{\log R} = \delta_R$; the square sensitivity (in transformed log space) to the load, $\alpha_{\log L}^2$, is $(\delta_L)^2/[(\delta_L)^2+(\delta_R)^2]$, and to the resistance, $\alpha_{\log R}^2$, is $(\delta_R)^2/[(\delta_L)^2+(\delta_R)^2]$. Since δ_L is larger than δ_R, $\alpha_{\log L}^2$ is larger than $\alpha_{\log R}^2$. Hence at the member failure level, the square sensitivity to the load is larger for the lognormal case than the Gaussian case. At the system level, the square sensitivity to the load is also larger for the lognormal case than the Gaussian case but the difference is smaller because the square sensitivity is already quite high in the Gaussian case.

Case #	ρ_c	r_x (ft.)	r_y (ft.)	r_z (ft.)	Nodal Force c.o.v.(δ)	P(MLM)	P(AFF)	P(MLS)	P(Sys. Failure)	System Factor
H.1	1.00	1	1	1	0.45	8.24e-3	13.3e-3	3.76e-3	5.36e-3	0.65
H.2	0.53	55	1000	55	0.56	9.72e-3	23.1e-3	5.78e-3	11.5e-3	1.18

Table 5. System Reliability Analyses with Lognormal Model

Results from system reliability analysis using lognormal model is summarized in Table 5. Cases H.1 and H.2 correspond to cases with perfect correlation among the nodal forces and correlation specified by the calibrated model, respectively. These two Cases, H.1 & H.2, correspond to the Gaussian Cases G.1 & G.2 in Table 4 (i.e., with same mean values, coefficients of variation and correlation structure for the random variables) respectively. As expected, in the lognormal case, the probability of failure increased both at the member and the system level compared to the corresponding Gaussian cases for both levels of correlation. In the Gaussian case with perfect correlation between the nodal forces (Case G.1), for the most likely member failure, the square sensitivity to the load, α_L^2, is 0.58 (also, $\alpha_R^2 = 1 - \alpha_L^2 = 0.42$), and for the system failure, the square sensitivity, α_L^2, to the load is 0.89. For the corresponding perfectly correlated lognormal case (Case H.1), the square sensitivity to the load, α_L^2, is 0.90 (also, $\alpha_R^2 = 1 - \alpha_L^2 =$ 0.10), and for the system failure, the square sensitivity to the load is 0.98. These sensitivities compare to the Gaussian as expected. Also, as expected, the system factors are much larger in the lognormal case, i.e., the failure probability of the most likely to fail member is close to the system failure probability.

To study the effects of lognormal modeling without the effects due to the obvious change in the absolute failure probability level, we carried out system reliability analyses for the lognormal model with the parameters adjusted such that the system failure probability for the perfectly correlated lognormal case (Case I.1, Table 6) is same as that of the corresponding Gaussian case (Case G.1, Table 4) i.e., about 1.9×10^{-4}. This was done by scaling down the mean nodal wave forces by approximately 35.5% and keeping all other mean values and the c.o.v.s the same. Case I.2 (Table 6) is the lognormal case with the same adjusted parameters as in Case I.1 and the correlation structure and the nodal force c.o.v. specified by the calibrated model.as in Case G.2 (Table 4). In general, the trends for the adjusted lognormal cases are the same for the as our earlier lognormal cases (Table 5) e.g., large system factors and high sensitivity to the load (in Case I.1, the square sensitivity to the load, α_L^2, is 0.90, and for the system failure, the square sensitivity to the load is 0.98).

Case #	ρ_c	r_x (ft.)	r_y (ft.)	r_z (ft.)	Nodal Force c.o.v.(δ)	P(MLM)	P(AFF)	P(MLS)	P(Sys. Failure)	System Factor
I.1	1.00	1	1	1	0.45	3.78e-4	6.57e-4	1.27e-4	1.93e-4	0.50
I.2	**0.53**	**55**	**1000**	**55**	**0.56**	**6.59e-4***	**1.73e-3**	**3.33e-4***	**7.09e-4**	**1.07**

* most likely to fail member or sequence is different from the base case

Table 6. System Reliability Analyses with Adjusted Lognormal Model

Because the load effect is more dominant in the lognormal case, one would expect to see greater changes in failure probabilities with decreasing correlation between the nodal forces (and the accompanying increase in the nodal force c. o.v.) compared to the Gaussian case. In comparing results from Table 4 and Table 5, we see the contrary. However, if we compare the adjusted lognormal results in Table 6, we see this effect. (Note that in the adjusted lognormal case the most likely to fail member is different for the $\rho_c = 0.53$ case from the $\rho_c = 1.0$ case. If we consider the same member for the $\rho_c = 0.53$ case that is most likely to fail in the $\rho_c = 1.0$ case, the failure probability is 5.9×10^{-4}). Note that even for the first member failure event where there is considerable increase in load sensitivity in the lognormal case compared to the Gaussian case, the failure probability does not increase significantly with decreasing correlation. This is the result of spatial load averaging involved in the internal member forces (in our truss model) which reduces the impact of the change in the nodal force c.o.v. arising from the change in correlation level. This is further explored below.

Case #	Distrb. Type	ρ_c	Sequence 1 (Tier 3 & 4) P(1st. mem.)	Sequence 1 (Tier 3 & 4) P(seq.)	Sequence 2 (Tier 4) P(1st. mem.)	Sequence 2 (Tier 4) P(seq.)	P(System Failure)
G.1	Normal	1.00	9.81e-4	1.09e-4	2.91e-4	0.49e-4	1.91e-4
G.2	Normal	0.53	21.7e-4	3.52e-4	9.42e-4	2.31e-4	6.99e-4
I.1	Lgnrm.	1.00	2.30e-4	1.27e-4	1.11e-4	0.83e-4	1.93e-4
I.2	Lgnrm.	0.53	6.59e-4	3.17e-4	5.05e-4	3.33e-4	7.09e-4

Table 7 Effect of Distribution Type on Sequence Failure Probability

Table 7 contains failure probabilities for two critical sequences, at two correlation levels for both the Gaussian and the adjusted lognormal model. Especially at the first-member failure, the square sensitivity to the load is much higher in the lognormal case, and hence a decrease in correlation (with the accompanying increase in the nodal force c.o.v.) causes a greater change in failure probability than in the Gaussian case. This is seen for the first member failures in both Sequences 1 & 2. Because the load sensitivity to the system failure is similar for Gaussian and lognormal models, the changes in the system failure probability (due to a similar change in the correlation level) are of same order for both models. In both the Gaussian and the lognormal case, the change in correlation has a greater impact on Sequence 2 (which has members from Tier 4 and therefore involves less load averaging) than on Sequence 1 (which has members primarily in Tier 3). It is also interesting to note that for the case of perfect correlation, the failure probabilities of the two sequences, 1 & 2, are closer to each other in the lognormal case than in the Gaussian case. This is expected because the problem is load dominated in the lognormal case, and the loads required to cause failure of Sequences 1 & 2 (with member capacities at their mean values) are similar. Also, in both models, for perfect correlation, Sequence 1 has higher probability of failure than Sequence 2. In the lognormal case (Table 7), because the failure probabilities for the two sequences at perfect correlation are similar, the greater increase of failure probability (with decrease in correlation) for Sequence 2 leads to Sequence 2 being more important than Sequence 1 at the calibrated correlation level. In contrast, in the Gaussian case, the difference in failure probability of the two sequences at the perfectly correlated level is larger and even though Sequence 2 has a larger increase in failure probability for the same decrease in correlation, Sequence 1 is still more important than Sequence 2 at the calibrated correlation level.

The same behavior is seen for the most likely to fail member for the lognormal model (Table 6). The most likely member to fail member in the perfectly correlated case (Case I.1) has many nodal loads influencing its internal member force, and therefore its failure probability is less sensitive to the change in the correlation between nodal forces. Hence when we decrease the correlation to the calibrated level, there is a smaller increase in failure probability of the most likely to fail member than in many other members and some other member becomes the most likely to fail member.

CONCLUSIONS

A framework to model the spatial correlation structure of nodal wave forces in system reliability analysis of offshore structures is presented.

Application of the framework to a fixed jacket structure modeled as a space truss shows that for a fixed base shear c.o.v., the correlation level among the nodal wave forces has to decrease markedly before such drop in the correlation level has a significant effect on the failure probabilities. This is because the internal member forces are influenced by multiple nodal loads which causes spatial averaging and lessens the impact of the decrease in correlation.

The correlation for our jacket structure was estimated from measured wave force and kinematics data. Although the correlation due to local uncertainties decays quite rapidly over space, the absolute correlation level remains high because of the high degree of common functional dependence of the nodal wave forces on the random wave height. For the estimated correlation, it appears that the more accurate spatial stochastic modeling of the nodal wave forces may be only of marginal benefit for truss type behavior of steel jacket structures. Nonetheless, depending on circumstances (e.g., with more accurate model considering failure in bending due to local loads), even this relatively high common correlation may affect the member failure probabilities, and the more accurate spatial stochastic model may be necessary.

The calibration procedure outlined here is quite crude; further analyses of the wave load data are needed before definite conclusions on the the common correlation and the rate of decay can be made.

ACKNOWLEDGEMENT

This work is a result of research sponsored in part by NOAA, National Sea Grant College Program, Department of Commerce, under grant number NA85AA-D-SG140, project number R/OE-5, through the California Sea Grant College Program, and in part by the California State Resources Agency and by the Office of Naval Research under grant number N00014-87-K-0475. The U.S. Government is authorized to reproduce and distribute for governmental purposes.

REFERENCES

1. Nordal, H., Cornell, C.A. and Karamchandani, A. (1987). "A Structural System Reliability Case Study of an Eight-leg Steel Jacket Offshore Production Platform." ***Proceedings of the Marine Structural Reliability Symposium***, October 1987, SNAME, USA.

2. Cornell, C.A., De, R.S., Karamchandani, A. and Bjerager, P. (1988). ***Enhanced Failure-Path Based Structural System Reliability.*** Offshore Structural System Reliability Joint Industry Project Report to Amoco Production Company (Proprietary until August, 1989). C. Allin Cornell, Inc., Portola Valley, California, USA.

3. Chan, H.Y. and Melchers, R.E. (1987). "Reliability of Complex Structures Under Wave Loads." ***5th International Conference on Application of Statistics and Probability in Soil and Structural Engineering***, ICASP, Vancouver, Canada.

4. Bjerager, P., Olesen, R. and Madsen, H.O. (1987). "System Reliability of Offshore Jacket Structures by Ideal Plastic Analysis." ***Proceedings of the Marine Structural Reliability Symposium***, October 1987, SNAME, USA.

5. Der Kiureghian, A. and Liu, P-L. (1986). "Structural Reliability Under Incomplete Probability Information." ***Journal of Engineering Mechanics*** , ASCE, Vol. 112, No. 1, January 1986.

6. De, R.S., Winterstein, S.R. and Bjerager, P. "Correlated Hermite Models in Offshore Structural Reliability." -- Proposed contribution, ***Fifth International Conference on Structural Safety and Reliability***, San Fransisco, 1989.

7. Heideman, J.C., Olsen, O.A. and Johansson, P.I. (1979). "Local Wave Force Coefficients." ***Proceedings of the Specialty Conference, Civil Engineering in the Oceans IV***, Volume II, September, 1979, ASCE.

8. Dean, R.G., Lo, J-M, Johansson, P.I. (1979). "Rare Wave Kinematics Vs. Design Practice." ***Proceedings of the Specialty Conference, Civil Engineering in the Oceans IV***, Volume II, September, 1979, ASCE.

9. Guenard, Y.F. (1984). ***Application of System Reliability Analysis to Offshore Structures***. Report #71, John A. Blume Earthquake Engineering Center, Stanford University, California, USA

10. Haring, R.E., Olsen, O.A.and Johansson, P.I. (1979). "Total Wave Force and Moment Vs. Design Practice." ***Proceedings of the Specialty Conference, Civil Engineering in the Oceans IV***, Volume II, September, 1979, ASCE.

11. Moses, F. (1981). ***Guidelines for Calibrating API RP2A for Reliability Based Design***, API-PRAC Project 80-22. Final Report to the American Petroleum Institute, Dallas, Texas, USA.

COLLAPSE PROBABILITY FOR ELASTO-PLASTIC BEAM STRUCTURES

M. V. Delmar, J. D. Sørensen & P. Thoft-Christensen
University of Aalborg
Sohngaardsholmsvej 57, DK-9000 Aalborg, Denmark

1 Introduction

In present design practice considerations on structural safety are usually concentrated on failure states corresponding to initial failure of an element. For a frame structure that means failure of a cross-section of the structure. Sufficient reliability with respect to this limit state is expected to be achieved by designing the structure on an element basis such that each member has at least a prescribed reliability. Most structures are redundant in the sense that failure of the first single element merely leads to redistribution of the load. The system reliability of the structure is in this paper defined as formation of a yield mechanism. System reliability aspects can be important in structural design and it is therefore relevant to consider limit state progressive failure in a probabilistic setting.

In reliability studies of progressive failures the analysis with respect to plastic collapse under proportional loading of linear elastic - ideal plastic structures has been undertaken in numerous works, but only a few authors have considered interaction of load effects in the yield condition. These works can be divided into two different formulations namely a total- and an incremental approach. The incremental approach has been used by Amdahl, Leira & Wu [1] and the total approach by Bjerager [2], Bådshauge & Bach-Gansmo [3], Ditlevsen [4], Høj & Loklindt [6], Murotsu, Okada & Matsuzaki [8] and Turner & Baker [12].

In this paper a new procedure to estimate the reliability of elasto-plastic beam structures based on the total approach is proposed. Interaction of different load effects is taken into account. The procedure consists of four main steps:

- Identification of the most significant sequences of failure elements giving collapse of the structure. The sequences are generated using a failure-tree approach.
- Identification of sets of fundamental mechanisms corresponding to the identified sequences of failed elements.
- Formulation of an optimization problem to determine the most significant collapse mechanisms by linear combination of fundamental mechanisms.
- The critical combined mechanisms are finally modelled as elements in a series system and an estimation of the probability of failure is obtained.

The procedure is formulated for general non-linear yield surfaces and is based on the fact, that multi degree-of-freedom mechanisms can be written as linear combinations of fundamental mechanisms. Due to the small number of fundamental mechanisms identified, the solution of the optimization problem, is very fast and efficient.

Finally the proposed procedure is illustrated by an example. The example considered is a space frame, where the load and the yield strength of the tubular elements are modelled as random variables.

2 Identification of Significant Failure Elements

In this section a general procedure to identify the most significant nodes (failure elements) leading to total collapse of the structure is presented. The following basic assumptions are used:

- The structural system is modelled with straight tree-dimensional beam elements each with two nodes 1 and 2.
- The external loads are applied to the nodes.
- The material is linear elastic-perfectly plastic.
- Second order geometrical effects are neglected.
- The influence of shear forces is neglected.
- Plastic deformations are confined to individual cross-sections as a yield hinge with no "spread length" along the structural member.

For each of the potential yield hinges (placed at the ends of each finite beam element) a failure element is defined. First a single finite element is considered. The nodal forces and the elastic nodal displacements in local coordinates for element t are denoted $\overline{r}_t$ and $\overline{v}_t^{\,e}$ and are related by

$$\overline{r}_t = \overline{\overline{k}}_t^{\,e} \; \overline{v}_t^{\,e} \tag{1}$$

where $\overline{\overline{k}}_t^{\,e}$ is the elastic stiffness matrix for element t.

A yield condition is defined at each node of the element (each failure element) and it is written

$$F_k(\overline{r}_t; Y_t) \; = \; 1 \;\; , \;\; k = 1,2 \tag{2}$$

where Y_t is the yield stress of element t. $F_k < 1$ indicates an elastic state, $F_k = 1$ indicates the plastic state and $F_k > 1$ is not possible. F_k is assumed to be convex. The nodal forces are related to the external loads by

$$r_{tj} = \sum_{m=1}^{l} a_{tmj}^{0} \; L_m \;\; , \;\; j = 1, \ldots, ndof \tag{3}$$

where $\overline{\overline{a}}_t^{\,0}$ contains coefficients of influence and $L_1, L_2, \ldots, L_l$ are the external loads. $ndof$, the number of degree-of-freedom, is equal to 12 for 3-dimensional beam elements.

The external loads $\overline{L}$ and the yield stresses $\overline{Y}$ are modelled as stochastic variables which are denoted $\overline{X}^T = [\overline{L}^T \overline{Y}^T]$. Corresponding to each potential yield hinge i a failure element is defined by the safety margin

$$M_i = 1 - F_i(\overline{r}_t; Y_t) \quad , \quad i = 1, \dots, n \tag{4}$$

where n is the number of failure elements. The probability of occurrence of yielding in the potential yield hinges can then be estimated using FORM (or SORM), see e.g. Thoft-Christensen & Murotsu [11] and Madsen, Krenk & Lind [7].

$$P_i = P(M_i \leq 0) \simeq \Phi(-\beta_i) \quad , \quad i = 1, \dots, n \tag{5}$$

where β_i is the reliability index corresponding to (4) and Φ is the standard normal distribution function.

When failure element i_1 has failed (yielding in the corresponding yield hinge) the behaviour of the structural system is generally controlled by the incremental changes in the response based on the usual approach of flow theory. For a piecewise linear yield surface the behaviour is controlled by the equations shown in the appendix. If the yield surface of the failed element i_1 is not piecewise linear it can be approximated by a piecewise linear yield surface. The linearization is performed at the mapping point $\overline{q}^*$† on the surface in normalised load effect space of the design point $\overline{x}^*$ corresponding to (4) and (5) calculated in normalised uncorrelated space. This approximation is expected to be insignificant as the purpose of the procedure in this section is only to identify the significant sequences of failure elements.

The yield condition of the failed element i_1 belonging to finite element t_1 is then approximated by

$$F_{i_1}(\overline{r}_{t_1}; Y_{t_1}) \simeq 1 + \overline{c}_{i_1}^T(\overline{r}_{t_1} - \overline{r}_{t_1}^*) = \overline{c}_{i_1}^T \, \overline{r}_{t_1} - d_{i_1} \tag{6}$$

where

$$c_{i_1 j} = \frac{\partial F_{i_1}(\overline{r}_{t_1}^*; Y_{t_1})}{\partial r_j} \quad \text{and} \quad d_{i_1} = \overline{c}_{i_1}^T \overline{r}_{t_1}^* - 1 \quad , \quad \overline{r}_{t_1} \in \Omega_{i_1} \tag{7}$$

Ω_{i_1} is the region for which (6) is valid (in the 2 dimensional case Ω_{i_1} is a quadrant in the normalised load effect space). The nodal forces in element t when element i_1 has failed are determined from

$$r_{tj}^{i_1} = \sum_{m=1}^{l} a_{tmj}^{i_1} \, L_m + b_{tj}^{i_1} \, L^{i_1} \quad , \quad j = 1, \dots, ndof \tag{8}$$

where $\overline{\overline{a}}_t^{\,i_1}$ contains coefficients of influence in element after failure element i_1 has failed. $\overline{b}_t^{\,i_1}$ contains coefficients of influence due to a unit load of the fictitious force L^{i_1} corresponding to failure of i_1, see appendix.

†* means design point value

The probability of simultaneous yielding in element i_1 and i $(i \neq i_1)$ can then be estimated from

$$P_i^{i_1} = P(M_{i_1} = 0 \cap \lambda_{i_1}^{i_1} \geq 0 \cap \{\bar{r}_{t_1} \in \Omega_{i_1}\} \cap_{j \neq i_1} \{M_j > 0\} \cap M_i^{i_1} \leq 0) \quad (9)$$

where $M_i^{i_1}$ is the safety margin for element i given failure of element i_1. $\lambda_{i_1}^{i_1}$ is a multiplier, see appendix. If $\lambda_{i_1}^{i_1} \geq 0$ no unloading takes place. The intersection $\cap_{j \neq i_1}$ is over all failure elements except i_1. $P_i^{i_1}$ can be estimated using FORM (or SORM).

If element $i_2(\neq i_1)$ in finite element t_2 now fails the same procedure can be used again and the probability of simultaneous yielding in element i_1, i_2 and i $(i \neq i_1, i_2)$ can be estimated from

$$P_i^{i_1,i_2} = P(M_{i_1} = 0 \cap \lambda_{i_1}^{i_1} \geq 0 \cap \{\bar{r}_{t_1} \in \Omega_{i_1}\} \cap_{j \neq i_1} \{M_j > 0\} \cap M_{i_2}^{i_1} = 0 \cap$$

$$\lambda_{i_1}^{i_1,i_2} \geq 0 \cap \lambda_{i_2}^{i_1,i_2} \geq 0 \cap \{\bar{r}_{t_2} \in \Omega_{i_1,i_2}\} \cap_{j \neq i_1,i_2} \{M_j^{i_1} > 0\} \cap M_i^{i_1,i_2} \leq 0) \quad (10)$$

where $M_i^{i_1,i_2}$ is the safety margin for element i given failure of element i_1 and i_2. $\lambda_{i_1}^{i_1,i_2}, \lambda_{i_2}^{i_1,i_2}$ are multipliers connected with the plastic deformations in failure element i_1 and i_2 given that the failure elements i_1 and i_2 have failed. Based on these probabilities the β-unzipping technique by Thoft-Christensen [11] or the Branch & Bound method by Murotsu[11], Guenard [5] can at least in some simple cases be used to identify the significant sequences of failure elements giving collapse. Collapse of the structural system is identified when the global stiffness matrix becomes singular.

The probabilities $P_i^{i_1,i_2,\ldots,i_n}$ can be estimated approximately in several ways, for example

$$P_i^{i_1,i_2,\ldots,i_n} = P(M_{i_1} = 0 \cap \lambda_{i_1}^{i_1} \geq 0 \cap M_{i_2}^{i_1} = 0 \cap \ldots \cap$$

$$\lambda_{i_n}^{i_1,i_2,\ldots,i_n} \geq 0 \cap M_i^{i_1,i_2,\ldots,i_n} \leq 0) \quad (11)$$

or more simply as

$$P_i^{i_1,i_2,\ldots,i_n} = P(M_{i_1} = 0 \cap M_{i_2}^{i_1} = 0 \cap \ldots \cap M_i^{i_1,i_2,\ldots,i_n} \leq 0) \quad (12)$$

or as calculated in most cases

$$P_i^{i_1,i_2,\ldots,i_n} = P(M_i^{i_1,i_2,\ldots,i_n} \leq 0) \quad (13)$$

The probabilities in (11), (12) and (13) can be estimated using FORM (or SORM). In the following sections eq. (13) is used to calculate these probabilities.

3 Identification of Fundamental Mechanisms

In this section a procedure to identify sets of fundamental mechanisms corresponding to identified sequences of failed elements giving collapse is presented.

Any failure mechanism is a linear combination of a set of independent mechanisms of one degree-of-freedom, called fundamental mechanisms. Let n be the number of internal degrees-of-freedom of the structure, i.e. the potential yield hinges identified multiplied by the degree-of-freedom in each hinge. Further, let r be the degree of redundancy of the structure. Then the number of independent fundamental mechanisms is $n - r$, Neal [9].

Once a sequence of failed elements is identified it may happen that the corresponding collapse mechanism have multi degrees-of-freedom. The mechanism is formed by using a failure-tree approach where combined load effects are taken into account by obeying the yield surface equation and the flow rule. Using this approach a certain ratio between mutual axial displacements and mutual moment rotations exists in each yield hinge in the sequence due to the modification procedure shown in appendix. It may occur that the structure will have a non-kinematically admissible strain rates field when the last failure element in the sequence is modified, because the modification is made according to the flow rule and the risk of a failure point placed in the singularity point at the yield surface is omitted, see appendix. It may also occur that the point on the yield surface, corresponding to the sequence of failure elements one level lower than the mechanism level, moves to a different point on a different yield surface defined in an other region. Therefore it is not possible to use the coefficients of influence calculated during the failure-tree approach as coefficients for the safety margin based on the principle of virtual work.

Using a technique proposed by Watwood [13], the multi degree-of-freedom mechanisms can be decomposed into independent one degree-of-freedom mechanisms.

4 Identification of Significant Failure Mechanisms

In this section the most significant collapse mechanisms are determined by combining the fundamental mechanisms.

The safety margin for the ith fundamental mechanism can, by the principle of virtual work, be written as

$$M_i = \sum_{m=i_1}^{i_n} W^i_{im} - \sum_{m=1}^{l} W^x_{im} \quad , \quad i = 1, \ldots, n-r \tag{14}$$

where W^i_{im} is the internal work in the potential yield hinge m. W^x_{im} is the external work done by the concentrated load L_m at yield hinge m.

The internal work W^i_{im} for yield hinge m in fundamental mechanism i can be calculated as

$$W^i_{im} = \Delta u_{im} N_{im} + \Delta\theta_{x_{im}} M_{x_{im}} + \Delta\theta_{y_{im}} M_{y_{im}} + \Delta\theta_{z_{im}} M_{z_{im}} \tag{15}$$

where

Δu_{im} is the mutual axial displacement.

$\Delta\theta_{j_{im}}$ is the mutual rotation about the j-axis, $j \in \{x, y, z\}$.

N_{im} is the axial force.

$M_{j_{im}}$ is the moment about the j-axis, $j \in \{x, y, z\}$.

The axial load capacity and yield moment capacity in yield hinge $i_1, \ldots, i_n$ can be written as

$$N_{p_m} = A_{p_m} Y_m \quad , \quad m = i_1, \ldots, i_n \tag{16}$$

$$M_{p_{jm}} = S_{p_{jm}} Y_m \quad , \quad m = i_1, \ldots, i_n, \quad j \in \{x, y, z\} \tag{17}$$

where Y_m models the yield stress. The plastic section modulus S_{p_j} and the cross-sectional area A_p are deterministic properties depending on the section geometry. Using the flow rule and the yield condition, the internal work W^i_{im} for yield hinge m can then be written

$$W^i_{im} = c_{im} Y_m, \; m \in 1, \ldots, i_n \tag{18}$$

where c_{im} depends on Δu_{im} and $\Delta\theta_{t_{im}}$ and the type of yield condition chosen for yield hinge m. For each yield hinge m, c_{im} is determined at the point on the yield surface, where a given strain velocity vector is outward normal to the yield surface.

The external work W^x_{im} can be written

$$W^x_{im} = d_{im} L_m \quad , \quad m = 1, \ldots, l \tag{19}$$

where d_{im} is the virtual displacement corresponding to the ith mechanism.

Using (18) and (19) equation (14) can be rewritten

$$M_i = \sum_{m=1}^{i_n} c_{im} Y_m - \sum_{m=1}^{l} d_{im} L_m \quad , \quad i = 1, \ldots, n - r \tag{20}$$

If a fictitious mechanism (a mechanism with no external work) is determined in the sequence of fundamental mechanisms, the coefficients of influence d_{im} are zero.

As mentioned in section 3, new failure mechanisms are made by linear combinations of fundamental mechanisms. An optimization problem from which the most significant failure mechanisms are determined is now formulated. The problem can be solved using a general optimization algorithm, e.g. the NLPQL algorithm, Schittkowski [10]. The optimization variables are denoted $\overline{z}$ and are real numbers used to combine the fundamental mechanisms. Each mechanism is described by a failure function $g(\overline{x}, \overline{z})$. Realisations $\overline{x}$ of $\overline{X}$ resulting in $g(\overline{x}, \overline{z}) \leq 0$ correspond to a failure state, while $g(\overline{x}, \overline{z}) > 0$ correspond to a safe state. In the first order reliability methods (FORM) a transformation $\overline{T}$ of the generally correlated and non-normally distributed variables $\overline{X}$ into standardised and normally distributed variables $\overline{U}$ is defined. Let $\overline{X} = \overline{T}(\overline{U})$. Then the failure function of general combined mechanism is written

$$g(\overline{T}(\overline{u}), \overline{z}) = \sum_{i=1}^{n-r} z_i \left(\sum_{m=l+1}^{i_n+l} c_{i(m-l)} T_m(\overline{u}) - \sum_{m=1}^{l} d_{im} T_m(\overline{u}) \right) \tag{21}$$

In order to determine the most significant collapse mechanisms corresponding to the failed sequence of nodes the following optimization problem is formulated

$$min_{\overline{z}} \ \beta(\overline{z}) \tag{22}$$

$$\text{s.t.} \ \ | \overline{z} | = 1$$

where $\beta(\overline{z})$ is determined from

$$\beta(\overline{z}) = min_{g(\overline{T}(\overline{u}), \overline{z})} \ (\overline{u}^T \ \overline{u})^{\frac{1}{2}} \tag{23}$$

The optimization problem (22) is generally non-convex. In order to identify the most significant local minima (and the corresponding mechanisms) one possibility is to use a large number of starting points. If s_j different start values of $\overline{z}$ for the optimization problem are used then no more than s_j different significant collapse mechanisms are generated as solutions to (22).

5 Estimation of System Failure Probability

In this section an estimate of the probability of failure of system collapse is calculated by modelling the identified significant mechanisms as elements in a series system.

A generalised reliability index β_s of the series system with the number of elements equal to the total number of significant collapse mechanisms $s = \sum s_j$ where the summation is over all different significant collapse mechanisms, can be determined from

$$\beta_s = -\Phi^{-1}(1 - \Phi_s(\overline{\beta}; \overline{\overline{\rho}})) \tag{24}$$

where $\overline{\beta} = (\beta_1, \ldots, \beta_s)$ are reliability indices of the mechanisms and $\overline{\overline{\rho}}$ is the correlation coefficient matrix determined from the linearized safety margins. Φ_s is the s-dimensional standard normal distribution function.

6 Example

To illustrate the procedure discussed in the previous sections the structure shown in figure 1 is analysed. The structure is inspired by an example analysed by Turner & Baker [12]. The space frame construction is a right angle bend made of tubular sections. The 5 potential yield hinges (failure elements) are shown in figure 1.

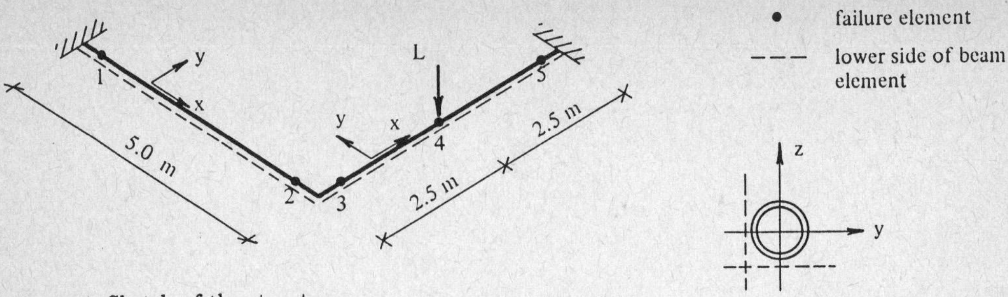

Figure 1. Sketch of the structure.

The beam is loaded by a single concentrated load L perpendicular to the plane of the structure. Let L be normally distributed with the expected value $E[L] = 60kN$ and the standard deviation $\sigma_L = 6kN$. The beam has a constant cross-section and is made of a material with $E = 2.1\ 10^8 kN/m^2$ and $G = 0.7\ 10^8 kN/m^2$. The cross-sectional area is $A = 4.59\ 10^{-3} m^2$. The following moments of inertia are chosen $I_x = 115.8\ 10^{-6} m^4$, $I_y = 57.9\ 10^{-6} m^4$ and $I_z = 57.9\ 10^{-6} m^4$. The yield stress Y is normally distributed with the expected value $E[Y] = 282.5\ 10^3 kN/m^2$ and the standard deviation $\sigma_Y = 28.25\ 10^3 kN/m^2$. The load and yield stresses are independent random variables.

First the reliability of the structure at level 1 is calculated (one plastic hinge to form). A non-linear safety margin is formulated for each failure element according to (4).

$$M_i = 1 - [(\frac{N_i}{N_{p_i}})^2 + (\frac{M_{x_i}}{M_{xp_i}})^2 + (\frac{M_{y_i}}{M_{yp_i}})^2 + (\frac{M_{z_i}}{M_{zp_i}})^2] \tag{25}$$

where $i = 1, \dots, 5$. The plastic axial load and bending moment capacities are determined by

$$\begin{aligned} N_{p_i} &= \pi\ d\ Y_i \\ M_{xp_i} &= 0.5\ \pi\ d^2\ Y_i \\ M_{yp_i} &= t\ d^2\ Y_i \\ M_{zp_i} &= t\ d^2\ Y_i \end{aligned} \tag{26}$$

$d = 0.3222m$ is the tube diameter, and $t = 0.0046m$ is the wall thickness. Y_i is the yield stress in simple tension. The 4 load effects included in the analysis are denoted N_i, M_{x_i}, M_{y_i} and M_{z_i}.

By linear elastic analysis the coefficients of influence $\overline{\overline{a}}_t^{\,0}$ and the reliability indices β_i are calculated. According to (13) and (5) the reliability indices at level 1 are

$$\beta_1 = 6.554, \quad \beta_2 = 9.089, \quad \beta_3 = 9.029, \quad \beta_4 = 7.319, \quad \beta_5 = 2.232 \tag{27}$$

For $\Delta\beta_1 = 1.0$, only one critical failure element is identified by the β-unzipping method, namely failure element no. 5. The non-linear safety margin for this failure element is linearized at the mapping point $\overline{q}^*$ in the normalised load effect space of the design point $\overline{x}^*$ calculated according to (4) and (5) in the normalised uncorrelated space. Load effects different from zero are illustrated in figure 2.

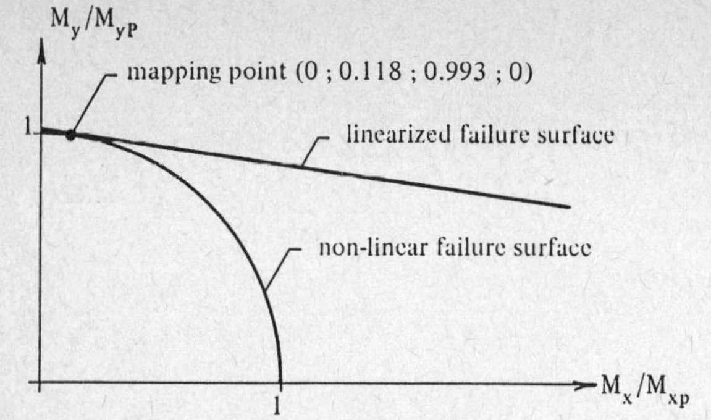

Figure 2. Linearization of failure element no. 5.

The linearized safety margin for failure element no. 5 is written on the same form as (4) using (6)

$$M_5 = 1 - \bar{c}_5^T \, \bar{r}_3 + d_5 \tag{28}$$

where failure element 5 is located in finite element 3 at end 2.

$$\bar{c}_5^T = [0,0,0,0,0,0, \frac{2\ sign(N_2)}{N_{p_2}^2}, 0, 0, \frac{2\ sign(M_{x_2})}{M_{xp_2}^2}, \frac{2\ sign(M_{y_2})}{M_{yp_2}^2}, \frac{2\ sign(M_{z_2})}{M_{zp_2}^2}]$$

$$\bar{r}_3 = [N_1, Q_{y_1}, Q_{z_1}, M_{x_1}, M_{y_1}, M_{z_1}, N_2, Q_{y_2}, Q_{z_2}, M_{x_2}, M_{y_2}, M_{z_2}]^T \tag{29}$$

where $\bar{c}_5$ is defined by (7) and d_5 is a constant.

The space frame is modified by inserting a yield hinge in failure element no. 5. Fictitious loads are added according to appendix. The modified structure is analysed at level 2 and the reliability indices are determinated according to (13) by the use of (5)

$$\beta_{1|5} = 5.661, \quad \beta_{2|5} = 9.086 \quad \beta_{3|5} = 9.014, \quad \beta_{4|5} = 6.751 \tag{30}$$

For $\Delta\beta_2 = 1.0$ only one critical failure element is identified, namely failure element no. 1. The same procedure as described above is used until a mechanism is obtained. For levels $> 2, \Delta\beta = 3.0$ is used. Reliability indices with failure assumed in 5 and 1 are

$$\beta_{2|5,1} = 1.941 \quad \beta_{3|5,1} = 5.554, \quad \beta_{4|5,1} = 6.534 \tag{31}$$

One critical failure element, no. 2 is identified. Reliability indices at level 4 are calculated assuming failure in 5, 1 and 2

$$\beta_{3|5,1,2} = -5.958, \quad \beta_{4|5,1,2} = -3.580 \tag{32}$$

Two critical failure elements are identified, namely failure elements no. 3 and 4. The global stiffness matrix is now singular and the identified failure tree at mechanism level is shown in figure 3.

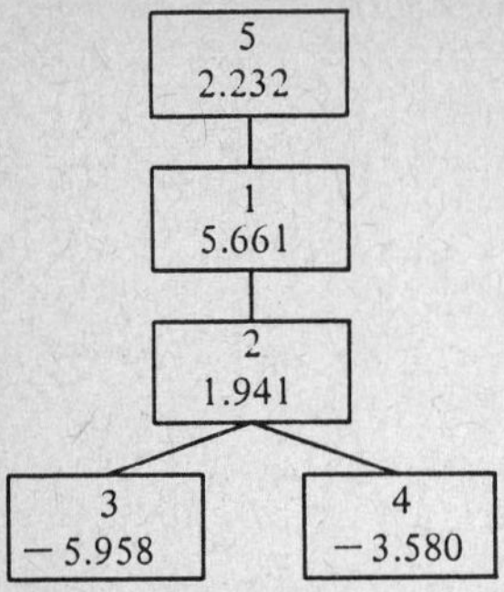

Figure 3. Failure tree at mechanism level.

Mech. No.	Yield hinge	Δu	$\Delta\theta_x$	$\Delta\theta_y$	$\Delta\theta_z$
1	1	0.0	0.0	0.0	−1.0
	2	0.0	0.0	0.0	1.0
	4	5.0	0.0	0.0	0.0
	5	0.0	0.0	0.0	0.0
2	1	2.5	0.0	0.0	0.0
	2	2.5	0.0	0.0	1.0
	4	0.0	0.0	0.0	−1.0
	5	0.0	0.0	0.0	0.0
3	1	0.0	−2.0	−1.0	0.0
	2	0.0	−2.0	1.0	0.0
	4	0.0	0.0	−2.0	0.0
	5	0.0	0.0	0.0	0.0
4	1	0.0	0.0	0.0	0.0
	2	0.0	0.0	−1.0	0.0
	4	0.0	1.0	0.0	0.0
	5	0.0	0.0	0.0	0.0
5	1	0.0	0.0	0.0	−1.0
	2	0.0	0.0	0.0	1.0
	4	0.0	0.0	0.0	0.0
	5	5.0	0.0	0.0	0.0
6	1	0.0	0.0	0.0	0.0
	2	0.0	0.0	0.0	1.0
	4	0.0	0.0	0.0	−2.0
	5	0.0	0.0	0.0	1.0
7	1	0.0	1.0	0.0	0.0
	2	0.0	1.0	0.0	0.0
	4	0.0	0.0	2.0	0.0
	5	0.0	0.0	−1.0	0.0
8	1	0.0	0.0	0.0	0.0
	2	0.0	0.0	−1.0	0.0
	4	0.0	0.0	0.0	0.0
	5	0.0	1.0	0.0	0.0

Table 1. Mutual displacements of fundamental mechanisms in sequence (5,1,2,4).

For each sequence in the failure tree an identification of a set of fundamental mechanisms is performed. For the sequence $(5, 1, 2, 4)$ seen in figure 3, 8 fundamental mechanisms are identified. n is equal to $14 = 6 + 2 \times 4$ due to the fact that two failure elements are in the same finite element and r is 6.

The mutual displacements for all fundamental mechanisms identified in the sequence are shown in table 1. The sign convention used for the mutual displacement is positive as elongation and for the mutual rotations about the axis as tension in the lower side of the beams. Twisting of the element is defined positive when it is a right-handed screw.

The only virtual displacement for the load L is obtained in fundamental mechanism no. 1 and d_{11} is equal to 2.50.

The fundamental mechanisms are combined and the NLPQL optimization algorithm by Schittkovski [10] is used for solving the optimization problem. The optimization variables $\overline{z}_1$ for the solution of the problem (22) are

$$\overline{z}_1 = (0.8527,\ 0.0,\ 0.4264,\ 0.2060,\ 0.0,\ 0.0,\ 0.00,\ 0.2204)^T \tag{33}$$

and the corresponding reliability index $\beta(\overline{z}_1) = 5.171$. This combined mechanism has the normalised mutual displacements as shown in table 2 with the virtual displacements of the load L equal to 2.50.

Yield hinge	Δu	$\Delta\theta_x$	$\Delta\theta_y$	$\Delta\theta_z$
1	0.0	0.0	−0.5	0.0
2	0.0	0.0	0.0	0.0
4	0.0	0.24	1.0	0.0
5	0.0	0.26	−1.0	0.0

Table 2. Normalised mutual displacements of combined mechanisms in sequence (5,1,2,4).

The same procedure is then applied to the second sequence (5,1,2,3) shown in figure 3. The corresponding reliability index is $\beta(\overline{z}_1) = 6.155$.

The generalised reliability index β_s for a series system with the two significant collapse mechanisms shown in figure 3 is then calculated with the correlation coefficient matrix

$$\overline{\overline{\rho}} = \begin{pmatrix} 1.000 & 0.997 \\ 0.997 & 1.000 \end{pmatrix} \tag{34}$$

The result is $\beta_s = 5.167$. (35)

The influence of different yield conditions and correlation between yield stresses for the sequence of failure elements in figure 3 is shown in table 3.

Yield func.	Y Fully correlated	Y Independent
Hyper Cube	5.805	8.041
Spherical	5.167	7.590
T & B	5.080	7.550
Simulation	5.091	7.561
Linear	3.855	5.492

Table 3. Reliability indices β_s for different yield functions. T & B refers to results from Turner & Baker [12].

The results from Turner & Baker [12] are based on a total approach where the spherical yield function (25) is used. The results from simulation are based on 1000 samples using direct simulation around the expected design point in the normalised and uncorrelated basic variable space.

7 Conclusion

A new numerical method for identification of the significant plastic collapse mechanisms is proposed. The method is based on the upper-bound theorem of plasticity. A formulation of the reliability with respect to plastic collapse of discretized linear elastic-ideal plastic structures is presented. For non-linear yield conditions an equivalent linear model is applied by linearizing the yield function at the design point.

The example indicates as expected that it is important to include the non-linear interaction of load effects in the yield condition. Further the results from using this new method coincides with simulation results.

8 References

[1] Amdahl, J., Leira, B. and Wu, Y.-L. : *On the Application of a Non linear Finite Element Formulation in Structural System Reliability.* Proceedings of the First IFIP WG 7.5 Working Conference, Aalborg, Denmark, May 6-8, 1987, Springer-Verlag, pp. 1-19.

[2] Bjerager, P. : *Plastic Systems Reliability by LP and FORM.* DCAMM Report No. 368, December, 1987, The Technical University of Denmark, Lyngby, Denmark, pp. 1-17.

[3] Bådshaug, O. and Bach-Gansmo, O. : *System Reliability Analysis of a Jacket Structure.* Proc. 4th International Conference on Structural Safety and Reliability, ICOSSAR, 1985, pp. II 613-617.

[4] Ditlevsen, O. : *Reliability Computations for Rigid Plastic Frames with General Yield Conditions.* Proceedings of the First IFIP WG 7.5 Working Conference, Aalborg, Denmark, May 6-8, 1987, Springer-Verlag, pp. 91-108.

[5] Guenard, Y. F. : *Application of System Reliability Analysis to Offshore Structures.* (Ph.D. Thesis). The John A. Blume Earthquake Engineering Center, Stanford University, California, U.S.A, Report No. 71, 1984.

[6] Høj, N. P. and Loklindt, E. : *Estimering af svigtsandsynlighed for rammekonstruktioner.* (In Danish). Report for M.Sc. No. 8302, June, 1983, Institute of Building Technology and Structural Engineering, University of Aalborg.

[7] Madsen, H. O., Krenk, S. and Lind, N. C. : *Methods of Structural Safety*, Prentice-Hall, Inc., Englewood Cliffs, New Jersey, 1986.

[8] Murotsu, Y., Okada, H. and Matsuzaki, S. : *Safety Margins for Reliability Analysis of Frame Structures.* The University of Osaka, Japan, Series A, Vol 32, No. 2, 1983.

[9] Neal, B. G. : *Plastic Methods of Structural Analysis*, Chapman & Hall Ltd, 1970.

[10] Schittkowski, K. : *NLPQL : A FORTRAN Subroutine Solving Constrained Non-Linear Programming Problems.* Annals of Operations Research, 1986.
[11] Thoft-Christensen, P. and Murotsu, Y. : *Application of Structural Systems Reliability Theory*, Springer-Verlag, Berlin Heidelberg New York Tokyo, 1986.
[12] Turner, R. C. and Baker, M. J. : *Structural System Reliability Analysis using Multi-Dimensional Limit State Criteria.* Proceedings of the First IFIP WG 7.5 Working Conference, Aalborg, Denmark, May 6-8, 1987, Springer-Verlag, pp. 433-457.
[13] Watwood, V. B. : *Mechanism Generation for Limit Analysis of Frames.* ASCE, Journal of the Structural Division, Vol. 109, No. ST1, January. 1979, pp.1-15.

Appendix

In this appendix the basic theory used in elasto-plastic analysis of beam elements with two nodes 1 and 2 is described.

First, a single finite element t is considered. The behaviour of the the element is described by a total formulation, where the relation between the nodal forces and the elastic nodal displacements is given by (1).

The plasticity condition, (2), is defined at each node of the element. If the flow rule is accepted then the total plastic displacements $\overline{v}_1^{\,p}$ and $\overline{v}_2^{\,p}$ in the element nodes 1 and 2 are given by

$$\overline{v}_1^{\,p} = \lambda_1 \frac{\partial F_1(\overline{r}_t^*; Y_t)}{\partial \overline{r}_t} = \lambda_1 \, \overline{c}_1 \tag{A1}$$

$$\overline{v}_2^{\,p} = \lambda_2 \frac{\partial F_2(\overline{r}_t^*; Y_t)}{\partial \overline{r}_t} = \lambda_2 \, \overline{c}_2 \tag{A2}$$

where λ_1 and λ_2 are multipliers. For example when section 1 is elastic, $\lambda_1 = 0$.

The total nodal displacement $\overline{v}_t$ is the sum of the elastic and the plastic displacements

$$\overline{v}_t = \overline{v}_t^{\,e} + \overline{v}_1^{\,p} + \overline{v}_2^{\,p} \tag{A3}$$

From (1) the nodal force $\overline{r}_t$ is expressed as

$$\overline{r}_t = \overline{\overline{k}}_t^{\,e} \left(\overline{v}_t \; - \; \overline{v}_1^{\,p} \; - \; \overline{v}_2^{\,p} \right) \tag{A4}$$

If the yield surface is linear or approximated by a linearized surface (evaluated at a point $\overline{q}_t^*$ on the yield surface) then the yield condition is given in the following form for failure element 1 according to (2) and (6)

$$F_1 = \bar{c}_1^T \bar{r}_t - d_1 = 1 \tag{A5}$$

where d_1 is a constant. If the original yield surface is linear d_1 is zero. Inserting (A4) into (A5) we get

$$\bar{c}_1^T \left(\bar{\bar{k}}_t^e \bar{v}_t - \lambda_1 \bar{\bar{k}}_t^e \bar{c}_1 - \lambda_2 \bar{\bar{k}}_t^e \bar{c}_2\right) = 1 + d_1 \tag{A6}$$

When both of the sections 1 and 2 are plastic it follows from (A6) that

$$\begin{pmatrix} \lambda_1 \\ \lambda_2 \end{pmatrix} = \bar{\bar{G}}^{-1} \bar{\bar{H}} \bar{v}_t - \bar{\bar{G}}^{-1} \begin{pmatrix} 1 + d_1 \\ 1 + d_2 \end{pmatrix} \tag{A7}$$

where

$$\bar{\bar{G}} = \begin{pmatrix} \bar{c}_1^T \bar{\bar{k}}_t^e \bar{c}_1 & \bar{c}_1^T \bar{\bar{k}}_t^e \bar{c}_2 \\ \bar{c}_2^T \bar{\bar{k}}_t^e \bar{c}_1 & \bar{c}_2^T \bar{\bar{k}}_t^e \bar{c}_2 \end{pmatrix} \quad \bar{\bar{H}} = \begin{pmatrix} \bar{c}_1^T \bar{\bar{k}}_t^e \\ \bar{c}_2^T \bar{\bar{k}}_t^e \end{pmatrix} \tag{A8}$$

The following resulting relation is obtained

$$\bar{r}_t = \bar{\bar{k}}_t^p \bar{v}_t + \bar{r}_t^p \tag{A9}$$

where the reduced element stiffness matrix is defined by

$$\bar{\bar{k}}_t^p = \bar{\bar{k}}_t^e - \bar{\bar{H}}^T \bar{\bar{G}}^{-1} \bar{\bar{H}} \tag{A10}$$

and the equivalent nodal force vector

$$\bar{r}_t^p = \bar{\bar{H}}^T \bar{\bar{G}}^{-1} \begin{pmatrix} 1 + d_1 \\ 1 + d_2 \end{pmatrix} \tag{A11}$$

when only one section is plastic, e.g. section 1, the relations are

$$\lambda_1 = \frac{\bar{c}_1^T \bar{\bar{k}}_t^e \bar{v}_t - (1 + d_1)}{\bar{c}_1^T \bar{\bar{k}}_t^e \bar{c}_1}, \quad \lambda_2 = 0 \tag{A12}$$

$$\overline{r}_t = \overline{\overline{k}}_t^{\,p}\, \overline{v}_t + \overline{r}_t^{\,p} \tag{A13}$$

where

$$\overline{\overline{k}}_t^{\,p} = \overline{\overline{k}}_t^{\,e} - \frac{\overline{\overline{k}}_t^{\,e}\, \overline{c}_1 \overline{c}_1^T \overline{\overline{k}}_t^{\,e}}{\overline{c}_1^T \overline{\overline{k}}_t^{\,e}\, \overline{c}_1} \tag{A14}$$

and

$$\overline{r}_t^{\,p} = \frac{\overline{\overline{k}}_t^{\,e}\, \overline{c}_1\, (1 + d_1)}{\overline{c}_1^T \overline{\overline{k}}_t^{\,e}\, \overline{c}_1} \tag{A15}$$

when no sections are plastic

$$\lambda_1 = 0, \quad \lambda_2 = 0 \tag{A16}$$

$$\overline{r}_t = \overline{\overline{k}}_t^{\,e}\, \overline{v}_t \tag{A17}$$

The physical meaning of the parameter λ_1 (or λ_2) is

$$\lambda_1 \begin{cases} > 0 & \textit{loading} \\ = 0 & \textit{neutral loading} \\ < 0 & \textit{unloading} \end{cases} \tag{A18}$$

For the total structure the following global analysis is performed. The equilibrium conditions are

$$\overline{R} = \overline{\overline{K}}^{\,p}\, \overline{V} \tag{A19}$$

where $\overline{\overline{K}}^{\,p}$ is given by

$$\overline{\overline{K}}^{\,p} = \sum_{t=1}^{NE} \overline{\overline{T}}_t^{\,T}\, \overline{\overline{k}}_t^{\,p}\, \overline{\overline{T}}_t \tag{A20}$$

NE is the number of elements, $\overline{\overline{k}}_t^{\,p}$ is given by (A10), (A14) or (A17) and $\overline{\overline{T}}_t$ is the geometric transformation matrix for element t. $\overline{V}$ is the total nodal displacement vector referred to the global coordinate system. $\overline{R}$ is given by

$$\overline{R} = \sum_{t=1}^{NE} \overline{\overline{T}}_t^{\,T} (\overline{r}_t - \overline{r}_t^{\,p}) \qquad (A21)$$

where $\overline{r}_t$ are the nodal forces in element t in the local coordinate system formed by the element nodal forces $\overline{r}$ and $\overline{r}_t^{\,p}$ is the equivalent nodal force vector referred to the local system given by (A11), (A15) or (A17).

STOCHASTIC WAVE LOADS FOR RELIABILITY ANALYSES OF JACKET STRUCTURES

M. Dogliani & R. Cazzulo
Registro Italiano Navale, C.P. 1195, I-16100 Genova, Italy

SUMMARY

This paper presents part of the results of our work within the Brite Project P1270, "Reliability Methods for Design and Operation of Offshore Structures". This project is funded through a 50% contribution from the Directorate General for Science, Research and Development of the Commission of the European Communities and another 50% contribution from eleven research partners and five industrial sponsors. The aim of this work has been to analyze some simplified regular wave models, which are more or less based on the hypothesis that the sea in nature can be modelled as a sufficiently narrow band process. The validity of these approaches is defined through the results obtained by a numerical simulation on a case study, trying to identify their different features for possible application in structural reliability analyses.

1 - INTRODUCTION

Jackets are among the more conventional offshore structures and their design is based on quite standard procedures. In particular, hydrodynamic loads are defined according to the following basic assumptions:

- hydrodynamic forces are assumed to be independent of tubular member flexibility and resonance phenomena are ignored*, thus jackets are considered "memory-less" structures;
- the dimensions of tubular members are assumed to be small with respect to wave height and length, thus hydrodynamic forces are evaluated via the well-known Morison equation;
- the wave kinematics is obtained from a periodic regular design wave with given characteristics (height, period, heading direction, current, ...).

Unfortunately, hydrodynamic loads obtained according to these quite notional assumptions are far from the "real" ones, which can be measured on a jacket in the

---

* except for vortex shedding effects, which are usually considered local effects and ignored in the global structural analysis.

sea. In fact, sources of uncertainties in the Morison equation can be found in the variability of the drag and added mass coefficients (which conversely depend on diameter, roughness, wave length, Reynold's and Koulegan Carpenter's numbers and on other parameters, such as marine growth and corrosion) as well as in the definition of the wave particle kinematics (Dogliani, Salza /1/). These problems have been discussed in depth in literature and a review was recently presented at the last ISSC (ISSC 88 - Com I.2 on "Environmental Forces" /2/).

Another point, rather more philosophical but not yet completely recognized is inherent in the use of a periodic regular wave itself. A regular wave is, in fact, unable to reproduce the "real" randomness in nature of the sea and this can lead to significant errors in the evaluation of hydrodynamic loads, too. This problem can easily be recognized by looking at the wave profile, which is typical of a storm, and the one, e.g. obtained in a model basin, for a regular periodic wave. The first produces an irregular behaviour, which holds also for the time and space series of velocities and accelerations, which cannot be reproduced by a regular periodic wave. The vertical distributions of velocities and accelerations are not very well fitted by the regular wave ones either (Figure 1).

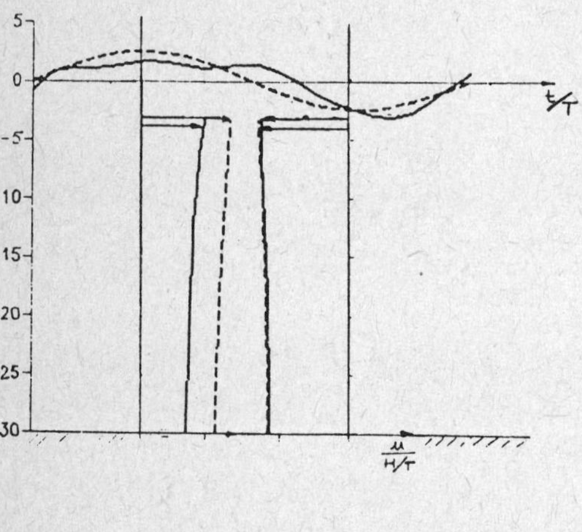

FIGURE 1 - Random and periodic wave profile and kinematics
----: results by regular periodic wave
——: results of random sea by numerical simulation

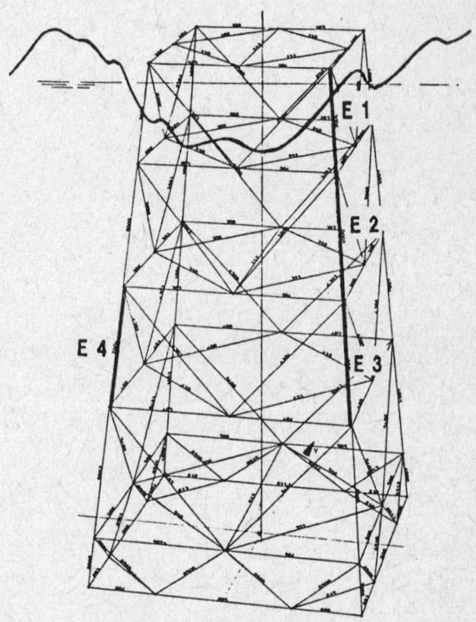

FIGURE 2 - Example jacket structure

Any mathematical idealization process induces systematic errors but perhaps these errors can be decreased by a more detailed model formulation (Ditlevsen - invited lecture on "Uncertainty and Structural Reliability" in /2/). In the case of the stochastic definition of hydrodynamic loads, this means trying to identify the suitable wave models which are able to retain, as far as possible, the randomness of the sea and to keep their model uncertainties within reasonable limits.

The correct way to solve this problem theoretically would be to consider a complete stochastic approach, by modelling the sea elevation and the corresponding water particle kinematics as non-uniform random fields. This leads to a definition of the Morison forces on tubular members of a jacket as being components of a vector random process, which is statistically fully described in terms of mean and second and higher order auto- and cross-correlation functions. According to this approach, structural component reliability can be evaluated by means of mean outcrossing techniques (Ditlevsen /3/), suitably applied in standard FORM procedures (Ferro, Cervetto /4/, Rackwitz et al /5/).

This has been shown to lead to good results, in the main hypothesis that each storm is assumed to be a stationary ergodic linear random process. Examples of this kind of analysis have been given for the reliability of ship (ISSC 85 - Com V.2 /6/) and submersible platform (ISSC 88 - Com V.1 /2/) components and it can be applied to jacket structures, too, provided that a suitable stochastic linearization technique for the non-linear drag force is used (Ferro et al /7/) but, in general, it is rather complicated to use when the number of force vector process components is high. This limits its applicability to the analysis of real jacket structures.

In fact, presently developed procedures for the system reliability analysis of jacket structures cannot avoid considering simple regular wave models for the definition of the hydrodynamic loads.
For example, when we studied the system reliability of jacket pile foundations, we considered a second order Stokes' wave train, with random height and period of given joint probability distribution, for different wave directions and crest positions (Cazzulo, Pittaluga, Ferro /8/).
Similar wave models, more refined for the definition of the distribution of the largest wave height and the associated period of the extreme sea state, have recently been proposed by Turner, Baker /9/ and Holm et. al. /10/.
Other system reliability procedures, such as that proposed by Thoft-Christensen and Morotsu /11/, require the use of very simple load paths, which, from the more optimistic point of view, are similar to the loads obtained by a regular harmonic wave, with some random parameters.

As stated by Guenard /12/, these approaches give an innacurate representation of the multidimensional stochastic nature of the loads (i.e. information on the statistical correlation of loads on different members) and they assume proportionality of loads from an initial value to the one corresponding to the collapse of the structure, converting the problem into a "time-independent" form. This conversion not only introduces in the system reliability analyses the same errors which are made in conventional design, but may lead to some inconsistencies in the general reliability scheme, with respect to the models chosen for the response and the collapse of the structure, i.e. when models of progressive collapse (such as the virtual distortion method /13/) are used.

These problems are well known to the members of Task III.2 of the Brite Project P1270 on the "Reliability methods for design and operation of offshore marine structures" /14/, whose main goal is "to treat the space/time variations typical of environmental loads on offshore structures in a way which is compatible with a practical procedure for system reliability analysis".

The RINA contribution to this task has been to try to identify the statistical properties of the errors made in the definition of hydrodynamic loads by modelling the sea, and the consequent velocity and acceleration fields, via some suitable periodic regular waves.

We have performed our study on an example jacket structure in 30 m water depth (Figure 2), subjected to a fully developed extreme sea state of 8 m significant wave height, 20 s peak period and 1 h of stationary duration time, comparing the hydrodynamic forces obtained by deterministic numerical simulation to the ones obtained by some regular wave approaches. Only the uniform equivalent forces on the four selected members (E1, E2, E3, E4) shown in Figure 2 and the component normal to each member and along the wave direction have been considered.

2 - WAVE LOADS BY NUMERICAL SIMULATION

The numerical simulation of the wave elevation and kinematics has been performed in the hypotheses of short-term statistics, for a long-crested sea, considered as a stationary ergodic linear Gaussian process and defined by a proper power spectrum. The simplest "deterministic approach" has been followed, which defines the wave elevation as the linear superposition of a finite number of Airy regular waves, with random phase and deterministic amplitudes, and conversely the wave kinematics as the superposition of the corresponding Airy velocity and acceleration terms. For this example, we have selected a hundred elementary Airy wave components. The sea-state conditions have been assumed according to Table 1.

Table 1 - Environmental data

Sea-depth	30 m
Significant wave height	8 m
Peak period	20 s
Wave direction	0 °
Power spectrum	JONSWAP
Peakedness parameter	3
Current speed	0 m/s
Duration time	1 h

The resulting simulation (for a total of 229 cycles) of the wave elevation in the origin, chosen as the center of the platform, is shown in Figure 3. According to this simulation, but it is quite a typical result, low and intermediate cycles are very irregular, whereas the highest wave cycles look more similar to regular waves. These irregularities in the time-scale hold also in space, considering different points over the x-axis. Moreover, the first highest waves have zero-crossing periods quite close to the peak period of the sea (see Table 2).

Table 2 - 1/10 highest waves

cycle	height	period	cycle	height	period
94	17.1	19.5	209	9.2	18.3
95	13.7	19.1	102	9.2	16.2
228	12.5	18.1	193	9.1	18.1
127	12.1	19.2	185	9.0	16.8
177	11.4	15.4	115	8.8	16.2
184	10.5	19.1	227	8.8	18.1
93	10.4	20.7	22	8.6	17.6
194	10.3	18.7	6	8.5	16.0
167	10.1	15.9	90	8.5	18.3
55	9.6	17.7	176	8.4	20.9
229	9.6	18.2	101	8.4	17.4

The numerical simulation of hydrodynamic forces on the jacket tubular members is straightforward, by applying the Morison equation. The randomness on drag and inertia coefficients as the other sources of uncertainties (marine growth, corrosion, ..) has been ignored, in order to take into account only the effects due to the actual water particle velocities and accelerations. Equivalent uniformly distributed forces have been considered as the average of those evaluated at each member's extremes. The equivalent uniformly distributed forces for piercing members have been approximated by the average of the forces evaluated at the piercing point and the submersed extreme.

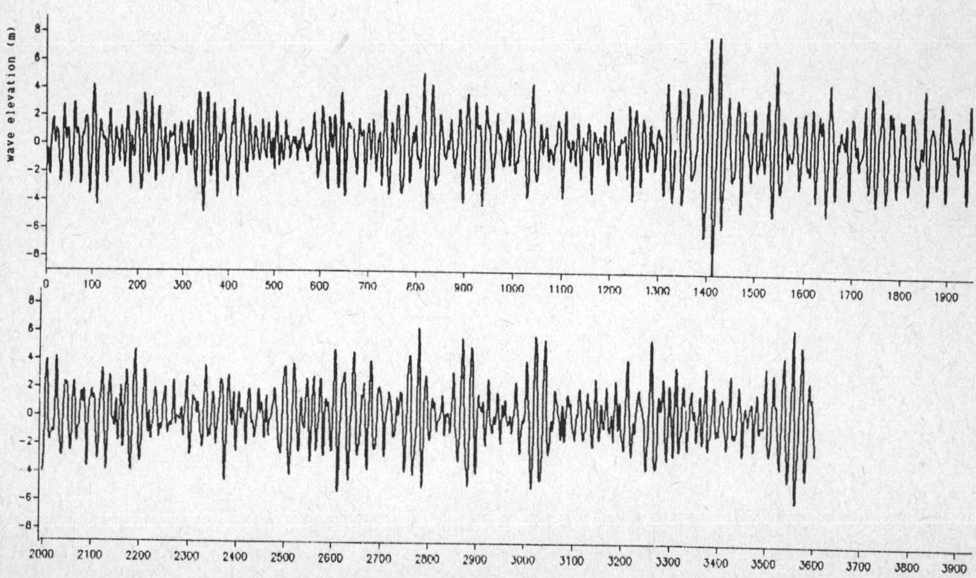

FIGURE 3 - Time series of wave elevation (1 hour storm)

3 - REGULAR WAVE APPROACHES

As discussed before, presently developed procedures for system reliability analyses cannot avoid to consider simplified wave models for the evaluation of the stochastic wave loads, which loose part of the randomness of the phenomenon. The simplest idea, which also has a certain connection with the conventional design of these structures, is to evaluate the Morison forces on the basis of the water particle velocities and accelerations obtained by a regular wave, with some suitable random parameters.

The definition of "regular wave" is not univocal. In this study we have compared three different models, which have different implications for their applicability in a reliability procedure. These models, which have a certain decreasing level of approximation, are summarized as follows:

A) "Narrow band approach", approximating the sea wave elevation by a narrow band process, with an infinitesimal bandwidth parameter $\varepsilon = O(0)$.
The consequence of this assumption is that all wave cycles have the same frequency ω_p and wave number K_p, while the wave amplitude A and phase φ vary so slowly that they can be assumed to be constant, in time and over a space, sufficiently wide to include the dimensions of the platform, in each cycle (Ochi /15/). The amplitudes of each cycle can be assumed as Rayleigh distributed, with properties depending on the variance of the sea spectrum, and the phases have a uniform distribution.

B) "Response surface approach", approximating the velocity and acceleration random fields as the product of some suitable time-independent shape functions (depending on the peak wave number K_p) and a scalar stochastic process (depending on the random wave amplitude A and period T). The wave number K_p is chosen as a function of the peak frequency ω_p, while (A/T) distribution can be obtained by considering a narrow but finite bandwidth sea spectrum.
This approach is similar to the one used in civil engineering for the design of tall buildings, e.g. with respect to wind loadings.

C) "Stochastic regular wave approach", dealing with a narrow but finite band spectrum, with small bandwidth parameter ε but not $= O(0)$.
According to this hypothesis, it is still reasonable to consider the wave amplitude and phase as random variables, slowly varying over time and space in each cycle but wave period and frequency randomly vary from cycle to cycle, too. In this case, the harmonic and hyperbolic shape functions in the velocity and acceleration formulas contain random terms, due to the wave number K, thus they cannot be defined in a suitable time-independent form.

The three approaches can also be seen as a different fitting of the numerical simulation results, defining the wave cycles by the zero-up-crossing periods of the wave elevation at the center of the platform, and applying for each cycle the transformation of time into a local phase variable ϑ, which varies from 0 to 2π, with origin at each instant of zero-up-crossing. In order to be consistent to the numerical simulation results, which have been obtained considering the sea as a Gaussian stochastic process, each wave cycle has been fitted by a single linear Airy wave, with amplitude equal to the half of the actual wave height (more refined wave models, such as the Stokes′ one, can be only compared by using non-linear numerical simulations). The differences of these approaches consist:

A) for the "narrow band approach", in assuming in all cycles a constant wave period Tp and wave number Kp, corresponding to the peak period of the sea;

B) for the "response surface approach", in varying the period T from cycle to cycle but in assuming a constant wave number Kp, equal to the peak one, thus violating the dispersion equation;

C) for the "stochastic regular wave approach", in varying the period T and the wave number K, too, in order to satisfy in each cycle the dispersion equation.

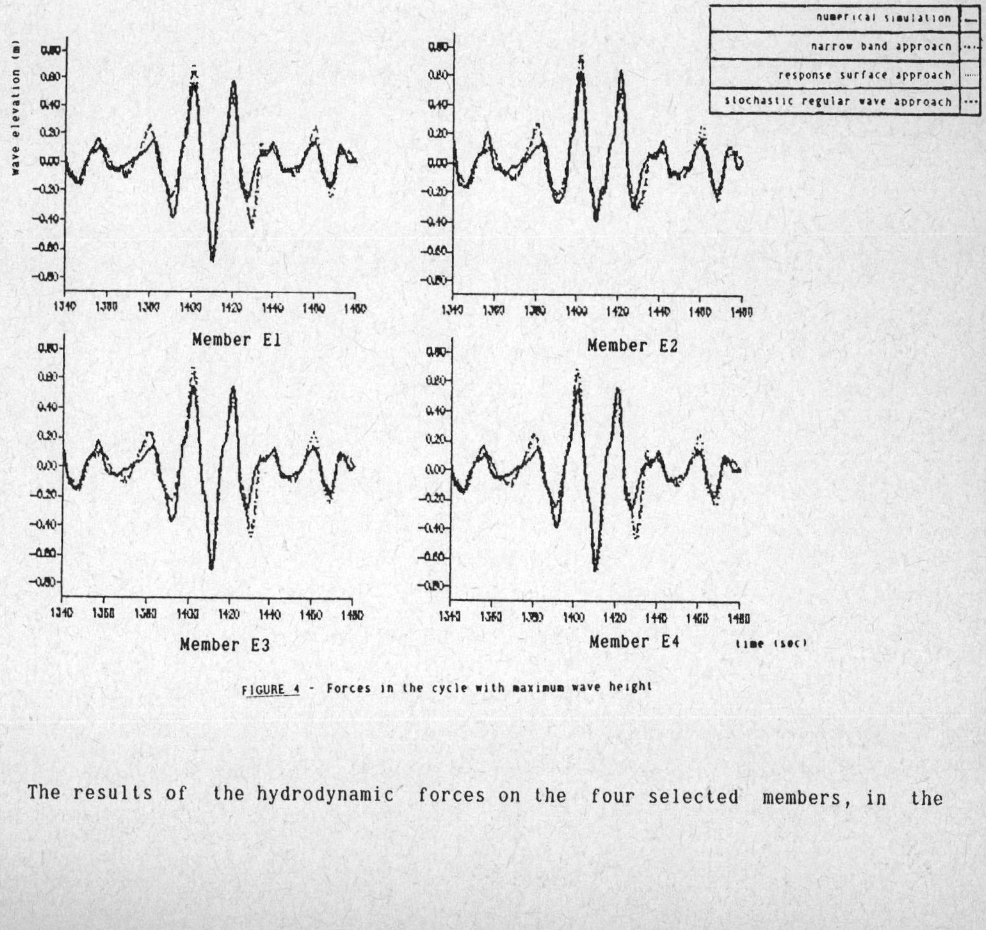

FIGURE 4 - Forces in the cycle with maximum wave height

The results of the hydrodynamic forces on the four selected members, in the period including the cycle with maximum wave height, are shown in Figure 4. The forces obtained by the regular wave approaches in these cycles seem similar to those obtained by numerical simulation and they only slightly overestimate the true force at the instant of maximum wave height. Greater differences occur at the immediately subsequent cycle, which is significantly non-symmetrical: in this case, the regular wave approaches underestimate the force obtained by numerical simulation. Some other differences can be found for smaller cycles, especially when the wave is in the trough position.

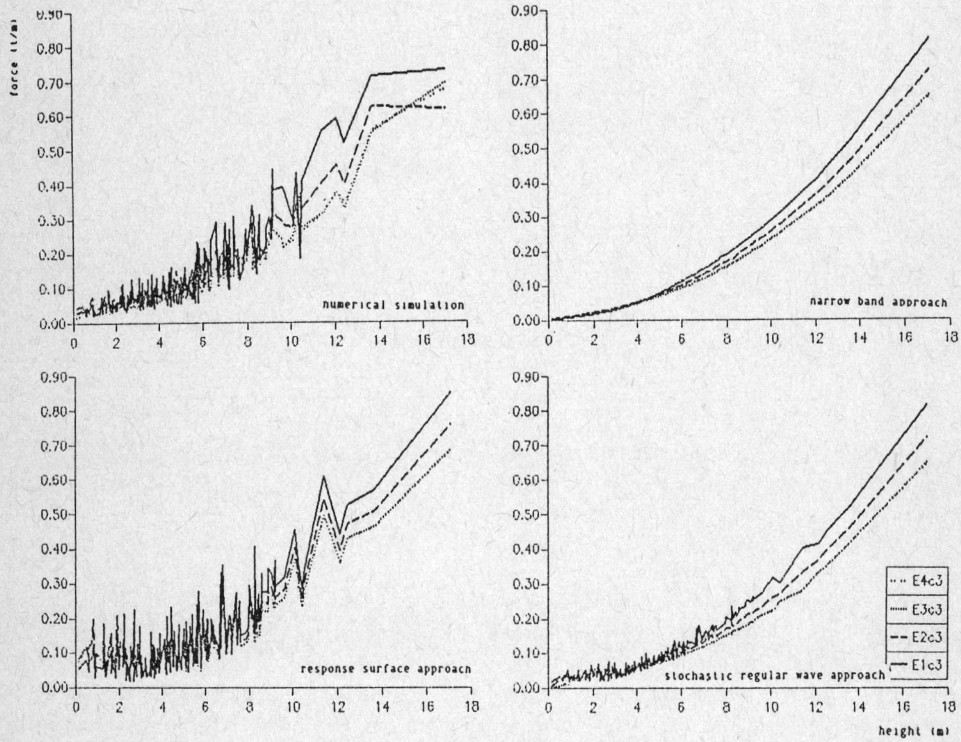

FIGURE 5 - Maximum forces versus wave height

The trend of these forces versus the wave height (Figure 5) looks similar to the numerical simulation results and the greatest differences are for low or intermediate cycles, which are less significant for the collapse of the structure. The narrow band and stochastic regular wave approaches are unable to reproduce the irregularities of small and intermediate cycles, whereas, on the other hand, the response surface approach seems to be more irregular.

The behaviour of the maximum forces versus the ratio H/T (Figure 6) is closer to what one would expect: the stochastic wave approach is more irregular than the other two and more similar to the randomness of the numerical simulation results.

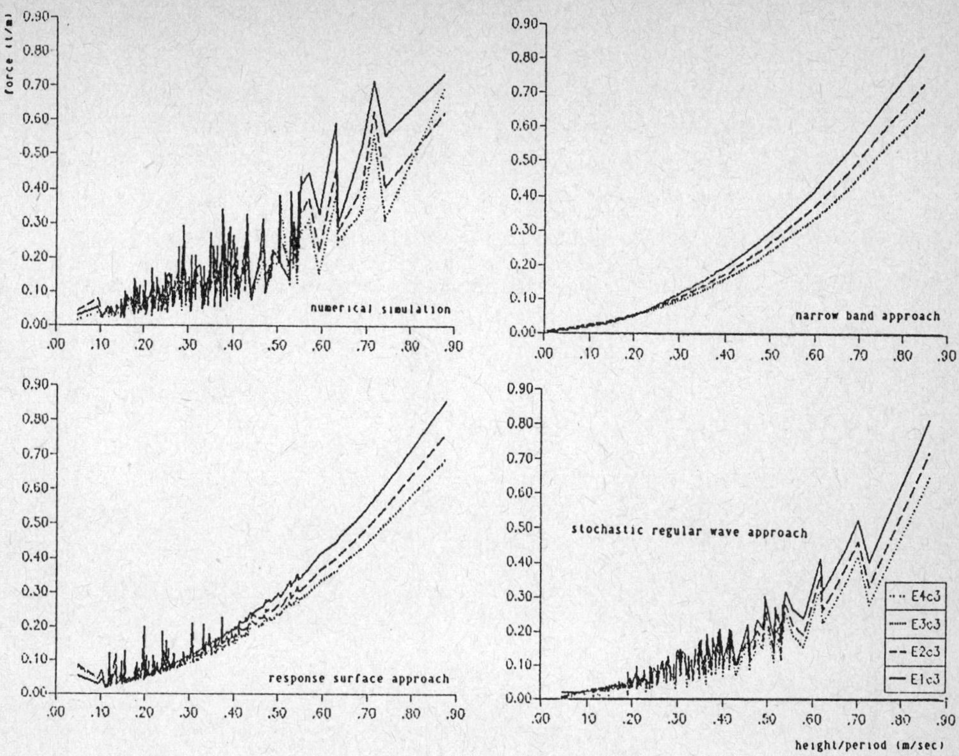

FIGURE 6 - Maximum forces versus ratio wave height over period

4 - COMPARISON OF ERRORS

These results shown that a certain correspondence exists between the forces obtained by the regular wave approaches and those obtained by numerical simulation but their errors can be noticeably high. The absolute errors between approximated and simulated results are conveniently defined with respect to the maximum force found in the whole simulations when, reasonably, the ultimate collapse of the structure occurs:

$$e = \frac{|F^R - F^S|}{\max F^S}$$

where, at a given time instant and for the generic force component, F^R is the force obtained by a regular wave approach, F^S is the one obtained by numerical simulation and $\max F^S$ is the maximum force found in the whole simulation.

The maximum errors found in the whole simulation on the four selected members by the three approximated approaches are shown in Table 3. Absolute errors given by the response surface approach are slightly higher than those obtained by the other two approaches; furthermore, the narrow band and the stochastic regular wave approaches produce about the same errors.

Table 3 - Maximum errors(%)	E1	E2	E3	E4
Narrow band approach	31	31	46	47
Response surface approach	35	36	51	52
Stochastic wave approach	31	31	45	45

The diagrams of absolute errors, defined with respect to the maximum forces in the whole simulation, against the ratio H/T (Figure 7) show that these errors for small and intermediate cycles are quite irregular but they always stay below the highest ones, because the forces at these cycles are considerably lower than the maximum forces obtained in the whole simulation.

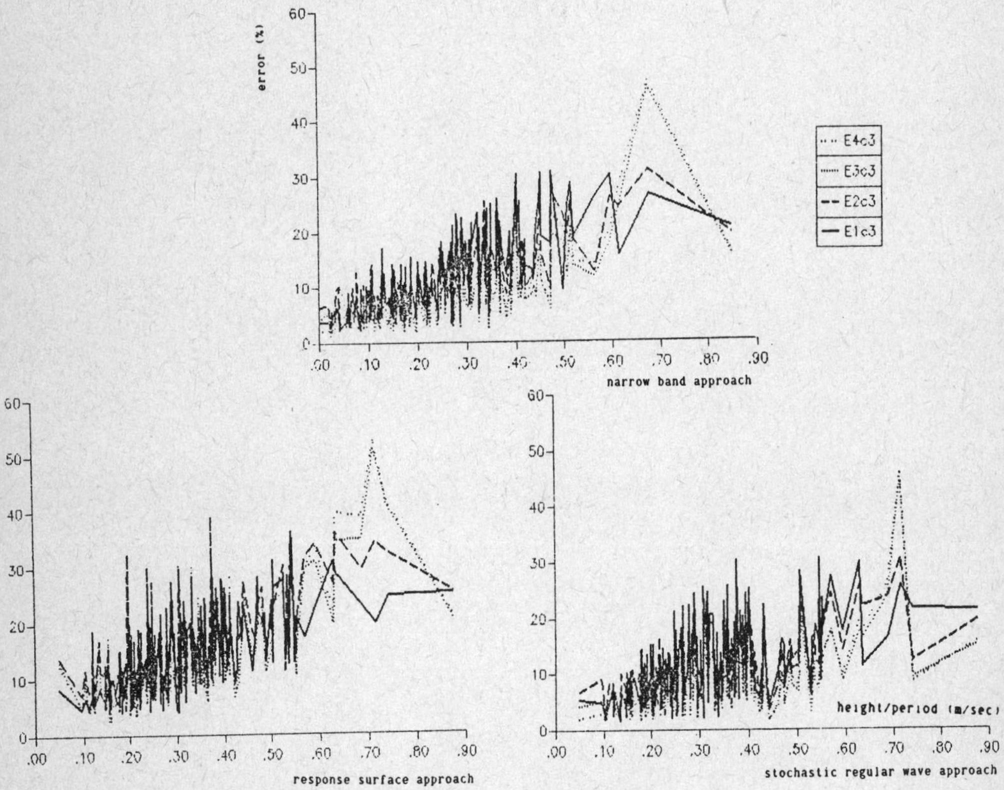

FIGURE 7 - Maximum errors versus ratio wave height over period

It is interesting to note that these absolute errors are lower than the relative errors, defined with respect to the maximum forces found in each cycle. The diagrams of the effective and absolute errors versus wave height for the stochastic regular wave approach (Figure 8) show that the effective errors are considerably higher (even 200%) than the absolute ones, for small and intermediate wave cycles.

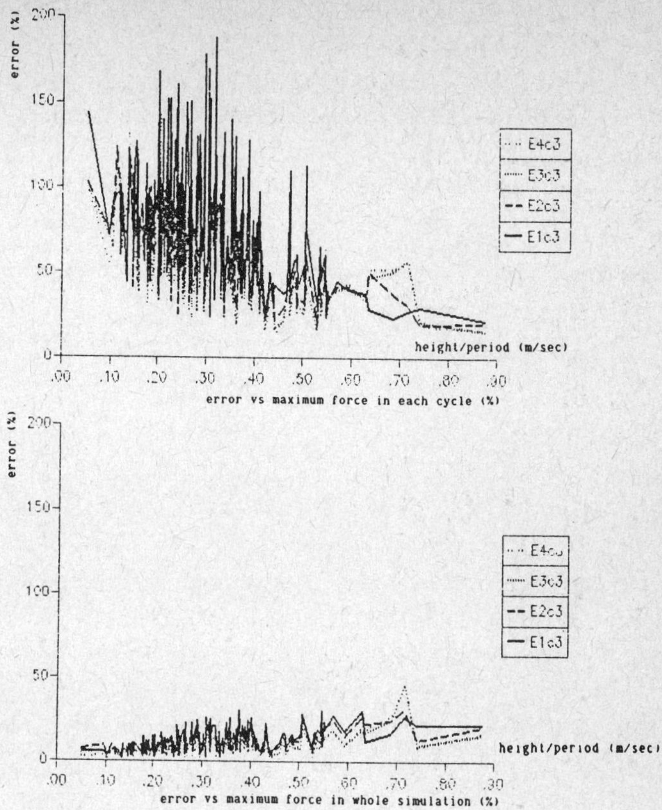

FIGURE 8 - Comparison of errors (stochastic regular wave approach)

5 - MEAN AND COVARIANCES

If independence of cycles is assumed, any cycle may be considered as a different realization of the same process and it is possible to define the statistics of hydrodynamic forces at each local phase in terms of mean and auto- and cross-covariance functions. The comparisons of mean and variances of forces at each local phase, obtained by the three regular wave approaches and by numerical simulation (shown in Figures 9 and 10), can be summarized as follows:

- for the simulation results, strangely, the variances of forces on always submersed members (E3 and E4) are slightly higher at local phase #5, when the wave, on average, is in crest positions at the center of the platform, than at local phase #15, corresponding to the wave trough.
  One possible explanation is that the chosen number of elementary wave components (100) is still insufficient to completely avoid undesired phase-locking effects. Obviously, the decrease of variance at high local phases is correct for the other two members (E1 and E2), because forces are decreased when the member is piercing or becomes zero when the member is above the water surface;

- The narrow band approach shows very good results for mean forces, with only a slight underestimate of their maximum, while gives the greatest differences for the variance of forces, especially for piercing members (E1 and E2), which are noticeably lower than those obtained by numerical simulation;

- The stochastic wave approach gives practically the same statistics, in terms of mean and averages, as the narrow band one;

- The response surface approach, on the other hand, generally overestimates absolute values of mean forces at all local phases and variances of forces on always submersed members (E3 and E4); whereas it seems to be close to simulation results when looking at variances of piercing members (E1 and E2).

Figure 11 shows the cross-covariance function between the members selected at opposite sides of the platform (E3 and E4) and for different local phase correlations. As the phase lags increase, the correlation between the forces evaluated at the opposite side of the platform decreases, but this behaviour is more evident for the stochastic wave approach. In fact, the response surface approach gives considerably higher correlations for all local phases, while the narrow band approach increases its errors when augmenting phase lags between forces on opposite members.

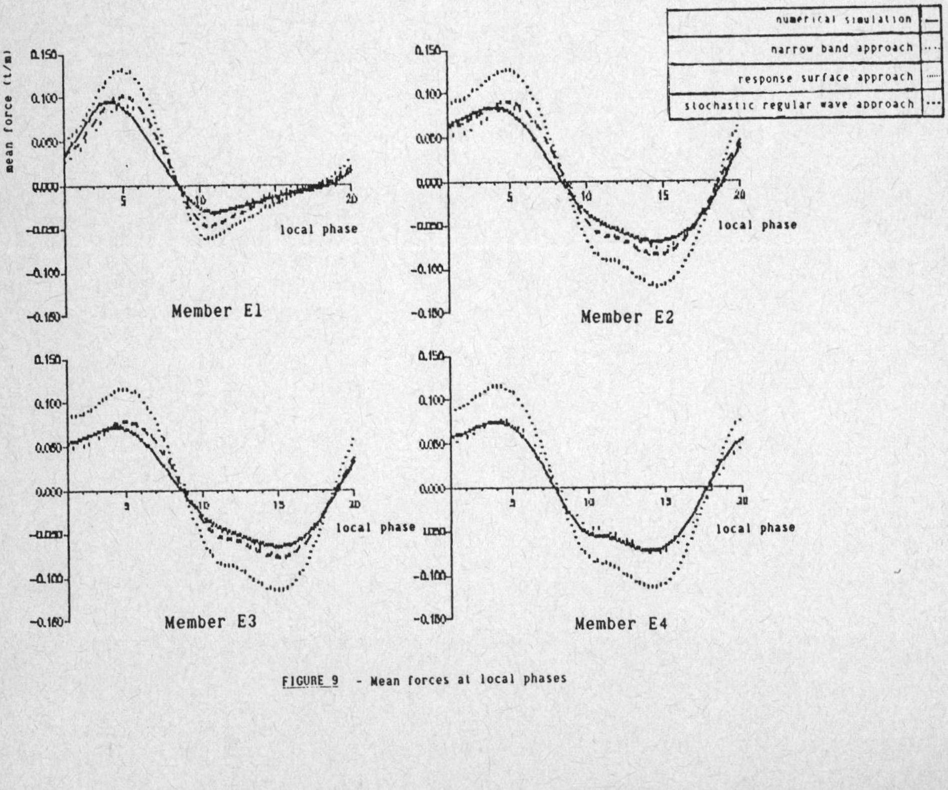

FIGURE 9 - Mean forces at local phases

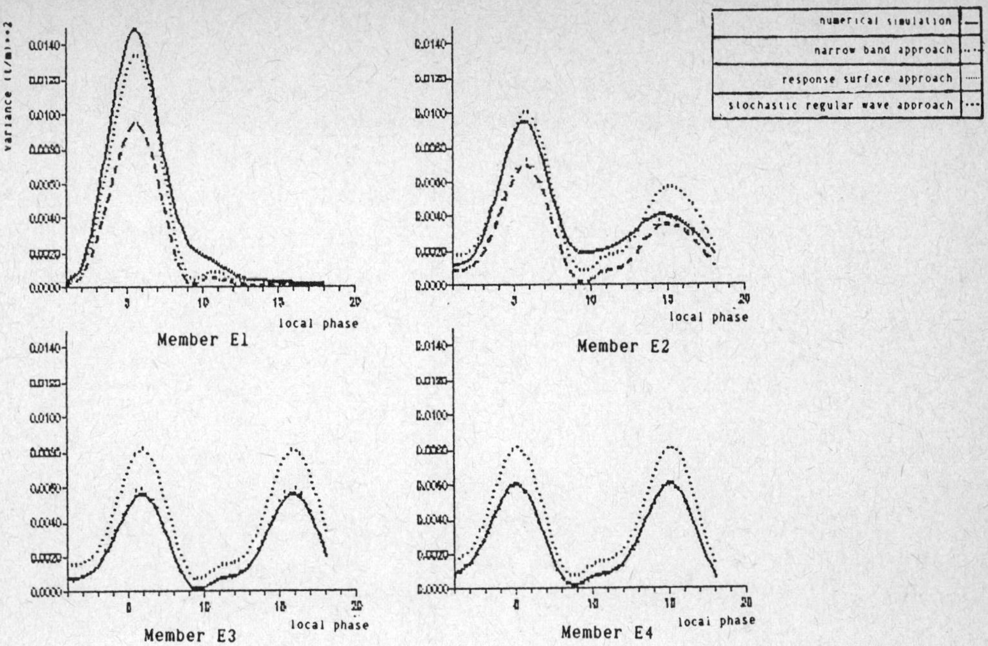

FIGURE 10 - Variances of forces at local phases

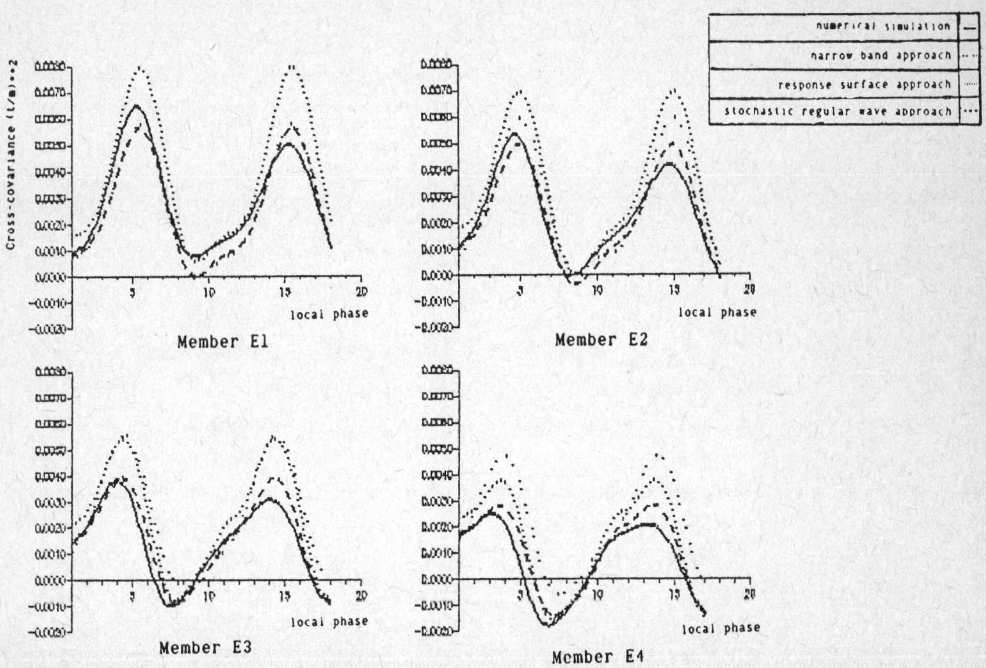

FIGURE 11 - Cross-covariances of forces at local phases (between members E3,E4 and local phases p,p-1,p-2,p-3)

6 - CONCLUSIONS

A suitable probabilistic model formulation should be based on certain hypotheses of regularity with respect to the natural physical fluctuations (Ditlevsen /2/). With reference to the above, all three regular wave models which have been considered in this study show good qualitative correspondence with "true" numerically simulated results but none of them can be judged as being able to reproduce quantitatively the effects of the real randomness of the sea.

The behavior of the maximum forces at each cycle, obtained by numerical simulation, is rather more irregular with respect to regular wave results, especially for small or intermediate cycles but the trend of simulated forces seems to be proportional to a power of wave height as regular wave ones are.

The statistics of hydrodynamic forces on local phases, obtained by assuming that each cycle is an independent realization of the same process, show some differences among the regular wave approaches. In particular, the simpler "narrow band approach" shows cross-covariances which are not very good with respect to simulation results, especially when augmenting the phase lag, and this is an indication of its limitations when it is applied to the analysis of large structures.

The so-called "response surface approach", which notionally defines the wave kinematics random fields as the product of some suitable time-independent shape functions and a scalar stochastic process, generally overestimates the mean and variances of forces on tubular members. This appproach is wrong also in assigning too much cross-covariance between the forces on fully-submersed members at opposite sides of the platform: this may lead to important errors, i.e. different bending or twisting effects, in the evaluation of the global response of the system. According to us, this shows an intrinsic limitation of the method, which artificially imposes the space variation of the wave kinematics as being independent of time.

Finally, the so-called "stochastic regular wave approach", which is in the line of the stochastic models proposed by Turner, Baker /9/ and other authors, seems to be the preferable one in view of applications in system reliability procedures (i.e. to a wide class of jacket structures). It has shown practically the same maximum errors, mean and variances as the narrow band approach but it is more sound in defining the cross-correlations between forces on opposite members.

These considerations are necessarily limited due to the restricted scope of the present study, which has only examined the results of a single deterministic numerical simulation on an example platform, for a specific sea state and an hour's duration.

However, we think that the most important conclusion is that the errors in evaluating the hydrodynamic forces by some regular wave approaches are very high: in certain conditions and for certain wave models they may be 30% or even 50% of the maximum forces which can actually be found during a storm in a structure. It is recommended that, in the general reliability scheme, proper model uncertainties are used, which are able to take into account the inability of regular wave models to follow the real randomness in nature of the sea.

A rigorous evaluation of these model uncertainties would require some systematic analyses on different classes of jackets and different selected sea states, possibly referring also to some systematic full scale measurements, which was beyond the scope of this work but would be possible for future developments.

REFERENCES

/1/ Dogliani M., Salza P., "Regular periodic waves: domains of application of analytic theories" (in Italian), Research Report 222, Registro Italiano Navale, April 1988

/2/ Proceedings of the 10th International Ship and Offshore Structures Congress, Lyngby, Denmark, August 1988

/3/ Ditlevsen O., "Gaussian outcrossing from safe convex polyhedrons", Journal of Engineering Mechanics, ASCE, Vol 109 No 1, 1983

/4/ Ferro G., Cervetto D., "Hull girder reliability", Ship Structure Symposium, Arlington, Virginia, October 1984

/5/ Hohenbichler M., Rackwitz R., "First-order concepts in system reliability", Structural Safety, Vol 1 No 3, April 1983

/6/ Proceedings of the 9th International Ship Structures Congress, Genova, Italy, September 1985

/7/ Ferro G., Pittaluga A., Costa Santos R., "Stochastic linearization in seakeeping analysis of complex offshore structures", 4th Offshore Mechanics and Arctic Engineering Conference, Dallas, Texas, February 1985

/8/ Cazzulo R., Pittaluga A., Ferro G., "Reliability of a jacket foundation system", 5th Offshore Mechanics and Arctic Engineering Conference, Tokyo, Japan, April 1986

/9/ Turner R.C., Baker M.J., "A wave model for the reliability analysis of offshore structures", London Center for Marine Technology, Department of Civil Engineering, Imperial College of Science and Technology, London, 1987

/10/ Holm C.A., Bjerager P., Olesen R., Madsen H.O., "System reliability of offshore jacket structures by elastic and plastic analysis", International Conference on Behaviour of Offshore Structures, Trondheim, June 1988

/11/ Thoft-Christensen P., Morotsu Y., "Application of structural systems reliability theory", Springer-Verlag, Heidelberg, 1986

/12/ Guenard Y., Goyet J., Labeyrie J., Remy B., "Structural safety evaluation of steel jacket platforms", Marine Structural Reliability Symposium, Arlington, Virginia, 1987

/13/ Holnicki-Szulc J., Gierlinski J.T., "Structural modifications simulated by virtual distortions", WS Atkins Engineering Science Report, 1988

/14/ BRITE Project P-1270, "Reliability methods for design and operation of off-shore and marine structures", Annex IV, Technical Contents, January 1986

/15/ Ochi M.K., "Stochastic analysis and probabilistic prediction of random seas", Advances in Hydroscience, Vol 13, Academic Press, 1982

EXCURSION PROBABILITIES OF NON-HOMOGENEOUS GAUSSIAN SCALAR FIELDS BASED ON MAXIMA CONSIDERATIONS

M. Faber* & R. Rackwitz**
***University of Aalborg, Sohngaardsholmsvej 57, DK-9000 Aalborg, Denmark**
****Technical University of Munich, Arcisstr. 21, D-8000 Munich 2, FRG**

Introduction

Many uncertain natural or technical phenomena are most realistically described by random fields. A typical example of a random field is the load effect in a floor slab which is loaded by a spatially distributed gravity load. Other examples of random fields include the sea–level around off–shore platforms, the distribution of material strength in a cross–section of a structural component, the concentration of detergents in ground water, the velocities or pressures in a turbulent medium or on surfaces, the ore content in geological deposits, etc. Several of their properties can be of interest in applications but their extremes are of special interest in reliability problems.

Random fields can be homogeneous or non–homogeneous. When the spatially distributed load in the first example is modeled by a homogeneous random field the response (i.e. the deflections or stresses) also becomes a random field and, hence, the extreme behavior of this field must be studied in order to quantify the reliability of the slab if it is considered to be in a failure state once certain response thresholds are exceeded at least once. If the loading field is varying only in space this field and the corresponding response field is a two–dimensional field. If the loading field is additionally a random function in time, the response must be modeled as a field in three–dimensions. If the response is a scalar (i.e. the vertical deflections) the random field is a scalar random field and, likewise, if the response is a vector (i.e. the principal stresses) the random field is a vector random field. Even though the loading random field may be homogeneous in the space coordinates the response random field can be inhomogeneous and, therefore, inhomogeneous random fields deserve special attention. Random fields can further be classified according to the nature of the state and the parameter space. Both states and parameters can be discrete or continuous. Among the various types of random fields Gaussian fields, continuous in state and parameter space, are studied most frequently not only because they have some nice analytical properties but primarily because they have so many applications. In this part I of a more general investigation the excursion probabilities of non–homogeneous scalar fields are studied using maxima considerations.

Review of concepts and theory for excursion probabilities

While there is a rich literature on excursion probabilities for simple stationary and non–stationary Gaussian scalar processes reporting relatively strong results and to some degree also for vector processes, much less is known for Gaussian fields. Relatively few attempts to formulate and solve the problem of the probability of (first) excursions of a random Gaussian field out of a given failure set, e.g. out of $F = \{g(x) \leq 0\}$ exist and those are for special cases with not necessarily much interest in structural reliability. In the following a short review of the present state of knowledge will first be given.

For scalar processes with differentiable sample paths many of the available results on excursion probabilities can be found in the books by Cramer/Leadbetter [8] and Leadbetter/Lindgren/Rootzen [15]. They are essentially based on the statistics of crossings of the process through given levels. Both the probabilistic characteristics of the process and its derivative, defined in the mean square sense, must be known. By modifying the formulation slightly, results have also been obtained for local maxima now requiring also the characteristics of the second–derivative process.

For vector processes, two alternative approaches have been proposed. On the one hand, a new scalar random function $Y(t) = g(\mathbf{X}(t))$ is formed so that the failure set becomes an open region on the real line, i.e. $F = \{y(t) \leq a \equiv g(x) = 0\}$. The excursion problem then can be solved by calculating the probability of a crossing of the level a or of a local maximum of the scalar random process Y(t) above the threshold a in a small time unit. Denote these probabilities by $\nu(a)$ and $\mu(a)$ or, more precisely, by the crossing rate and the rate of local maxima, respectively. Further, assume that the point processes of crossings and of local maxima are regular point processes so that the property of additivity holds and denote by $E[N(B)]$ and $E[M(B)]$ their mean numbers obtained by integration over the parameter domain B. It is then clear that since $\nu(a) \leq \mu(a)$ and, correspondingly, $E[N(B)] \leq E[M(B)]$, any approach based on local maxima is less efficient than the upcrossing approach. Remember also that second–order derivative statistics must be known for the local maximum approach which sometimes are not easy to assess. Yet, for extreme excursions there is only a small difference and asymptotically both quantities converge to each other. This is a general statement valid also for random fields.

The same concept of scalarization would carry over to the multi–parameter case. However, the random process Y(t) or the scalar random field $Y(\mathbf{t})$ where $\mathbf{t}$ is a vector parameter (denoting e.g. time and space coordinates) then, in general, is no more Gaussian and the properties of the non–Gaussian (process) field are extremely difficult to obtain. The alternative formulation to determine crossing rates in the original Gaussian space for a generally curved failure surface ∂F proved to be more successful for vector processes. Clearly, a corresponding approach based on local maxima is no more straightforward because the notion of a maximum has no obvious equivalent in the vectorial case.

For random fields on the other hand the conventional notion of a crossing of a random field into some domain is not useful and must be replaced by the notion of an excursion set. It can contain several local maxima, local minima and the corresponding saddle points. The excursion set may not even be simply connected. It can have one or more isolated minima. Nevertheless, for extreme excursions there should be a tendency towards simply connected excursions under suitable regularity conditions for the random field and sufficient smoothness of the failure surface. Letting those excursions becoming more and more rare for homogeneous scalar fields with growing "levels" a, there should be only one maximum associated with each excursion set and this is the case which initially found certain solutions (see [1] and [2]) the strongest of which are of asymptotic nature. For later reference the result in [2] is given below. If $X(\mathbf{t})$ is a homogeneous m–dimensional, zero mean Gaussian field satisfying certain regularity conditions, the mean number of maxima above the level a for $a \to \infty$ and $\mathbf{t} \in B$ with B a given domain is:

$$E[M(B)] \sim L(B)\,(2\pi)^{-1/2(m+1)}\,\det(\Sigma_{\dot{X}})^{1/2}\,\sigma^{-2(m-1/2)}a^{m-1}\exp[-1/2(a/\sigma)^2] \qquad (1)$$

where L(B) is the Lebesque measure (volume) of B and $\Sigma_{\dot{X}}$ the covariance matrix of $\dot{X}$. For example, let B be given as $B = \{0 \leq t_i \leq T_i\}$, then $L(B) = \Pi_{i=1}^{m} T_i$. Another important result concerning the limit distribution of the maxima of scalar fields in a hypercube, i.e. the Gumbel distribution, is due to Bickel/Rosenblatt [3] which parallels the corresponding result in the scalar case.

Improvements of these results rest almost entirely on a better description of the characteristic of an excursion set. A detailed account of some of the attempts reported in the literature can be found in [4]. Based on certain considerations in part summarized in [1] Bolotin [5] developed such a refined excursion characteristic which essentially is based on the curvature properties of the excursion set of a homogeneous field defining local maxima, minima and the associated saddle points and some topological considerations due to Morse [9]. In a series of papers Adler and Hasofer (for example, [4,6,7]) developed under mild conditions but homogeneity of the field another excursion characteristic which again used topological elements and whose most important probabilistic quantity, the mean number in a set B, can be given explicitly without approximation for any level a. It is a generalization of the notion of an upcrossing in the one–dimensional case and it is a more powerful definition of an excursion characteristic than Bolotin's. For high levels a it converges to the asymptotic result quoted above. A fairly exhaustive discussion of these results is contained in the monograph by Adler [17] where many other important results and references can be found.

For non–homogeneous scalar fields virtually no result is known to the authors. Apparently, this is due to the fact that an analytical evaluation of an appropriately modified formula for the mean number of excursion characteristics becomes hopelessly cumbersome if possible at all in analytic form. This appears to be true for simple local maxima and for both Bolotin's and Adler's definition of refined excursion characteristics. Furthermore, there seems to exist no result for vector fields which, from an engineering point of view, is especially interesting as indicated before at an example.

In this note maxima considerations for excursions of scalar, homogeneous random fields into certain domains are generalized to the non–homogeneous case. Practical numerical schemes are proposed which are based on concepts of asymptotic analysis and which are considered as the main new contributions.

Definitions

Define the vectors $\mathbf{X}$, $\dot{\mathbf{X}}$ and $\ddot{\mathbf{X}}$. $\mathbf{X}$ is assumed to be a n–dimensional vector for given parameter vector $\mathbf{t}$. The dimension of the parameter vector is m. Therefore, $\dot{\mathbf{X}}$ can be represented by a $r = n \cdot m$ –vector and $\ddot{\mathbf{X}}$ is r^2–dimensional. If $\mathbf{X}$ is Gaussian so is $\dot{\mathbf{X}}$ and $\ddot{\mathbf{X}}$ with distribution function

$$\begin{bmatrix} X \\ \dot{X} \\ \ddot{X} \end{bmatrix} \sim N_s \begin{bmatrix} m_X & \Sigma_{XX} \, \Sigma_{X\dot{X}} \, \Sigma_{X\ddot{X}} \\ m_{\dot{X}} \; ; & \Sigma_{\dot{X}X} \, \Sigma_{\dot{X}\dot{X}} \, \Sigma_{\dot{X}\ddot{X}} \\ m_{\ddot{X}} & \Sigma_{\ddot{X}X} \, \Sigma_{\ddot{X}\dot{X}} \, \Sigma_{\ddot{X}\ddot{X}} \end{bmatrix} \tag{2}$$

with $s = n + r + r^2$ and where:

$$\Sigma_{XX} = \{\mathrm{Cov}[X_i(t_1), X_j(t_2)];\ 1 \le i,j \le n\}$$
$$\Sigma_{X\dot{X}} = \{\mathrm{Cov}[X_i(t_1), \dot{X}_j(t_2)];\ 1 \le i \le n,\ n+1 \le j \le nm\}$$
$$\Sigma_{X\ddot{X}} = \{\mathrm{Cov}[X_i(t_1), \ddot{X}_j(t_2)];\ 1 \le i \le n,\ nm+1 \le j \le nm(nm+1)/2\}$$
$$\Sigma_{\dot{X}X} = \{\mathrm{Cov}[\dot{X}_i(t_1), X_j(t_2)];\ n+1 \le i \le nm,\ 1 \le j \le n\}$$
$$\Sigma_{\dot{X}\dot{X}} = \{\mathrm{Cov}[\dot{X}_i(t_1), \dot{X}_j(t_2)];\ n+1 \le i \le nm,\ n+1 \le j \le nm\}$$
$$\Sigma_{\dot{X}\ddot{X}} = \{\mathrm{Cov}[\dot{X}_i(t_1), \ddot{X}_j(t_2)];\ n+1 \le i \le nm,\ nm+1 \le j \le nm(nm+1)/2\}$$
$$\Sigma_{\ddot{X}X} = \{\mathrm{Cov}[\ddot{X}_i(t_1), X_j(t_2)];\ nm+1 \le i \le nm(nm+1)/2,\ 1 \le j \le n\}$$
$$\Sigma_{\ddot{X}\dot{X}} = \{\mathrm{Cov}[\ddot{X}_i(t_1), \dot{X}_j(t_2)];\ nm+1 \le i \le nm(nm+1)/2,\ n+1 \le j \le nm\}$$
$$\Sigma_{\ddot{X}\ddot{X}} = \{\mathrm{Cov}[\ddot{X}_i(t_1), \ddot{X}_j(t_2)];\ nm+1 \le i \le nm(nm+1)/2,\ nm+1 \le j \le nm(nm+1)/2\}$$

Note that due to symmetry of covariance matrices it suffices to determine the diagonal and the upper or lower triangle. Further, since the partial (mean square) derivatives of a quantity with respect to the parameters indexed by i and j and j and i are identical the vector $\ddot{\mathbf{X}}$ can be reduced to a vector of $(m^2 + m)/2$ dimensions.

Here, we consider only the scalar case $\mathbf{X} = X$ $(n = 1)$. For simplicity of notation and in accordance with the procedures adopted for scalar processes [8], it is convenient to introduce a centralized field :

$$Z(t) = X(t) - m_X(t) \qquad (3a)$$

$$\dot{Z}_i(t) = \dot{X}_i(t) - m_{\dot{X}_i}(t) \qquad (3b)$$

$$\ddot{Z}_{ij}(t) = \ddot{X}_{ij}(t) - m_{\ddot{X}_{ij}}(t) \qquad (3c)$$

with threshold function

$$\alpha(t) = a(t) - m_X(t) \qquad (3d)$$

and derivatives

$$\dot{\alpha}_i(t) = \dot{a}_i(t) - \dot{m}_{Xi}(t) \qquad (3e)$$

$$\ddot{\alpha}_{ij}(t) = \ddot{a}_{ij}(t) - \ddot{m}_{Xij}(t) \qquad (3f)$$

with obvious notation. Obvious simplifications result when $Z(t)$ is a homogeneous field but only the threshold function is parameter–variant or $Z(t)$ is non–homogeneous but the threshold function is constant.

Alternatively, the field

$$Z(t) = X(t) - a(t) \qquad (4a)$$

can be considered where the maxima of the original field above $a(t)$ correspond to maxima of $Z(t)$ above zero. The means of the components of $(Z(t),\dot{Z}(t),\ddot{Z}(t))$ of this field are

$$m_Z(t) = m_X(t) - a(t) \qquad (4b)$$

$$m_{\dot{Z}_i}(t) = m_{\dot{X}_i}(t) - \dot{a}_i(t) \qquad (4c)$$

$$m_{\ddot{Z}_{ij}}(t) = m_{\ddot{X}_{ij}}(t) - \ddot{a}_{ij}(t) \qquad (4d)$$

while the covariance matrix of $(Z(t),\dot{Z}(t),\ddot{Z}(t))$ equals the one in eq. (2).

Excursion probabilities for scalar fields based on local maxima considerations

We now generalize Bolotin's derivation for the rate of local maxima above a level a in [5] for the non–homogeneous case. Define by B $(B \subset \mathbb{R}^n\}$ a certain domain in the parameter space with ∂B its boundary and by

$$A = \{\mathbf{t} \in B \backslash \partial B;\ Z(\mathbf{t}) \geq \alpha\} \tag{5}$$

the excursion set. The case where excursions also can occur on the boundary ∂B of B is discussed later. For brevity of notation, reference to the parameter vector is now temporarily omitted.

The mean rate of local maxima above α in an excursion set is defined by [5]:

$$\mu_{\max}(\alpha; \Delta\mathbf{t}) = \lim_{\Delta\mathbf{t}\to 0} \frac{P(\alpha\ ;\ \Delta\mathbf{t})}{\Delta\mathbf{t}} = \lim_{\Delta\mathbf{t}\to 0} \frac{P(M(\Delta\mathbf{t}) > 0)}{\Delta\mathbf{t}} \tag{6}$$

with

$$P(\alpha; \Delta\mathbf{t}) = P(Z_{\max} > \alpha; \Delta\mathbf{t}) = P\begin{bmatrix} \{Z > \alpha\} \cap \\ \{\dot{Z} \in \Delta\dot{z}\} \cap \\ \{H \in \mathbb{H}_-\} \cap \\ \{\mathbf{t} \in \Delta\mathbf{t}\} \end{bmatrix} \tag{7}$$

The field obeys certain regularity conditions:

a: Z and a are at least twice differentiable

b: $P((Z > \alpha)\cap(\bigcap_{i=1}^{m} \dot{Z} = 0)\cap(\det(H) = 0)) = 0$

c: $P((Z > \alpha)\cap(\bigcap_{i=1}^{m-1} \dot{Z} = 0)\cap(\det(H) = 0)\cap(Z(\mathbf{t}) \in \partial B) = 0$

d: The point process $M(\tau)$ of local maxima is assumed to be a regular process, i.e. with $P(M(\Delta\mathbf{t}) > 0) = o(\Delta\mathbf{t})$ when $\Delta\mathbf{t}\to 0$

The first condition is obvious. The second condition requires that at a maximum the field has non–vanishing curvatures. Therefore, for example, plane parallel waves are excluded from consideration. The third usually imposed condition namely that the probability of maxima on the boundary ∂B is negligible will be discussed in some detail later as mentioned before. The forth condition ensures that the mean number of local maxima in B can be obtained by integration.

The derivations are particularly simple if excursions above zero for the field as defined by eqs. (4) are considered. Then, $\Delta\dot{Z}$ in eq. (7) is a volume element containing $\dot{Z} = \mathbf{0}$ and H is the Hessian matrix of second derivatives of Z. Clearly, at a maximum this matrix must belong to the set of negative definite matrices $\mathbb{H}_-$. Neglecting higher order terms the volume of $\Delta\dot{\mathbf{z}}$ can be written as

$$\Delta\dot{\mathbf{z}} = |\det(H)|\ \Delta\mathbf{t} \tag{8}$$

with

$$\Delta\dot{z}_j = \sum_{k=1}^{n} h_{jk}\, \Delta t_k$$

Given the joint density $\varphi(z,\dot{z},\ddot{z};t)$ of Z, $\dot{Z}$ and $\ddot{Z}$ the probability in eq. (7) can be written as:

$$P(Z_{max} \geq 0;\ \Delta t) = \int_{\ddot{z}\in \mathbf{I\!I}_-} d\ddot{z} \int_{\Delta\dot{z}} d\dot{z} \int_0^{\infty} \varphi(z,\dot{z},\ddot{z};t)\ dz \qquad (9)$$

For $\Delta\dot{z}$ small and a slowly varying integrand the integration over $\Delta\dot{z}$ by applying the mean value theorem of integral theory yields:

$$P(Z_{max} \geq 0;\ \Delta t) = \Delta t \int_{\ddot{z}\in \mathbf{I\!I}_-} d\ddot{z} \int_0^{\infty} \varphi(z,0,\ddot{z};t)\ |\det(H)|\ dz \qquad (10)$$

and for the rate of maxima one obtains:

$$\mu_{max}(0) = \int_0^{\infty} \int_{\ddot{z}\in \mathbf{I\!I}_-} \varphi(z,0,\ddot{z};t)\ |\det(H)|\ d\ddot{z}\ dz \qquad (11)$$

The expected number of maxima, therefore, is:

$$E[M(B)] = \int_B \int_0^{\infty} \int_{\ddot{z}\in \mathbf{I\!I}_-} \varphi(z,0,\ddot{z};t)\ |\det(H)|\ d\ddot{z}\ dz\ dt \qquad (12)$$

For a sufficiently mixing random field the reliability function then becomes [8]:

$$R(B) = P(\text{no excursion in } B) \sim \exp[-E[M(B)]] \qquad (13)$$

Similar formulae can be derived for the field defined by eqs. (3). They are also given here because they will allow generalizations to vector fields. $\Delta\dot{z}$ is now a parallelepiped containing $\dot{z} = \dot{a}$. H is given by

$$H = H_{\dot{Z}} - H_{\dot{a}} \qquad (14)$$

As before $\Delta\dot{z}$ is written as

$$\Delta\dot{z} = |\det(H)|\ \Delta t \qquad (15)$$

with

$$\Delta \dot{z}_j = \sum_{k=1}^{n} h_{jk} \Delta t_k$$

Given the joint density $\varphi(z,\dot{z},\ddot{z};t)$ of Z, $\dot{Z}$ and $\ddot{Z}$ the probability in eq. (7) can be written as:

$$P(Z_{\max} \geq \alpha;\, \Delta t) = \int_{\ddot{z} \in \mathbb{I}_-} d\ddot{z} \int_{\Delta \dot{z}} d\dot{z} \int_{\alpha}^{\infty} \varphi(z,\dot{a},\ddot{z};t)\, dz$$

$$= \Delta t \int_{\ddot{z} \in \mathbb{I}_-} d\ddot{z} \int_{\alpha}^{\infty} \varphi(z,\dot{a},\ddot{z};t)\, |\det(H)|\, dz \tag{16}$$

Therefore,

$$\mu_{\max}(\alpha;t) = \int_{\alpha}^{\infty} \int_{\ddot{z} \in \mathbb{I}_-} \varphi(z,\dot{a},\ddot{z};t)\, |\det(H)|\, d\ddot{z}\, dz \tag{17}$$

and the expected number of maxima is

$$E[M(B)] = \int_{B} \int_{\alpha}^{\infty} \int_{\ddot{z} \in \mathbb{I}_-} \varphi(z,\dot{a},\ddot{z};t)\, |\det(H)|\, d\ddot{z}\, dz\, \mathbf{dt} \tag{18}$$

with

$$\varphi_s(\zeta;t) = (2\pi)^{-s/2} \det(R(t))^{-1/2} \exp[-1/2\, (\zeta^T R(t)^{-1} \zeta)]$$

where $\zeta = (z,\dot{a},\ddot{z})^T$ and $s = 1 + m + (m^2 + m)/2$ and $R(t)$ is the covariance matrix for the elements of ζ.

Bolotin met serious difficulties when evaluating eq. (11) or (17) numerically even in a special homogeneous case and we shall focus in the remainder of this paper primarily on this task.

Before discussing the numerical evaluation of the derived formulae it is necessary to study the case of excursions on the boundary. Whereas it is natural to assume that in the homogeneous case the number of excursions on the boundary ∂B is negligible as compared with the number in the interior of B the same may not be true for non–homogeneous fields and/or varying thresholds. An example where excursions on the boundary might not be neglected is a plate with fixed edges under spatially varying loading. The formal derivation of the rates of excursions on the boundary is relatively straightforward as will be seen. Its numerical evaluation, however, will require some effort.

Using the additivity of rates of excursions (regularity property of the point process of local maxima) one can modify eq. (12) as:

$$E[M(B)] = \int_B \mu_{max}(F;t)\, dt + \int_{\partial B} \tilde{\mu}_{max}(F;t)\, ds(t) \tag{19}$$

ds(**t**) means surface integration and $\tilde{\mu}_{max}$ (F:t) is the rate of local maxima on the boundary. Let $t_m = \psi_m^{-1}(\ldots,t_1,t_2,\ldots t_{m-1})$ be a suitable parameterization so that one can write:

$$\int_{\partial B} \tilde{\mu}_{max}(F;t_1,t_2,\ldots t_m)\, ds(\mathbf{t})$$

$$= \int_B \tilde{\mu}_{max}(F;t_1,t_2,\ldots,\psi_m^{-1}(\ldots,t_1,t_2,\ldots t_{m-1}))$$

$$\Big(1 + \sum_{i=1}^{n-1} \big(\frac{\partial \psi_m}{\partial t_i}\big)^2\Big)^{1/2} dt_1\, dt_2 \ldots dt_{m-1} \tag{20}$$

It is easy to see that the rate $\tilde{\mu}_{max}(F;\, t_1,t_2,\ldots,t_{m-1})$ equals the one in eq. (12) but multiplied with an additional factor and with the last dimension of $\dot{z}$ and $\ddot{z}$ omitted. Unfortunately, the function $\psi^{-1}(.)$, in general, cannot be given explicitly and we will make no attempt to investigate possible numerical solutions but rather comment briefly on the subject in a qualitative manner.

In accordance with [4] and [5] it is reasonable to assume that the probability of excursions on the boundary is negligible small whenever the probability of excursions with respect to the parameter vector **t** is largest in the interior of B. Therefore, the second term in eq. (19) can be neglected as usual.

For the case where the maximum point of the integrand of eq. (12) is on the boundary ∂B one obviously has to distinguish between two situations. If in the neighborhood of the maximum point of the integrand in the second integral of eq. (19) the non–homogeneous, deterministic part of Z(**t**) is only "slowly" varying, the contribution of the surface integration in that neighborhood is supposed to be negligible small as compared to the volume integration in that neighborhood. Consequently, the second part in eq. (19) is negligible. If, however, the non–homogeneous, deterministic part of Z(**t**) is not slowly varying as, for example, in the abovementioned case of a response field in the corners of rectangular plates under spatially varying loading, the surface integration part may not be negligible. The assessment of "slowness" of variations of the field near the maximum point of the integrand is beyond the scope of this study and should in any case be rather difficult.

Evaluation of the integrals by Laplace's method

Eqs. (12) is a simple volume integral as well as eq. (18). For a time–invariant scalar field in the plane the dimension of the corresponding integral is 6, and for each new dimension of the field the dimension of the integral increases drastically, i.e for a tree–dimensional random field the dimension of the integral is 10, for a four–dimensional field (three space coordinates and one time coordinate) it is 15. In general, this integral must be evaluated numerically and the numerical effort becomes considerable for larger dimensions of **t** as just exemplified. That numerical integration for scalar fields is required already in the two–dimensional case is shown in [5] for a special, simple case. Since we are interested in rare occurrences of such excursion characteristics it is however justified to apply asymptotic concepts which essentially reduce the integration problem to a problem of mathematical programming and some simple algebra.

Denote by $\mathbf{y} = (\mathbf{x},\mathbf{t})^T$ the vector of integration variables. Integrals of the type

$$I(\lambda) = \int_D h(\mathbf{y}) \exp[\lambda f(\mathbf{y})] \, d\mathbf{y} \tag{21}$$

then have certain asymptotic solutions for $\lambda \to \infty$ depending on the behavior of the functions $h(\mathbf{y})$ and $f(\mathbf{y})$ and the boundary of the domain D. D is assumed to be a bounded simply connected domain in the n–dimensional Euclidean space containing the origin. $f(\mathbf{y})$ is assumed to have a maximum at $\mathbf{y}^*$ either in the interior of D or on its boundary ∂D. Also, $f(\mathbf{y})$ and $h(\mathbf{y})$ are sufficiently smooth functions around $\mathbf{y}^*$ and $h(\mathbf{y}^*) \neq 0$. In particular, $f(\mathbf{y})$ and $h(\mathbf{y})$ are at least twice differentiable in $\mathbf{y}^*$. Without loss of generality it is assumed that $\mathbf{y}^* = \mathbf{0}$ since this can always be achieved by a translation of the coordinate system. The basic idea to obtain asymptotic approximations for the integral is to show that with $\lambda \to \infty$ only a small neighborhood of $\mathbf{y}^*$ contributes to the integral which implies that the function $f(\mathbf{y})$ and possibly the boundary of D can be expanded into a Taylor series of which only the first non–vanishing terms are retained.

We distinguish several cases for which it is shown in [10] and [11] that the following asymptotic approximations for $\lambda \to \infty$ hold:

I: $\mathbf{y}^* = \mathbf{0}$ is an interior point of D [10]:

$$I(\lambda) = \int_D h(\mathbf{y}) \exp[\lambda f(\mathbf{y})] \, d\mathbf{y}$$

$$\sim h(\mathbf{y}^*) \exp[\lambda f(\mathbf{y}^*)] \left(\frac{2\pi}{\lambda}\right)^{n/2} \frac{1}{(|\det(S(\mathbf{y}^*))|)^{1/2}}$$

where the Hessian matrix $S(\mathbf{y}^*) = \{\frac{\partial^2 f(\mathbf{y}^*)}{\partial y_i \, \partial y_j};\ i,j=1,\dots,n\}$ of $f(\mathbf{y})$ at point $\mathbf{y}^*$ is negative definite.

II: $\mathbf{y}^* = \mathbf{0}$ is on the boundary of D but $\nabla f(\mathbf{y}^*) \neq 0$ and ∂D is not differentiable in $\mathbf{y}^*$. D is represented as $D = \bigcap_{i=1}^{k} D_i$ ($k \in \{1,\dots,n\}$) with $D_i = \{g_i(\mathbf{y}) \leq 0\}$ and the functions $g_i(\mathbf{y})$ are twice differentiable in a neighborhood of $\mathbf{y}^*$ [11]:

$$I(\lambda) = \int_D h(y) \exp[\lambda f(y)]\, dy$$

$$\sim h(y^*) \exp[\lambda f(y^*)] \frac{(2\pi)^{(n-\kappa)/2}}{\lambda^{(n+\kappa)/2}}$$

$$\frac{1}{\det(A)} \left(\prod_{i=1}^{\kappa} |\gamma_i|^{-1}\right) \frac{1}{(|\det(S(y^*))|)^{1/2}}$$

where

$g_i(y) = 0$ for $i = 1,\ldots,\kappa$ ($\kappa \in \{1,\ldots,k\}$)

$a_i = \nabla g_i(y^*)$ $(i=1,\ldots,\kappa)$ are linearly independent and there is $a_{ij} = 0$ for $i=1,\ldots,\kappa$ and $j=\kappa+1,\ldots,n$

$A = (a_1, a_2, \ldots, a_\kappa)$

$\nabla f(y^*) = \sum_{i=1}^{\kappa} \gamma_i a_i$ with $\gamma_i < 0$

$S(y^*) = \{s_{ij};\ i,j = \kappa+1,\ldots,n\}$ with

$$s_{ij} = \frac{\partial^2 f(y^*)}{\partial y_i\, \partial y_j} - \sum_{q=1}^{\kappa} \gamma_q \frac{\partial^2 g_q(y^*)}{\partial y_i\, \partial y_j}$$ and where $\det(S(y^*)) = 1$ for $\kappa = n$

The special case were y^* is on the boundary of D but $\nabla f(y^*) \neq 0$ and ∂D is at least twice differentiable in y^* leads to the formula [10,11]:

$$I(\lambda) = h(y^*) \exp[\lambda f(y^*)] \frac{(2\pi)^{(n-1)/2}}{\lambda^{(n+1)/2}} \frac{1}{\|y^*\|} \frac{1}{(|\det(S(y^*))|)^{1/2}}$$

where $S(y^*) = \{s_{ij};\ i,j = 2,\ldots,n\}$ with

$$s_{ij} = \frac{\partial^2 f(y^*)}{\partial y_i \partial y_j} - \frac{\|y^*\|}{\|\nabla g(y^*)\|} \frac{\partial^2 g_1(y^*)}{\partial y_i\, \partial y_j}$$

In order to use these results for the integrations in eq. (12), (18) we perform a central scaling of the integration domain, i.e. we determine the integral

$$I(\beta D) = \int_{\beta D} r(y) \exp[s(y)]\, dy \tag{22}$$

Making the substitution $y = \beta x$ leads to

$$I(\beta D) = \beta^n \int_D r(\beta x) \exp[s(\beta x)]\, dx$$

$$= \beta^n \int_D \exp[s(\beta x) + \ln r(\beta x)]\, dx \tag{23}$$

It is seen that the foregoing results apply with:

$$h(\mathbf{x}) = 1 \tag{24}$$

$$f(\mathbf{x}) = s(\mathbf{x}) + \ln r(\mathbf{x}) \tag{25}$$

provided that $f(\beta\mathbf{y})$ has a maximum in D and increases monotonically with β. Note that the maximum point of $f(\beta\mathbf{y})$ may explicitly depend on β in this case.

The mean number of excursion characteristics in B

Eq. (18) is repeated here:

$$E[M(B)] = \int_B \int_a^{\infty} \int_{\ddot{z}\in\mathbb{L}} \varphi_s(z,\dot{a},\ddot{z};t)\ |\det(H)|\ d\ddot{z}\ dz\ dt \tag{26}$$

It is seen that this integral is, in fact, of the type (23).

Using the notation $\zeta = (z,\dot{z},\ddot{z})^T = (z,\dot{a},\ddot{z})^T$ we have:

$$s(\beta \begin{bmatrix}\zeta\\ \mathbf{t}\end{bmatrix}) = -\tfrac{1}{2}\beta^2\ \zeta^T\ R^{-1}(\beta t)\ \zeta$$

$$r(\beta \begin{bmatrix}\zeta\\ \mathbf{t}\end{bmatrix}) = \beta^m\ |\det(H)|\ (\det(R(\beta t)))^{-1/2}$$

and

$$f(\beta \begin{bmatrix}\zeta\\ \mathbf{t}\end{bmatrix}) = \beta^2\ [(-\tfrac{1}{2}(\zeta^T\ R^{-1}(\beta t)\ \zeta) + \frac{1}{\beta^2}\ln\{\beta^m\ |\det(H)|\ (\det(R(\beta t)))^{-1/2}\}]$$

where $\lambda = \beta^2$. For large λ the second term is negligible as compared to the first term. Assuming that the integration domain is already a large domain allows to set $\lambda = \beta^2 = 1$. This leads to:

$$E[M(B)] \sim (2\pi)^{-s/2} \int_B \int_a^{\infty} \int_{\ddot{z}\in\mathbb{L}} \exp[f(\zeta,t)]\ d\ddot{z}\ dz\ dt \tag{27}$$

with

$$f\left(\begin{bmatrix}\zeta\\ t\end{bmatrix}\right) = [(-\tfrac{1}{2}(\zeta^T R^{-1}(t)\,\zeta) + \ln\{|\det(H)|\,(\det(R(t)))^{-1/2}\}] \tag{28}$$

and the critical point $(\zeta^*, t^*)^T$ must be found by minimizing the object function

$$G(\zeta,t) = (\zeta^T R^{-1}(t)\,\zeta) + \ln\{|\det(H)|\,(\det(R(t)))^{-1/2}\} \tag{29}$$

for $(\zeta,t)^T$ under the constraints

$$D = \{\zeta \geq a \cap \bigcap_{i=1}^{\frac{m^2+m}{2}} (\omega_i > 0) \cap (t \in B)\} \tag{30}$$

where the ω_i's are the eigenvalues of H.

Example

For illustration of the developments in the above and in order to investigate its accuracy a simple example is investigated in the following. Consider a zero mean unit variance homogeneous scalar gaussian random field in two dimensions $Z(t_1,s_1)$ with covariance function $R_{ZZ}(t,s)$ given by

$$R_{ZZ}(t,s) = \exp[-(\frac{t^2+s^2}{h^2})]$$

where h represents a parameter determining the correlation length. Of interest is the expected number of maxima of this field above the threshold a, which here is chosen as

$$a(t_1,s_1) = a_0 - t_1^{\alpha} \cdot (1-t_1)^{\beta} \cdot s_1^{\delta} \cdot (1-s_1)^{\gamma} \cdot \lambda$$

where a_0 represents a stationary term and $\alpha,\beta,\delta,\gamma$ and λ are shape constants determining the non–homogeneous part of the threshold function. With this threshold function it is possible to make comparisons of the results with the analytical result by Hasofer[2] for homogeneous scalar Gaussian fields.

The numerical results for eq. (12) or (18) are given first. The accuracy of asymptotic approximation to eq. (18) is then tested by an integral evaluation by importance sampling. Finally, the results are compared with the result in eq. (1) by assuming that the field is homogeneous over the domain B with properties as determined in the maximum point for eq. (12) or (18).

For this example $f(\zeta,\tau)$ is written as

$$f(\zeta,\tau) = (-\tfrac{1}{2}\zeta^T R^{-1}(t)\zeta) \cdot \ln\{|\det(H)|(\det(R(t)))^{-1/2}\}$$

with ζ,τ and R(t) given by

$$\zeta^T = (z,0,0,\frac{\partial^2 z}{\partial t_1^2},\frac{\partial^2 z}{\partial s_1^2},\frac{\partial^2 z}{\partial t_1 \partial s_1}), \qquad \tau^T = (t_1,s_1)$$

$$R(t) = R = \begin{bmatrix} 1 & 0 & 0 & -\frac{2}{h^2} & -\frac{2}{h^2} & 0 \\ & \frac{2}{h^2} & 0 & 0 & 0 & 0 \\ & & \frac{2}{h^2} & 0 & 0 & 0 \\ & & & \frac{12}{h^4} & \frac{4}{h^4} & 0 \\ & \text{Symmetric} & & & \frac{12}{h^4} & 0 \\ & & & & & \frac{4}{h^4} \end{bmatrix}$$

with $h = 1$, $\alpha = \beta = \delta = \gamma = 1$ and $\lambda = 1 \cdot 10^{-3}$ and with index parameter set $t_1 x s_1$, given by $[0,1]x[0,1]$. The minimum point is obtained by a suitable optimization algorithm in an interior point of the index set for all positive values of a_0 with only one active constraint namely $z = a$. This means that the formula IV is valid. E[M(B)] is plotted as a function of a_0 in figure 1 (curve 1) together with the stationary results of Hasofer (curve 2).

Importance sampling

In order to justify the asymptotic approximation for the expected number of maxima above a certain threshold the integral of equations (12) or (18) is estimated using direct importance sampling. The direct importance sampling method is a variance reducing simulation technique which can be used to estimate general integrals (see i.e. Rubinstein [16])

$$I(x) = \int_D f(x)\, dx \tag{31}$$

where f(x) is a function and D is some integration domain bounded by g(x) such that $\{x \in D | g(x) \leq 0\}$. This integral can be rewritten as

$$I(x) = \int_{\mathbb{R}^n} 1_{g(x)\leq 0}\, f(x)\, dx$$

where $\mathbf{1}_{g(x)\leq 0}$ is defined by

$$\mathbf{1}_{g(x)\leq 0} = \begin{cases} 1 \,;\, g(x) \leq 0 \\ 0 \,;\, g(x) > 0 \end{cases}$$

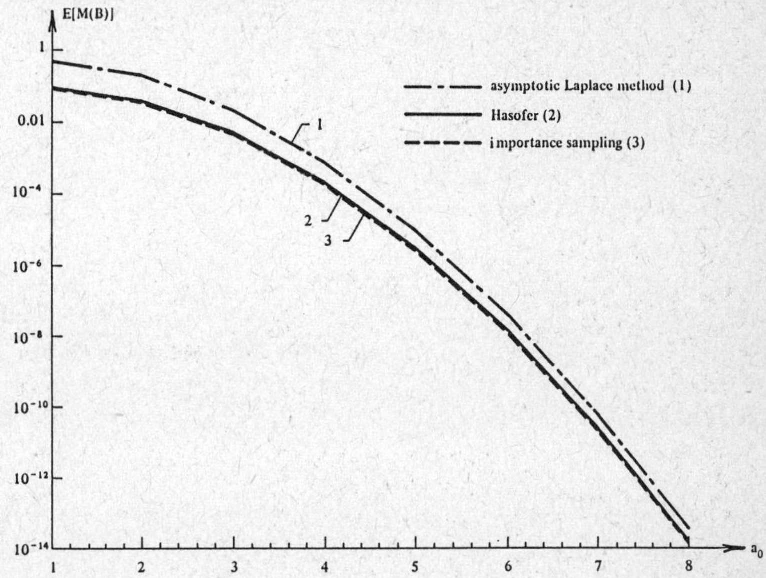

Figure 1. Plots of the asymptotic approximation of the integral (1) and the Hasofer results (2) as function of the threshold value a_0. (3) represents the integral evaluated by importance sampling.

Furthermore it is seen that the introduction of a sampling function h(x) leaves the integral unaltered

$$I(x) = \int_{\mathbb{R}^n} \mathbf{1}_{g(x)\leq 0} \frac{f(x)}{h(x)} \cdot h(x)\, dx \tag{32}$$

Now it is obvious that when h(x) is chosen to be a sampling density function then the integral can be considered as an expectation

$$I(x) = E[\mathbf{1}_{g(x)\leq 0} \frac{f(X)}{h(X)}] \tag{33}$$

where the variable X is distributed according to h(x) and which can be estimated by simulation. The choice of the sampling function h(x) is important for the reduction of the variance of the estimate of the integral. In this example the sampling is performed using a sampling function with all the field variables independently Gaussian distributed and the index parameter variables uniformly distributed.

The mean values of the sampling variables are chosen as the values of the optimum point identified previously. The standard deviations are chosen to be one.

Again for varying level a_0 the expected number of maxima, but here obtained by simulation, are plotted in figure 1 as curve 3. It is observed that the simulated results are rather close to the results according to eq. (1). The asymptotic approximation of the integral for the expected number of maxima still must be considered as very good especially in view of the fact that simulation requires a by orders of magnitude greater numerical effort.

For a number of interesting applications the optimum point will be located on the boundary of the index set. This is, for example, the case when the moment response random field of uniformly loaded, restrained floor slabs is considered. In order to investigate the accuracy of the asymptotic

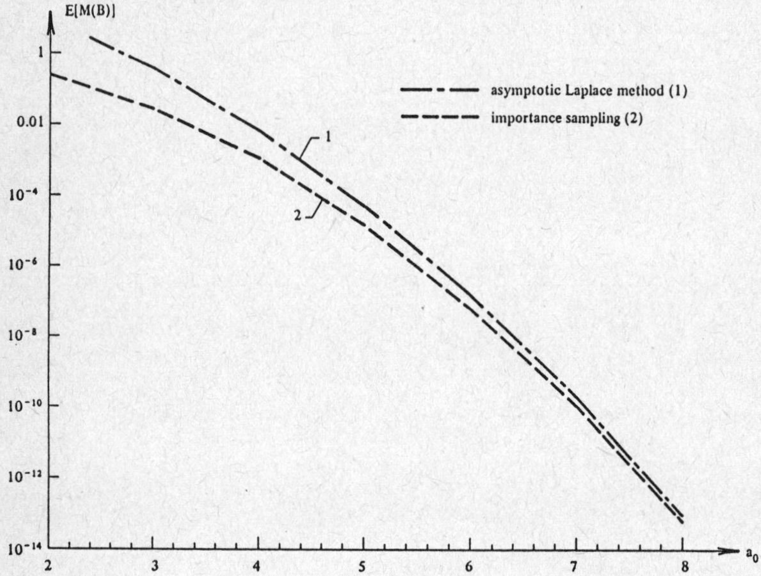

Figure 2: (1) Plot of asymptotic Laplace method results (2) Plot of results obtained by importance sampling

Laplace method in these cases a comparison is made with results obtained by importance sampling as described before.

A maximum on the boundary of the index set is obtained by adjusting the constants in the threshold function. Now we have $\alpha = 0$, $\beta = 6$, $\delta = \gamma = \lambda = 1$. $\beta = 6$ assures that the maximum will be steep with respect to the corresponding index parameter which would resemble the case of the floor slab.

As before the results of the asymptotic Laplace method and the importance sampling are presented as functions of a_0. Plots of the functions are presented in figure 2. It is seen that the results of the two different approaches are again very similar for the range of a_0 presented in figure 2.

Summary and Conclusions

The concepts and results available for the determination of excursion probabilities for Gaussian processes and fields are reviewed. A formula for the mean number of maxima of non–homogeneous Gaussian fields in arbitrary, simply connected parameter sets is derived. The resulting multidimensional integral can be calculated by applying concepts of asymptotic analysis according to Laplace. The resulting approximation is shown to be fairly accurate for high threshold surfaces as demonstrated by a comparison with another asymptotic result for the stationary case and by an integral evaluation by an importance sampling scheme. It should, however, be mentioned that refined descriptions of excursion sets exist, for example, due to Bolotin and Adler whose application should yield more accurate excursion probabilities. Those are not considered herein.

References

[1] Belyaev, Y., K., Point processes and first passage problems, Proc. 6th Berkeley Symp. Math. Statist. Prob., 3, Univ. of California Press, 1972, pp. 1–17

[2] Hasofer, A.M., The Mean Number of Maxima above High Levels in Gaussian Random Fields, Journ. Appl. Prob., 13, 1976, pp. 377–379

[3] Bickel, P., Rosenblatt, M., Two–dimensional random fields, in: Multivariate Analysis III, Ed. P. R. Krishnaiah, Harcourt, Brace, Jovanavich, New York

[4] Adler, R. J., On generalizing the notion of upcrossings to random fields, Adv. Appl. Prob., 8, 1976, pp. 789–805

[5] Bolotin, V.V., Wahrscheinlichkeitsmethoden zur Berechnung von Konstruktionen, VEB Verlag für Bauwesen, Berlin, 1981

[6] Adler, R. J., Hasofer, A. M., Level crossings for random fields, Ann. Prob., 4, 1, 1976, pp. 1–12

[7] Adler, R. J., Excursions above a fixed level by n–dimensional random fields, J. Appl. Prob., 13, 1976, pp. 276–289

[8] Cramer, H., Leadbetter, M.R., Stationary and Related Stochastic Processes, Wiley, New York, 1967

[9] Morse, M., Cairns, S., Critical point theory in global analysis and differential topology, Academic Press, New York, 1969

[10] Bleistein, N., Handelsman, R.A., Asymptotic Expansions of Integrals, Holt, Rinehart and Winston, New York, 1975

[11] Breitung, K., Hohenbichler, M., Asymptotic approximations for multivariate integrals with an application to multinormal probabilities, manuscript, Munich, 1986

[12] Lindgren, G., Extreme Value and Crossings for the χ^2–Process and Other Functions of Multidimensional Gaussian Processes with Reliability Applications, Adv. Appl. Prob., Vol. 12, 1980, pp. 746–774

[13] Breitung, K., Asymptotic Approximations for Multinormal Integrals, Journ. of the Eng. Mech. Div., Vol. 110, No.3, 1984, pp. 357–366

[14] Veneziano, D., Grigoriu, M., Cornell, C.A., Vector–Process Models for System Reliability, Journ. of Eng. Mech. Div., ASCE, Vol. 103, EM 3, 1977, pp. 441–460

[15] Leadbetter, M. R., Lindgren, G., Rootzen, H., Extremes and Related Properties of Random Sequences and Processes, Springer Verlag, New York, 1983

[16] Rubinstein, R.Y., Simulation and the Monte Carlo Method, John Wiley & Sons, New York, 1981

[17] Adler, R.J., The Geometry of Random Fields, John Wiley & Sons, New York, 1981

OPTIMIZATION OF STRUCTURAL SYSTEMS UNDER RESERVE AND RESIDUAL RELIABILITY REQUIREMENTS

Dan M. Frangopol & Gongkang Fu
Dept. of Civil, Environmental and Architectural Engineering
University of Colorado at Boulder, Boulder, Colorado 80309-0428, USA

1 Introduction

Most researchers interested in optimal structural designs based on system reliability requirements consider only the reserve capacity of the systems [5]. The reserve capacity is quantified in the sense of probability by structural system reliability, along with the assumption that system collapse is initiated from the intact state. However, damages will exist in virtually all structural systems [2]. The system residual reliability, referring to the system reliability of structures with damages, should be also included in the optimization process. This is because optimizations considering only the reserve reliability requirement may result in a dramatic reduction in residual capacities of these systems induced by global or partial loss of component capacities.

This paper emphasizes the importance of considering *both* system reserve and system residual reliability requirements in the optimization process. Such an optimization process is performed in a vector (multiobjective) optimization environment. The concepts of reserve and residual reliability are discussed. A solution searching strategy is presented for vector optimization. This strategy is especially suitable for vector optimization of structural systems based on reliabilities. These concepts are illustrated by a numerical example.

2 Reserve and Residual Reliabilities of Structural Systems

The system reserve reliability R_{rsv} is herein defined as the complement of system collapse probability with respect to intact state $P_{f(int)}$:

$$R_{rsv} = 1 - P_{f(int)} \tag{1}$$

where $P_{f(int)}$ is computed by considering failure modes g_i $(i = 1, 2, \cdots, m)$

$$P_{f(int)} = Prob\,[\; any\ g_i \leq 0\;] \tag{2}$$

in which *Prob* denotes probability.

The m failure modes g_i are found by assuming that the failure sequences are initiated from the intact state of the system.

The system residual reliability R_{rsd} is, similarly, defined as the complement of system collapse probability with respect to damage states $P_{f(dmg)}$:

$$R_{rsd} = 1 - P_{f(dmg)} \tag{3}$$

where $P_{f(dmg)}$ is to be obtained by considering failure modes h_j $(j = 1, 2, \cdots, n)$

$$P_{f(dmg)} = Prob\ [\ any\ h_j \leq 0\ ] \tag{4}$$

The n failure modes h_j are found based on the assumption that the failure sequences are initiated from given damage states of the system.

Certain levels of both system reserve and system residual reliabilities are usually required in structural designs. The former addresses the risk of extreme overloadings and the latter deals with the risk of resistance reductions due to, for example, unpredictable accidents. It is practically important to consider both of them in optimal structural design searchings, although overwhelmingly only the system reserve reliabilty has been included in such activities [5].

3 Definition of Problem and Solution

The vector structural optimization under requirements of system reserve and residual reliabilities is formulated as follows:

$$min\ \ [V\ (\mathbf{x}),\ P_{f(int)}(\mathbf{x}),\ P_{f(dmg)}(\mathbf{x})] \tag{5a}$$

$$for\ \ \mathbf{x} \in X = \{\mathbf{x} \mid \mathbf{Y}_1(\mathbf{x}) \leq 0;\ \mathbf{Y}_2(\mathbf{x}) = 0\} \tag{5b}$$

where $\mathbf{x}$ is the design variable vector to be determined; V, $P_{f(int)}$ and $P_{f(dmg)}$ are objective functions to be minimized, representing material volume, system probability of collapse initiated from the intact state, and system probability of collapse initiated from given damage states, respectively; $\mathbf{Y}_1(\mathbf{x}) \leq 0$ and $\mathbf{Y}_2(\mathbf{x}) = 0$ are constraints on the design variable vector $\mathbf{x}$, and serve to define the feasible solution set X.

The solution $\mathbf{x}^*$ to problem (5a-b) is defined as the non-inferior solution. This is a solution under which none of the objectives can be further reduced without increasing at least one of the others. Obviously this kind of solution is not a single point in the design variable space and therefore not unique. Such solutions constitute a solution set which is referred to as Pareto solution set [2,3,8,9,11]. The corresponding objectives evaluated at those Pareto solution points are called Pareto optimal objectives thereafter.

In the next section a strategy to find the Pareto solution set to the vector structural system optimization problem (5a-b) and associated Pareto optimal objectives is presented. Further preferences on those objectives are needed in order to choose one of the solutions as the final design.

4 Pareto Solution Searching Strategy

Approaches to find the complete set of Pareto solution to a multiobjective optimization problem have been presented in many papers [3,4,5,6,8,9,11], among others. This section will, however, present an original three-step strategy of searching the solutions to the structural optimization problem defined by eqns.(5a-b).

Step 1: Choosing Ranges on Upper Limits of System Failure Probabilities to be Considered.

In the problems of structural optimization considering system reliabilities as objectives it is not always necessary to find the complete solution set but only the part which is sensible for further consideration in final decision-making. It is often possible to predetermine the acceptable upper limit for structural failure probabilities. The Pareto solution set to be found will be reduced in size by choosing the size of Pareto optimal objectives as described.

Step 2: Solving Problems of Biobjective Optimization Decomposed from the Original Problem.

For a problem with several objectives it is often very informative to investigate first the interactive influence between any two of the objectives without considering the others. These investigations can be performed in a series of biobjective optimizations. It is obviously possible to have various pairs of objectives in this step. The number of such pairs depends on the interest and experience of the analyst, or simply depends on what the decision maker desires to know. Technically, the Pareto set solution with respect to these two objectives can be found by the so called ε-constraint method [3,8,9,11]. This method converts one of the two objectives to a constraint and conducts a series of the resulted single objective optimizations by giving different levels of the constraint transferred from the objective. The results of these biobjective optimizations are desired to have basic insights to the problem and are often helpful to understand the results produced in the next step.

Step 3: Solving the Problem of Multiobjective Optimization.

This can be done by following the ε - constraint method, i.e. keeping one of the objectives and transferring the rest of them into constraints, and then optimizing the single objective based on a group of organized levels of the transformed constraints.

The three-step solution searching technique described above constitutes an organized strategy suitable for vector structural optimization based on system reliabilities. Its basic idea is to decompose the original problem into a series of simpler problems.

5 Bridge Optimization Example

A truss bridge structural optimization is presented here for illustration (Fig.1). The material to be used for the uniaxial components is assumed ductile (elastic-perfectly plastic). The design variables are chosen as the cross-sectional areas of these uniaxial components **A** , namely:

$$\mathbf{x} = \mathbf{A} \tag{6}$$

Considering the symmetric vertical (vehicle) load S as shown in Fig. 1, symmetric components are required. In order to illustrate the basic philosophy in a simpler way the components are classified in four deterministic design variables (i.e, cross-sectional areas) as follows: vertical members (A_1), lower chord members (A_2), diagonal members (A_3), and upper chord members (A_4). They are indicated also in Fig.1. Based on this classification, the total material volume of the truss bridge system V (in cm^3) is computed as follows:

$$V = 2100 \cdot A_1 + 2100 \cdot A_2 + 5525 \cdot A_3 + 2265 \cdot A_4 \tag{7}$$

Table 1 lists the random variables (i.e., load S and resistances R_1, R_2, R_3, R_4 of the four type of components) considered in this problem and their statistical parameters; $[\bar{\sigma}_y]_i$ is the mean value of yield stress limit of component i. The superscripts "+" and "−" are used to indicate the tension and compression capacities, respectively. The latter are reduced considering buckling effects. All the random variables are assumed normal distributed and independent of each other:

$$\rho_{(R_i,R_j)} = \rho_{(R_k,S)} = 0 \quad (i,j,k = 1,2,3,4) \tag{8}$$

where ρ is correlation coefficient.

Table 2 contains the failure mode expressions for system failure probabilities regarding intact and damage states. The damage states here refer to those states with one component lost. These failure modes are found by the incremental loading method [10]. The failure probabilities are evaluated by Ditlevsen's upper bound [1].

This problem is solved by following the strategy described in the previous section.

Step 1: Setting ranges on upper limits for the failure probabilities to be considered

$$10^{-9} = P^L_{f(int)} \leq P_{f(int)} \leq P^U_{f(int)} = 10^{-5} \tag{9}$$

$$10^{-2.5} = P^L_{f(dmg)} \leq P_{f(dmg)} \leq P^U_{f(dmg)} = 10^{-0.5} \tag{10}$$

in which the superscripts "L" and "U" indicate lower and upper bounds on the upper limit of the failure probabilities considered.

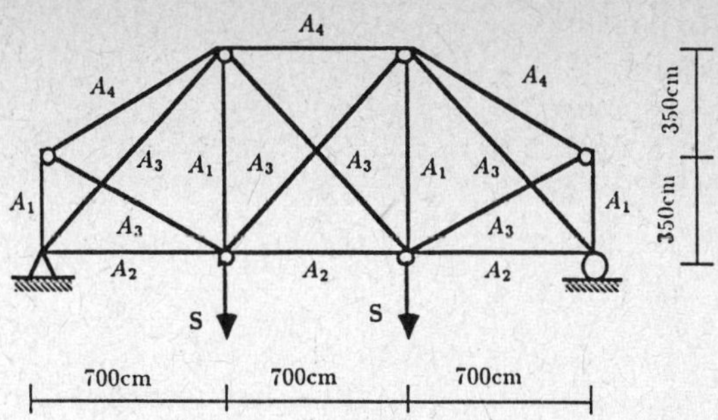

Fig. 1. Truss Bridge Example

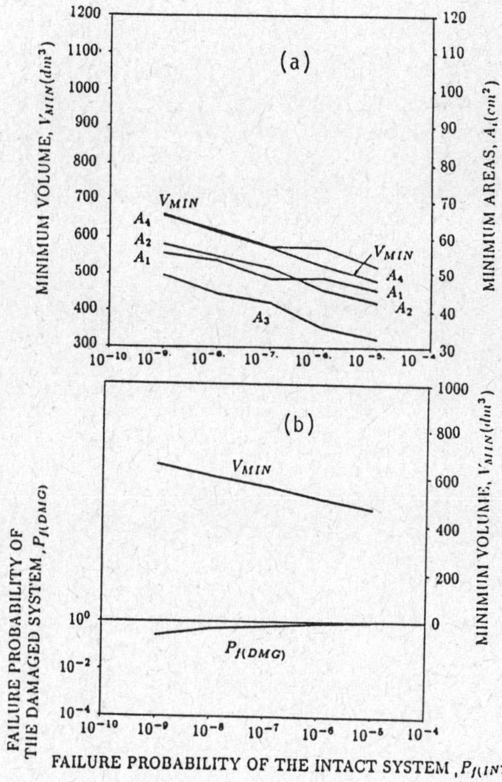

Fig. 2 **Biobjective Optimization of a Truss Bridge** [see eqn. (11); V is denoted V_{MIN}]: **(a) Associated Optimum Areas; (b) Associated Failure Probability of the Damaged Bridge.**

Table 1: Basic Variables

Random Variable	Mean	Coefficient of Variation
R_1	$[\bar{\sigma}_y]_1\ A_1$	11%
R_2	$[\bar{\sigma}_y]_2\ A_2$	11%
R_3	$[\bar{\sigma}_y]_3\ A_3$	11%
R_4	$[\bar{\sigma}_y]_4\ A_4$	11%
S	320 kN	30%

E (Young's Modulus) $=$ 20,000 kN/cm^2

$[\bar{\sigma}_y]_i^+ \;=\; 25kN/cm^2 \qquad (i = 1,2,3,4)$

$[\bar{\sigma}_y]_i^- \;=\; 12.5kN/cm^2 \qquad (i = 1,2,3,4)$

$A_i\ (i = 1,2,3,4)$ — Design Variables

$\rho(_{R_i,R_j}) \;=\; \rho(_{R_k,S}) \;=\; 0 \quad (i,j,k \;=\; 1,2,3,4)$

Table 2: Failure Modes for Reliability Evaluations

1. Failure Modes with Respect to Intact State, $P_{f(int)}$

$$g_1 = 1.0000\ R_1^- + 1.0000\ R_2^+ - S$$

$$g_2 = 1.0000\ R_1^- + 0.7071\ R_3^- - S$$

$$g_3 = 1.0000\ R_2^+ + 0.8944\ R_3^+ - S$$

$$g_4 = 1.0000\ R_2^+ + 0.8944\ R_4^- - S$$

$$g_5 = 0.8944\ R_3^+ + 0.7071\ R_3^- - S$$

$$g_6 = 0.7071\ R_3^- + 0.8944\ R_4^- - S$$

$$g_7 = 1.0000\ R_2^+ + 0.7071\ R_3^+ - S$$

$$g_8 = 0.5000\ R_2^+ + 0.5000\ R_4^- - S$$

$$g_9 = 0.5000\ R_1^- + 1.0000\ R_1^+ + 0.7071\ R_3^+ - S$$

$$g_{10} = 0.2500\ R_1^- + 0.5000\ R_1^+ + 0.5000\ R_4^- - S$$

$$g_{11} = 1.1543\ R_3^+ + 1.0000\ R_1^+ - S$$

$$g_{12} = 0.2236\ R_3^- + 0.5000\ R_1^+ + 0.5000\ R_4^- - S$$

$$g_{13} = 0.4472\ R_4^- + 1.0000\ R_1^+ + 0.7071\ R_3^+ - S$$

$$g_{14} = 0.7236\ R_4^- + 0.5000\ R_1^+ - S$$

2. Failure Modes with Respect to Damage States, $P_{f(dmg)}$ (with One Component Lost)

$$h_1 = 1.0000\ R_1^- - S$$

$$h_2 = 0.5000\ R_2^+ - S$$

$$h_3 = 0.7071\ R_3^- - S$$

$$h_4 = 0.5000\ R_4^- - S$$

$$h_5 = 1.0000\ R_1^+ + 0.7071\ R_3^+ - S$$

$$h_6 = 0.5000\ R_1^- + 0.7071\ R_3^+ - S$$

$$h_7 = 0.5000\ R_1^- + 1.0000\ R_1^+ - S$$

$$h_8 = 0.5000\ R_1^+ + 0.5000\ R_4^- - S$$

$$h_9 = 0.2500\ R_1^- + 0.5000\ R_4^- - S$$

$$h_{10} = 0.2500\ R_1^- + 0.5000\ R_1^+ - S$$

$$h_{11} = 1.1543\ R_3^+ - S$$

$$h_{12} = 0.5000\ R_1^+ + 0.5000\ R_4^- - S$$

$$h_{13} = 0.2236\ R_3^+ + 0.5000\ R_4^- - S$$

$$h_{14} = 0.2236\ R_3^+ + 0.5000\ R_1^+ - S$$

$$h_{15} = 1.0000\ R_1^+ + 0.7071\ R_3^+ - S$$

$$h_{16} = 0.4472\ R_4^- + 0.7071\ R_3^+ - S$$

$$h_{17} = 0.4472\ R_4^- + 1.0000\ R_1^+ - S$$

$$h_{18} = 0.5000\ R_1^+ - S$$

$$h_{19} = 0.7236\ R_4^- - S$$

Step 2: Solving two biobjective optimization problems

$$\min \; [\, V,\; P_{f(int)} \,];\; Subject\ to\ A_i \;\geq\; 25cm^2\ (i = 1,2,3,4) \tag{11}$$

$$\min \; [\, V,\; P_{f(dmg)} \,];\; Subject\ to\ A_i \;\geq\; 25cm^2\ (i = 1,2,3,4) \tag{12}$$

These problems are solved by using the ε-constraint method [3,8,9,11].

Step 3: Solving the following problem using different values of c and d to provide the decision space

$$\min V \tag{13a}$$

subject to

$$A_i \;\geq\; 25cm^2\ (i = 1,2,3,4) \tag{13b}$$

and

$$P_{f(int)} \;\leq\; c\ \ and\ \ P_{f(dmg)} \;\leq\; d \tag{13c}$$

where

$$c \in [\, P^L_{f(int)},\, P^U_{f(int)}]\ \ and\ \ d \in [\, P^L_{f(dmg)}, P^U_{f(dmg)} \,] \tag{13d}$$

Figs. 2(a) and 2(b) depict the results of the first biobjective optimization problem defined in eqn. (11), namely V vs. $P_{f(int)}$. Fig. 2(a) contains the Pareto solution and optimized objectives. It shows that the upper chord (A_4) is required to be relatively strengthened to reach the optimal designs because components composing the upper chord are in compression. It can be observed in Fig. 2(b) that the damage probability associated with the nonoptimized objective $P_{f(dmg)}$ is very high and varies between 0.1 and 1.0. This indicates that the system residual reliability is reduced by the optimization meeting only the requirement of system reserve reliability.

Figs. 3(a) and 3(b) display the other pair of biobjective optimization, i.e. V vs. $P_{f(dmg)}$. By comparing Figs. 2(a) and 3(a), it is observed that the Pareto optimal solution and resulted material volume are more sensitive to $P_{f(dmg)}$ in the given consideration range. Fig. 3(a) also shows that the upper chord (A_4) is still required to have the largest area.

Fig. 4 shows the decision space obtained by solving the problem defined by eqns. (13a-d). It provides Pareto optimal objectives to the decision maker for the final decision. It is easy to recognize that V and $P_{f(dmg)}$ are dominant figures for the final decision making. In fact, this is already indicated by the behaviors of the objectives in Figs. 2(a) and 3(a).

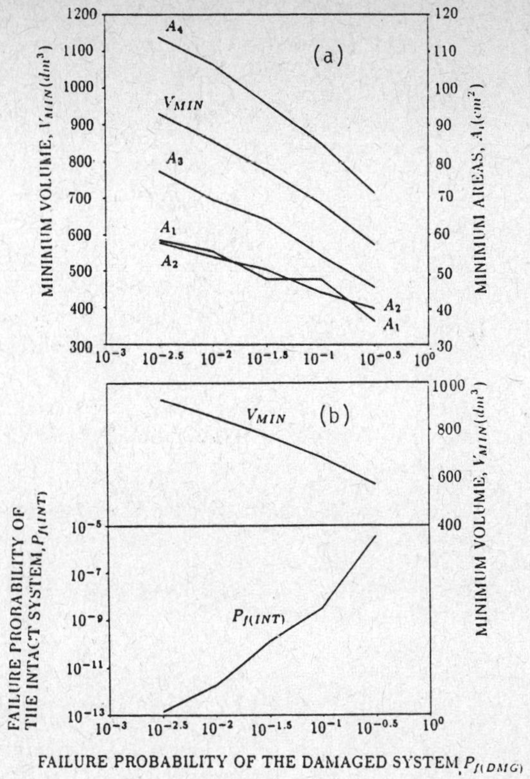

Fig. 3 **Biobjective Optimization of a Truss Bridge** [see eqn. (12); V is denoted V_{MIN}]: **(a) Associated Optimum Areas; (b) Associated Failure Probability of the Intact Bridge.**

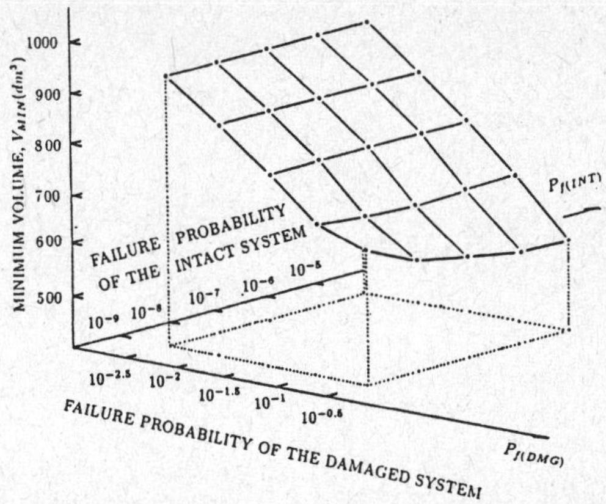

Fig. 4 **Three-Objective Decision Space of a Truss Bridge** [see eqns. (13a-d); V is denoted V_{MIN}].

The similar tendencies of material distribution among the components for the two suboptimization problems defined by eqns. (11) and (12) [see Figs. 2(a) and 3(a)], imply that one of the system failure probabilities need not be considered in determining the final design. Comparing these two figures quantitatively leads to the recognition of the dominant objective (i.e., $P_{f(dmg)}$) since it requires much higher total material volume in the range of consideration.

6 Summary and Conclusions

This paper deals with optimal structural designs under requirements of both system reserve and system residual reliabilities. This problem is solved in a framework of vector (multiobjective) optimization where the system failure probabilities are included in the objective function vector. A strategy to search for Pareto optimal solution and objectives to the problem is presented, which is especially suitable for multiobjective structural optimization based on reliabilities. A truss bridge optimization example is presented for illustration purposes. The following conclusions are reached:

1. Considering only the system reserve reliability in structural optimization may lead to higher risk with respect to the system residual reliability; and

2. In the numerical example presented, the interaction between V and $P_{f(dmg)}$ is dominant. Consequently, the final design can be determined based only on the trade-offs between V and $P_{f(dmg)}$, and $P_{f(int)}$ will be automatically kept at a satisfactorily low level.

Finally, it is noted that work is in progress [7] to incorporate a system redundancy measure in the vector optimization algorithm for both truss and frame structures.

7 Acknowledgement

This research is supported by the U.S. National Science Foundation under Grants No. MSM-8618108 and ECE-8609894 to the University of Colorado at Boulder. Special thanks are due to Ove Ditlevsen for insightful questions during the presentation of this paper.

References

[1] Ditlevsen, O., "Narrow Reliability Bounds for Structural Systems," Journal of Structural Mechanics, Vol. 7, 1979.

[2] Frangopol, D.M., Nakib, R. and Fu, G., "Bridge Reliability Evaluation Using 3-D Analysis and Damage Scenarios," *Probabilistic Methods in Civil Engineering*, Ed. P.D. Spanos, ASCE, New York, 1988.

[3] Frangopol, D.M. and Klisinski, M., "Vector Optimization of Structural Systems," Proceedings ASCE Structures Congress '89, San Francisco, California, May, 1989 (in print).

[4] Fu, G. and Frangopol, D.M., "Reliability-Based Multiobjective Structural Optimization, Phase 2: Applications to Frame Systems," Structural Research Series No. 88-01, Department of Civil, Environmental and Architectural Engineering, University of Colorado, Boulder, Colorado, January, 1988.

[5] Fu, G. and Frangopol, D.M., "Optimal Structures with Respect to System Reliabilities of Intact and Postintact States," Structural Research Series No. 88-11, Department of Civil, Environmental and Architectural Engineering, University of Colorado, Boulder, Colorado, June, 1988.

[6] Fu, G. and Frangopol, D.M., "Multicriterion Reliability-Based Optimization of Structural Systems," *Probabilistic Methods in Civil Engineering*, Ed. P.D. Spanos, ASCE, New York, 1988.

[7] Fu, G. and Frangopol, D.M., " System Reliability and Redundancy in a Multiobjective Optimization Framework," Preprints NSF Workshop on Research Needs for Applications of System Reliability Concepts and Techniques in Structural Analysis, Design and Optimization, Ed. D.M. Frangopol, Boulder, Colorado, September, 1988.

[8] Heimes, Y.Y., Hall, W.A. and Freedman, H.T., *Multiobjective Optimization in Water Resources Systems*, Elsevier Scientific Publishing Company, 1975.

[9] Hwang, C.L. and Masud, A.S.M., *Multiple Objective Decision Making - Methods and Applications*, Springer - Verlag, 1979.

[10] Moses, F., "System Reliability Developments in Structural Engineering," Structural Safety, Vol. 1, 1982.

[11] Osyczka, A., *Multicriterion Optimization in Engineering*, Ellis Horwood Limited, 1984.

STRUCTURAL RELIABILITY ANALYSIS OF FLOATING OFFSHORE STRUCTURE

Joo-Sung Lee
Department of Naval Architecture and Ocean Engineering
University of Glasgow, Glasgow G12 8QQ, UK

ABSTRACT

Present paper is concerned with the system reliability analysis of a floating offshore structure using the developed algorithm called ' Extended Incremental Load Method ', which is an extended approach of the conventional incremental load method, and has been developed in order to extend its applicability to the system reliability analysis of a structure under multi-loadings and to more realistically take into account the post-failure component behaviour.

The modified equation of safety margin is proposed to use the existing strength formula for principle components of a floating offshore structure, in which the concept of the First-Order Second Moment Method is used.

The application to a simple frame structure and a Tension Leg Platform have been carried out and their results show the validity of the proposed approach.

1. INTRODUCTION

Since the begining of the 1970's, the reliability analysis has been applied to the design of marine structures [1]. New design codes have been developed based on reliability analysis [2] and some are in developing stage. During the last decade the need for system reliability analysis has been emphasised, and many studies on this subject have been performed. Most of them, however, deal with discrete structures such as a frame structure for which the possible failure modes can be identified relatively easily. In the case of a continuous stiffened plate structure such as a floating platform, it may not be so simple and easy as in the case of a discrete structure to identify the possible failure modes and to define the redistribution of load effects following the failure of each component.

Up to the present the application of system reliability analysis is a relatively new area. One of the important tasks in structural system reliability analysis is to identify the failure modes. It is not usually practical to include all possible failure modes in evaluating the system reliability of a complex structure. The modes to contribute mainly to the system failure must be defined to estimate the system safety level. These are often referred to as the most likely [3] or the most important or the most significant failure modes [4,10,11], or the stochastically dominant failure modes [5]. In this paper these will be referred to as the most important failure modes. During the last decade several methods have been developed. After

Moses and Stahl [6] suggested the incremental load method, several approximate methods have rapidly been developed, which originally stemmed from the incremental load method of which the failure path approach is most popular. In the failure path approach, since the less important modes are are neglected, the result may lie on the unconservative side, but it is more efficient from the computational efforts' point of view than the other methods[19]. Above all, this approach can take into account the non-linear behaviour by using the two-state model as in Fig. 1. Because of this, most research on the system reliability analysis has been done by the failure path approach. Table 1 shows the comparison of two typical failure path approaches [10,11] : the incremental load method (ILM) and the element replacement method (ERM).

A major aim of the present study is to introduce the system reliability algorithm called 'Extended Incremental Load Method' which has been developed based on the conventional incremental load method [4,6] and the application to a continuous structural such as a TLP. As seen in Table 1, one of the major limitations of the conventional incremental load method is that its application is restricted to a single load pattern. Moses and Rashed [7] presented the applications to the multi-loading cases for the ductile system by incrementing one load and keeping the rest fixed at their final value. The formulation, however, is neither clear nor consistent because all loads should probably be incremented proportionally until failure occurs. The present method is an extended approach of the conventional incremental load method and has been developed to extend its applicability to the system reliability analysis of a structure under multi-loadings. At present only a few works are concerned with the system reliability of floating platforms [8,9,10,11]. Murotsu et al [9] presented the system reliability analysis of semi-submersibles using the linearised failure equation. Component failure was based on the plastic collapse analysis. It is well recognised that buckling failure may be dominant in such a structure.

Since in the incremental load method the load factor up to the particular failure stage can be obtained, it can be used to predict the deformation of the failed component, in a general sense, based on the mean value of the load. This method may have potential to more realistically take into account the post-ultimate strength than other methods and can also be easily applied to evaluating and assessing the system safety with the pre-defined failure modes for the sensitivity study.

In the incremental load method the utilisation ratio generally represents the relationship between component strength and load increments, which may be a complex expression, such as the interaction formula for combinded loading [4]. The existing strength formula developed for the principle components of offshore structure[2,12-14], such as column and pontoon, can be used for this purpose. In this study the modified safety margin equation is employed to use the strength formula.

2. PRESENT APPROACH OF SYSTEM RELIABILITY ANALYSIS

In this section the deriving procedure of the safety margin equation following the present method called ' Extended Incremental Load Method ' is introduced, and the proposed safety margin equation and the reduced element stiffness matrix are described. The identifying procedure of the most important failure modes is also briefly described.

2.1 Derivation of Safety Margin Equation

The deriving procedure of the safety margin equation for the multi-loading case is similar to that of the conventional incremental load method for a single loading case, except that the contribution factor defined below for each loading is included [10,11]. When L loading is acting on a structure of which j components $r_1, r_2, \ldots, r_j$ have failed, the utilisation equation for each loading may be expressed as Eq.(1) similar to the single loading case [4]. For the l th loading :

$$\begin{Bmatrix} R_1 \\ R_2 \\ \cdot \\ \cdot \\ R_j \end{Bmatrix} = \begin{bmatrix} a_{11}^{(l)} & & & & \\ a_{21}^{(l)} & a_{22}^{(l)} & & & \\ \cdot & \cdot & \cdot & & \\ \cdot & \cdot & & \cdot & \\ a_{j1}^{(l)} & \cdot & \cdot & \cdot & a_{jj}^{(l)} \end{bmatrix} \begin{Bmatrix} P_1^{(l)} \\ P_2^{(l)} \\ \cdot \\ \cdot \\ P_j^{(l)} \end{Bmatrix} \tag{1}$$

or

$$\{R_e\} = [a^{(l)}]\,\{P^{(l)}\} \tag{2}$$

where $[a^{(l)}]$ is the utilisation matrix of which elements are defined as the proportion of component strength utilised in the particular load increment and represent the relationship between component strength and load increment, and is a triangular matrix of which all elements in the upper triangle are zero. $\{R_e\}$ and $\{P^{(l)}\}$ are the resistance vector consisting of the strength of failed components and the load increment vector respectively.The superscript (l) represents the terms related to the l th loading. The load increment vector is obtained by solving Eq.(2)

$$\{P^{(l)}\} = [a^{(l)}]^{-1}\,\{R_e\} \tag{3}$$

System resistance is expressed as the sum of each load increments

$$R_{sys} = P_1 + P_2 + \ldots + P_j = \sum_{k=1}^{j} C_k^{(l)'} R_k \tag{4}$$

where $C_k^{(l)'}$ is the sum of the k th column of $[\,a^{(l)}\,]^{-1}$

$$C_k^{(l)'} = \sum_{i=k}^{j} a_{ik}^{(l)^{-1}} \tag{5}$$

The resistance coefficients for l th load can be obtained by normalising the coefficients by the final one. The resistance coefficient $C_k^{(l)}$ corresponding to resistance R_k for the l th loading is

$$C_k^{(l)} = \sum_{k=1}^{j} (a_{ik}^{(l)^{-1}} / C_j^{'(l)}) \qquad k = 1, 2, \ldots, j-1 \tag{6}$$

and

$$C_j^{(l)} = 1.0 \tag{7}$$

When the contribution of resistance for each loading to the system resistance is represented by the contribution factor, $CF^{(l)}$, which is defined here as the relative proportion of utilisation of the l th loading, the resultant resistance coefficients are expressed as a sum of contributions of resistance for each loading to the system resistance.

$$C_k = \sum_{l=1}^{L} C_k^{(l)} . CF^{(l)} \qquad k = 1, 2, \ldots, j-1 \tag{8}$$

Hence, the resistance terms in the equation of safety margin can be expressed as

$$C_1 R_1 + C_2 R_2 + \ldots + C_j R_j \tag{9}$$

where C_j is 1.0. The loading terms can be easily obtained as the sum of the product of the utilisation ratio, $a_{jj}^{(l)}$ and load, $P^{(l)}$.

$$-(a_{jj}^{(1)} P^{(1)} + a_{jj}^{(2)} P^{(2)} + \ldots + a_{jj}^{(l)} P^{(l)}) \tag{10}$$

With Eq.(9) and (10), the safety margin equation for the m th failure mode becomes

$$Z_m = \sum_{k=1}^{j} C_{mk} . R_k - \sum_{l=1}^{L} B_{ml} . P^{(l)} \tag{11}$$

where C_{mk} and B_{ml} are resistance and loading coefficients for m th failure mode respectively, and $B_{ml} = a_{jj}^{(l)}$. The details of deriving the safety margin equation following the above procedure are described in Ref.[15] for a simple structure.

As described in the previous section the utilisation ratio represents the utilised strength of a component, and this can be obtained from the complex strength formula. Conceptually the strength formula can be expressed as

$$F(\mathbf{R}, \mathbf{Q}) = 1 - G(\mathbf{R}, \mathbf{Q}) \tag{12-a}$$

This expression represents the failure surface of a component. When the modelling error, X_M, defined as Eq.(13) [13,16], is introduced as a random variable, Eq.(12-a) can be rewritten as

$$F(X_M, \mathbf{R}, \mathbf{Q}) = X_M - G(\mathbf{R}, \mathbf{Q}) \tag{12-b}$$

$$X_M = \frac{\text{actual behaviour}}{\text{predicted behaviour}} \tag{13}$$

Mean of X_M is referred to as the bias and its coefficient of variation V_{X_M} referred to as the modelling uncertainty which should be no greater than about 13 % for good strength models as recommended by Faulkner et al [16]. Guenard et al [17] also proposed the same idea with the mean bias of the strength model being hopefully close to 1 and its probability density function representing the model uncertainty.

In Eq.(12-a) and (12-b), **R** and **Q** are the random variable vectors related to the component strength, such as the geometric and material properties, and related to the load effects such as axial force and bending moment. The component reliability index is obtained from

$$\beta = -\Phi^{-1}\{\ F(X_M, \mathbf{R}, \mathbf{Q}) \leq 0\ \} \tag{14}$$

In order to use the strength formula for the present purpose, the safety margin equation may be expressed conceptually as a modified form of Eq.(11) for the m th failure mode.

$$Z_m = X_M + \sum_{k=1}^{j-1} G_k(R_k, R_j) - \sum_{l=1}^{L} G_l(Q, R_j) \tag{15}$$

where the first summation term is the contribution of the strength of the failed components to the system resistance and the second term that of loadings.

Expression of the safety margin equation as Eq.(15) may be a feasible way to use the strength formula in the system reliability analysis. The mean and the variance of each coefficient in Eq.(15) are obtained by using the concept of the First-Order Second Moment Method. Let **X** be a vector of the random variable representing the resistance and load effects, i.e., $\mathbf{X} = \{ X_M, \mathbf{R}, \mathbf{Q} \}$ and the mean and the variance of the coefficient is obtained from Eq.(16) [18].

$$\mu_{G_l} \cong G_l(\underline{X}) \tag{16-a}$$

$$\sigma_{G_l}^2 \cong \sum_i \{\frac{dG_l}{dX_i}\}^2 . \sigma_{X_i}^2 + \sum_i \sum_j \{\frac{dG_l}{dX_i}\} . \{\frac{dG_l}{dX_j}\} \sigma_{X_i} \sigma_{X_j} \rho_{ij} \tag{16-b}$$

where $\underline{X}$ is the mean value vector of the random variables, $(X_1, X_2, \ldots, X_i, \ldots)$, σ_{X_i} is the standard deviation of a random variable, X_i and ρ_{ij} is the correlation coefficient between X_i and X_j. The partial derivatives of the function $G_l(X)$ in Eq.(16-b) are numerically evaluated at mean values of the random variables.

2.2 Reduced Element Stiffness Matrix

One problem in the system reliability analysis is to derive the reduced stiffness matrix of a component after failure. The strength formula has usually the form of the non-linear interaction equation, and up to the present there is no way to consider the non-linear interaction failure surface [18]. In this study the concept of plasticity flow theory is employed to derive the reduced stiffness matrix using the strength formula, generally given as Eq.(12).

When an element remains in an elastic state, the relation between the nodal displacement and the nodal force vector, $\{u\}$ and $\{f\}$ is given by

$$[\, k_0 \,]\,\{u\} = \{f\} \tag{17}$$

where $[\, k_0 \,]$ is the elastic stiffness matrix of an element. Let $\{\, a_i \,\}$ and $\{\, a_j \,\}$ be the flow vectors for node i and j respectively, the flow vector of an element is

$$\{ a \} = \begin{Bmatrix} \{ a_i \} \\ \{ a_j \} \end{Bmatrix} \tag{18}$$

and components of the flow vector of a failed element is given by

$$a_k = \frac{dF}{df_k'} \tag{19}$$

where F is the interaction equation as a strength formula, i.e. failure surface equation, and f_k' is the nodal force normalised by the ultimate strength when each loading is applied separately.

The explicit form of the reduced stiffness matrix $[k_r]$, when one or both nodes of an element have failed, is given as follows.

(1) When one node, i has failed.

$$[k_r] = [k_o] - \frac{[k_o] \{ a_i \} \{ a_i \}^T [k_o]}{\{ a_i \}^T [k_o] \{ a_i \}} \tag{20}$$

(2) When both nodes, i and j have failed.

$$[k_r] = [k_o] - [H]^T [G]^{-1} [H] \tag{21}$$

where $[G_{ij}] = \{ a_i \}^T [k_o] \{ a_j \}$ and $[H_i] = \{ a_i \}^T [k_o]$

2.3 Identification of the Important Failure Modes

Identification of the failure modes is one of the most important parts of the failure path approach. In complex structure the number of potential failure modes is usually very large but, as is well recognised, only a few of them are important in evaluating the probability of system failure.

The idea of identifying the most important failure modes in the present study is, in some sense, similar to the Branch and Bound method [3] and is towards reducing the computational time. The identifying procedure is composed of two procedures: Searching Procedure to select the most important failure mode and Discarding Procedure to discard the relatively less important modes in which are adopted the similar idea of bounding technique in Ref.[5] and similar criteria in Ref.[4]. The details are described in Ref.[10]. According to the procedure, failure paths are likely to be obtained in decreasing order of failure probability, or in increasing order of corresponding reliability indices, i.e.

$$P_{f_1} > P_{f_2} > \ldots > P_{f_m} > \ldots \quad \text{or} \quad \beta_1 < \beta_2 < \ldots . < \beta_m < \ldots \tag{22}$$

where P_{f_m} and β_m are the failure probability of the m th path and the corresponding reliability index.

In the procedure of identifying the failure modes, the collapse of structure is defined as the occurrence of the singularity in the structural stiffness matrix. Practically, the occurrence of the structural collapse may be judged from

$$\frac{\text{Det [K] of Current Failure State}}{\text{Det [K] of Intact State}} \leq \varepsilon_{det} \tag{23}$$

where ε_{det} is the prescribed small number of which value may be related to the number of failed components in the particular mode and is given from experience. As it decreases, the system reliability evaluated naturally increases because more components are included in the path. In evaluating the bounds of the system failure probability Ditlevsen's bounds, called narrow bounds, are calculated [20].

2.4 Example of Simple Structure : Plane Frame Model

The plane frame model in Fig. 2 has 8 possible hinges as components. Component failure is assumed when the bending moment at a particular hinge reaches the plastic bending moment, and hence the strength formula can be expressed simply as Eq.(24).

$$F(X_M, R, Q) = X_M - \frac{Q}{R} \tag{24}$$

where Q is the bending moment as the load effect, and R is the plastic bending moment as the strength. Here modelling error, X_M, is assumed to be deterministic having a mean value of 1.0, and all resistance and loading terms in Eq.(15) are assumed to be normal. Table 2 shows results when using the safety margin equation (11) and (15). The 2 nd column is a reliability index of each failure mode when using Eq.(11) as a safety margin equation and exactly the same as those in Ref.[21]. Comparing the results by two types of the safety margin equation Eq.(15) usually gives a lower reliability index than Eq.(11), but the difference is less than 1% and can be ignored.

Results of the simple frame structure justify the validity and the applicability of the present system reliability analysis method and the proposed safety margin equation (15) to a structure under multi-loading. It can be said that using Eq.(15) may be one possible way of using the strength formulae directly in the system reliability analysis towards its results based design.

3. APPLICATION TO TLP STRUCTURE

The Tension Leg Platform is a relatively new concept of platform applied in deep water. At present there are two examples : Hutton TLP and Jolliet TLWP [22]. The structural system of TLPs is a typical continuous structure and its main components such as columns, bracing, pontoons and decks have high redundancy, especially around their joints. It is also very sensitive to weight and its development might therefore depend on how to reduce its weight while retaining sufficient strength [23].

3.1 Environmental Load and Load Effects

Generally environmental loading is categorised into three part [24], i.e. static loading due to dead load and static pressure, quasi-static load due to current, wind, wave drift and quasi-static motion, and dynamic load due to wave and dynamic motion. All loads are included in the present analysis except the load due to wave drift. A Morison type approach [25] is adopted to estimate the hydro-dynamic loading [10,11]. The three categories of load effects are obtained by modelling the TLP structure into the finite element type space frame (Fig. 4) and the load effects, such as axial force and bending moment, are calculated using the beam theory. Boundary conditions are imposed to minimise restraining of the structure. The coefficient of variation of static, quasi-static and dynamic load effect are assumed to be 10%, 20% and 10% respectively.

3.2. Results and Discussion

The Hutton TLP[26] is chosen for the present study (Fig. 3). It is assumed that component failure is restricted within the components on columns and pontoons since deck structure may be stiffer than column and pontoon, and joints are sufficiently stiff so that they can transfer forces from each other and the components on elements around the joints do not fail. Component failure is assumed to occur under the axial load and bending moments. Hence, the effects of shear action are ignored.

The reliability analysis result may differ depending on the distribution type of a random variable such as the strength modelling error. In this study, when using Eq.(15) as the equation of safety margin, the distribution of the strength modelling error, X_M and resistance coefficients and loading coefficients are assumed to be log-normal. It is further assumed that the three catagories of loading, static, quasi-static and dynamic, give the same load effects. COV of important variables are listed in Table 3 [10]. Mean bias of the strength modelling error is given by 0.99 for the cylindrical components and 1.00 for the rectangular box girders. Among them, because at present there is no available test result of a rectangular box-girder under the combined axial compression and bi-axial bending, the modelling error is postulated based on the fact that its failure is closely related to the failure of a stiffened panel in the weakest portion in the section [14].

The evaluated system safety level may differ from the strength formula used in the analysis. In this study the relatively simple interaction Eq.(25) [12] is used.

$$\left(\frac{p}{p_u}\right)^m + \left(\frac{\sigma_x}{\sigma_{xu}}\right)^n = 1 \tag{25}$$

where p is the pressure and σ_x is the total axial stress resulting from axial force, bending moments and pressure, and p_u is the ultimate pressure, and σ_{xu} is the ultimate axial compression. In the above equation, m = 2 and n = 1 is used.

The system reliability analysis of the Hutton TLP model has been performed under the design environmental loading condition [26] with the assumed component behaviour of ductile when wave angle, χ, is 0° (following sea), 45° (quarter sea) and 90° (beam sea) (Fig. 3). Results of the system reliability analysis are summarised in Table 4 (the 3 rd row in Table 4), in which the failure probability is the average value of lower and upper bounds. The total average of the system reliability index is the value corresponding to the average failure probability obtained under the assumption that the same probability of each wave angle is given.

As mentioned in the previous section the incremental load method can also be easily applied to evaluating and assessing the system safety with the pre-defined failure modes for the sensitivity study. In Table 4 the results when V_{X_M} of the cylindrical components has different values , i.e., $V_{X_M} = 5$ % and 13 %, keeping X_M at the same value with the pre-defined failure modes obtained when $V_{X_M} = 10$ % through the probabilistic searching procedure are also includeded (see the 2 nd and 4 th row in Table 4). From Table 4 total average of β_{sys} is increased by 9 % by reducing V_{X_M} by 5 % and decreased by 9 % by increasing V_{X_M} by 3 %. Component reliability indices of the component which is on Column 2 just above pontoon (shaded part in Fig. 3 (2)) are listed in Table 5 for three COV's. As seen in Table 5 the component safety level is much sensitive to change in V_{X_M} as the system safety level. These results say that developing the strength formula with low value of the modelling uncertainty may be one of the easiest ways to raise up the system safety, and therefore to achieve the weight saving with retaining the same level of β_{sys}.

Fig. 5 shows a typical example of deformed shape of the mode having the lowest path reliability for each wave direction at each section in Fig. 4.

When using the incremental load method the load factor, λ_i can be obtained, from which the mean load can be predicted up to the particular failure stage, i. This can be easily extended to estimate the mean value of collapse load from Eq.(26).

$$[\text{ Collapse Load }] = \lambda_T * [\text{ Mean Applied Load }] \qquad (26)$$

where λ_T refers to the total load factor when collapse occurs and may be related to the reserve strength index (RI) [27]. The relation between the total load factor (λ_T) and the path reliability (β_{path}) of the Hutton TLP models is shown in Fig.6 (4 modes for each wave direction). As β_{path} increases, λ_T usually increases, although the path of higher β_{path} does not always have the higher load factor. The average values of λ_T for for each wave direction in Fig. 6 are 2.11, 3.05 and 2.74 respectively and the total average is 2.63. Because of the assumptions in the analysis, the total load factor of a real structure may be lower than these values, but the high value of load factors also support the high redundancy of a continuous structure such as a TLP.

4 CONCLUSIONS

In this paper an 'Extended Incremental Load Method' has been introduced, which is an extended approach of the conventional incremental load method to extend its applicability to the system reliability analysis of a structural system under the multi-loading. A modified form of the safety margin equation is proposed in order to use the strength formula in which mean values and COV's of the resistance and loading coefficients are obtained by using the First-Order Second-Moment method.

The results of simple frame structure justify the validity and the applicability of the present method of system reliability analysis and the proposed safety margin equation (15) to a structure under multi-loading. Hence, it can be said that using the proposed modified safety margin equation may be a possible way to directly use the existing strength formulae.

The modelling uncertainty (V_{X_M}) may give much influence not only on the system reliability but also on the component reliability, and therefore it should be kept as low as possible to achieve the economic benefit in the design with the mean bias being kept close to unity [16].

As far as the present results are concerned, TLP structural system shows a high reserve strength, i.e., a high redundancy, which can be supported by the values of the total load factors (Fig. 6).

The results of the present study are obtained under the assumption of ductile behaviour of a component. The system reliability under actual behaviour may be lower than those values. In the incremental load method the load factor up to the particular failure stage can be obtained, and hence the deformation of a failed component can be predicted in the mean sense using the load factor. It can be, therefore said that the present method may be one possible approach of more practically taking into account the post-failure behaviour of a failed component in a structure under multi-loading.

ACKNOWLEDGEMENT

This work has been carried out in the Department of Naval Architecture and Ocean Engineering at the University of Glasgow under the supervision of Professor D. Faulkner, head of the Department, and the author's many thanks are due to him.

REFERENCES

[1] Mansour, A.E. and Faulkner, D., "On Applying the Statistical Approach to Extreme Sea Loads and Ship Hull Strength", Trans. RINA, Vol.115 (1973) pp.277-314

[2] ABS-CONOCO, "Model Code for Structural Design of Tension Leg Platforms", CONOCO-ABS TLP Rule Case Committee, Final Report, ABS, New York (1984)

[3] Guenard, Y.F., "Application of System Reliability Analysis to Offshore Structures", John A. Blume Earthquake Engineering Centre, Stanford Univ., U.S.A., Report No.71 (1984)

[4] Moses,F., "System Reliability Developments in Structural Engineering", Structural Safety, Vol.1 (1982) pp.3-13

[5] Murotsu, Y., "Reliability Analysis of Framed Structure Through Automatic Generation of Failure Modes", Reliability Theory and its Application in Structural and Soil Mechanics, P.Thoft-Christensen ed., Martinus Nijhoff Pub. (1983), pp.525-540

[6] Moses, F. and Stahl, B., "Reliability Analysis Format for Offshore Structures", Proc. 10 th Offshore Technology Conference, OTC 3046 (1978), pp.29-38

[7] Moses, F. and Rashedi, M.R., "The Application of System Reliability to Structural Safety", Proc. 4 th Intl. Conf. on Applications of Statics and Probability in Soil and Structural Engineering, Univ. de Fierenze, Florence, Italy, G.Augusti et al ed., Vol.1 (1983), pp.573-584

[8] Amdahl, J., Taby, J. and Granli, T., "Progressive Collapse Analysis of Mobile Platforms", Proc. PRADS'87 (1987), pp.1060-1072

[9] Murotsu, Y., Okada, H., Ikeda, Y. and Matsuzaki, S., "On the System Reliability of Semi-submersible Platform", Proc. PRADS'87 (1987), pp.752-764

[10] Lee, J.S. and Faulkner, D., "System Reliability Analysis of Structural System", Dept. of Naval Arch. and Ocean Engineering, Univ. of Glasgow Report, NAOE-88-33 (1988)

[11] Lee, J.S. and Faulkner, D., "Reliability Analysis of TLP Structural Systems", 8 th Intl. Conf. on Offshore Mech. and Arctic Engg., The Hague, The Netherlands (1989) (paper submitted)

[12] Faulkner, D., Chen, Y.N. and de Oliveira, J.G., "Limit State Design Criteria for Stiffened Cylinders of Offshore Structures", ASME 4th National Congress of Pressure Vessels and Piping Technology, Portland, Or. (1983), pp.1-11

[13] Faulkner, D. and Warwick, D.M., "Predicting the Strength of Welded Stiffened Cylinders", Proc. Intnl. Symp. Developments in Deep Waters, RINA, London (1986)

[14] Lee, J.S., "Pre- and Post-Ultimate Behaviour Analysis and Derivation of Strength Model of Rectangular Box Girder", Dept. of Naval Arch. and Ocean Engineering, Univ. of Glasgow Report, NAOE-87-27 (1987)

[15] Lee, J.S., "Basic Study on the System Reliability Analysis of Floating Offshore Structures', Progressive Report, Dept. of Naval Arch. and Ocean Engineering, Univ. of Glasgow (1987) (unpublished)

[16] Faulkner, D., Guedes Soares, C. and Warwick, D.M., "Modelling Requirements for Structural Design and Assessment", Proc. 3rd Intl Conf. on Integrity of Offshore Structures, Elsevier Applied Science, Univ. of Glasgow, D.Faulkner et al ed., Paper No.2 (1987), pp.25-54

[17] Guenard, Y., Goyet, J., Labeyrie, J. and Remy, B., "Structural Safety Evaluation of Steel Jacket Platforms", Proc. Marine Structural Reliability Symposium, The Ship Structure Committee and SNAME, Arlington, VA (1987), pp.169-183

[18] Thoft-Christensen, P. and Baker, M.J. ,"Structural Reliability Theory and its Applications", Springer-Verlag (1982)

[19] Karamchandani, A., "Structural System Reliability Analysis Method", John A. Blume Earthquake Engineering Centre, Stanford Univ., U.S.A., Report No.83 (1987)

[20] Ditlevsen, O., "Narrow Reliability Bounds for Structural Systems", J. of Struct. Mech., Vol.7, No.4 (1979), pp.453-472

[21] Thoft-Christensen, P. and Murotsu, Y., "Application of Structural Systems Reliability Theory", Springer-Verlag (1986)

[22] Faulkner, D., "Tension Leg Platforms - From Hutton to Jolliet" or "Deep Water without Deep Pockets", Dept. of Naval Arch. and Ocean Engineering, Univ. of Glasgow Report, NAOE-88-26 (1988)

[23] Chung, J.S., "Offshore and Arctic Frontier : Structures, Ocean Mining", Mechanical Engineering (1985), pp.55-63

[24] Chen, Y.N., Liu, D. and Shin, Y.S., "Probabilistic Analysis of Environmental Loading and Motion of a Tension Leg Platform for Reliability-Based Design", Centenary Conference on Marine

Safety, Univ. of Glasgow, Paper No.17 (1983)
[25] Morison, J.R., O'Brien, M.P., Johnson, J.W. and Schaaf, S.A., "The Force Exerted by Surface Waves on Piles", J. of Petroleum Technology, AIME, Vol.189 (1950), pp.149-154
[26] Ellis, N., Tetlow, J., Anderson, D. and Woodhead, L., "Hutton TLP Vessel - Structural Configuration and Design Features", Proc. 14 th Offshore Technology Conf., OTC 4427 (1982), pp.557-571
[27] de Oliveira, J.G. and Zimmer, R.A., "Redundancy Consideration in the Structural Design of Flating Offshore Platforms", The Role of Design, Inspection and Redundancy in Marine Structural Reliability, D.Faulkner et al ed., National Academy Press (1984), pp.293-327

Table 1 Comparison between ILM and ERM

	ILM	ERM
behaviour of a failed component	can follow the post- ultimate behaviour	failed components are removed from the structure and their strengths are replaced by the equivalent forces acting on the structure
deformation	deformation of failed components can be allowed	deformation of the failed components are restricted
merit	post-ultimate behaviour of failedcomponents can be considered more realistically	applicable to multi-loading case
limitation	applicable only to single loading cases	post-ultimate behaviour of failed components are idealised as a two-state model : ductile, semi-brittle and brittle

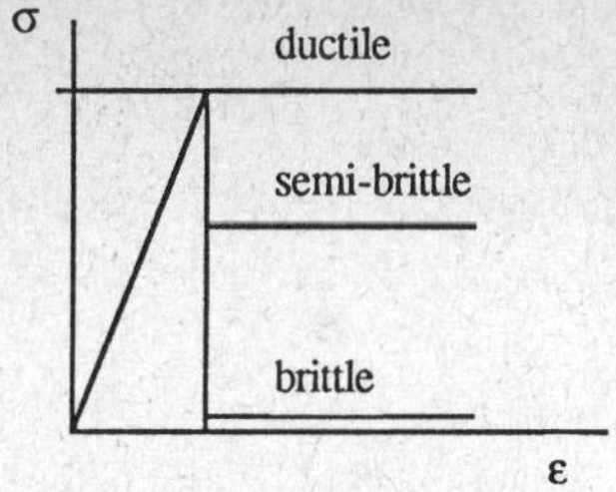

Fig. 1 Two-State Model of Non-linear Behaviour

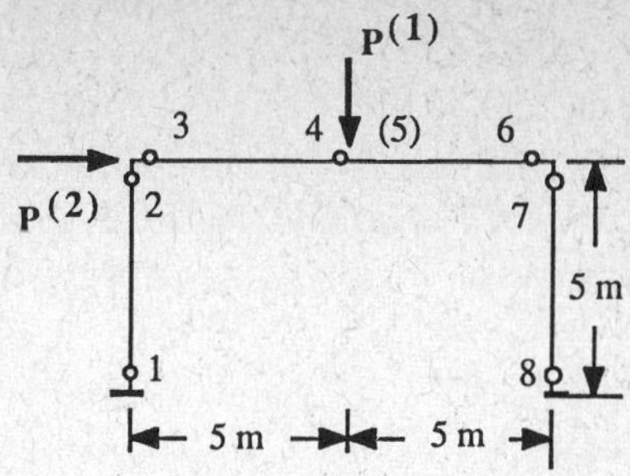

comp.	A_k	I_k	R_k
1, 2	4.0×10^{-3}	3.58×10^{-5}	0.075
3, 4	4.0×10^{-3}	4.77×10^{-5}	0.101
5, 6	4.0×10^{-3}	4.77×10^{-5}	0.101
7, 8	4.0×10^{-3}	3.58×10^{-5}	0.075

COV of $R_k = 0.05$, COV of $P^{(1)}$ & $P^{(2)} = 0.3$

A_k = Cross sectional area (m^2)
I_k = Moment of inertia (m^4)
R_k = Mean value of reference strength (= plastic bending moment, MN)
Mean of yield stress = 276 MPa
$P^{(1)} = 0.02$ MN, $P^{(2)} = 0.04$ MN

Fig. 2 Plane Frame Model (from Ref.[21])

Table 2 Comparison between Two Safety Margin Equations for the Frame Model

Failure Paths	Eq. (11)	Eq. (15)
7-4-2	2.48	2.46
7-4-8-2	2.48	2.46
7-4-3	2.88	2.88
7-4-8-3	2.88	2.88
7-4-8-1	2.91	2.90

Table 3 Coefficients of Variations of Random Variables

variables	COV
E	0.04
σ_y	0.08
Geometric parameters	0.04
Load effect : dynamic	0.1
static	0.1
quasi-static	0.1
Strength modelling error :	
cylinder	0.1
box girder	0.1

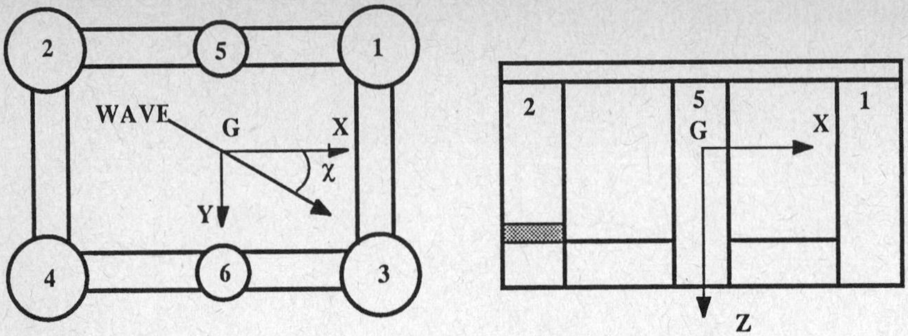

Fig. 3 Hutton TLP

Table 4 System Reliability Indices

V_{X_M}	$\chi = 0$	$\chi = 45$	$\chi = 90$	total average
0.05	6.58	7.61	7.02	6.74
0.10	6.00	6.77	6.83	6.18
0.13	5.43	6.19	5.85	5.61

Table 5 Component Reliability Indices

V_{X_M}	$\chi = 0$	$\chi = 45$	$\chi = 90$
0.05	5.19	7.00	8.08
0.10	3.78	5.31	6.39
0.13	3.13	4.45	5.47

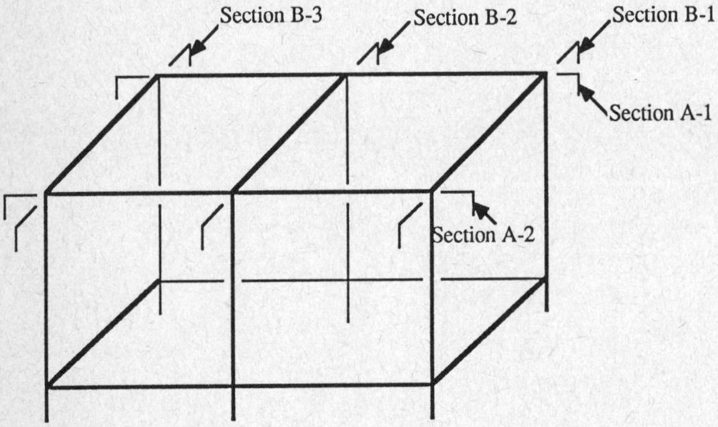

Fig. 4 Space Frame Model

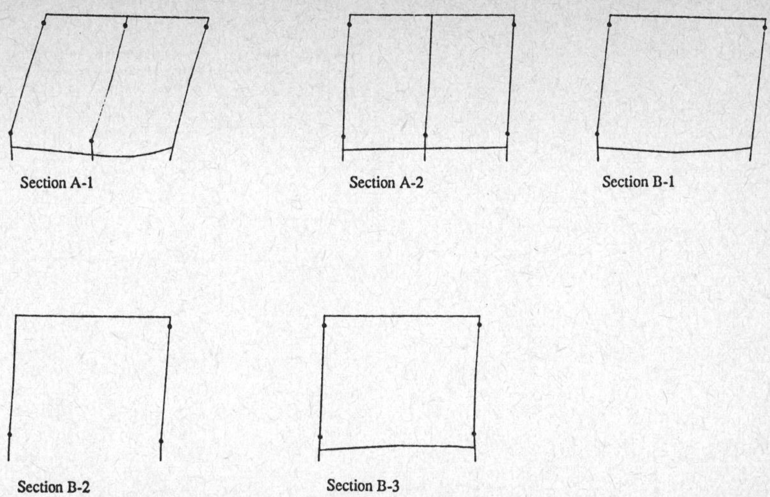

(1) $\chi = 0$ deg.

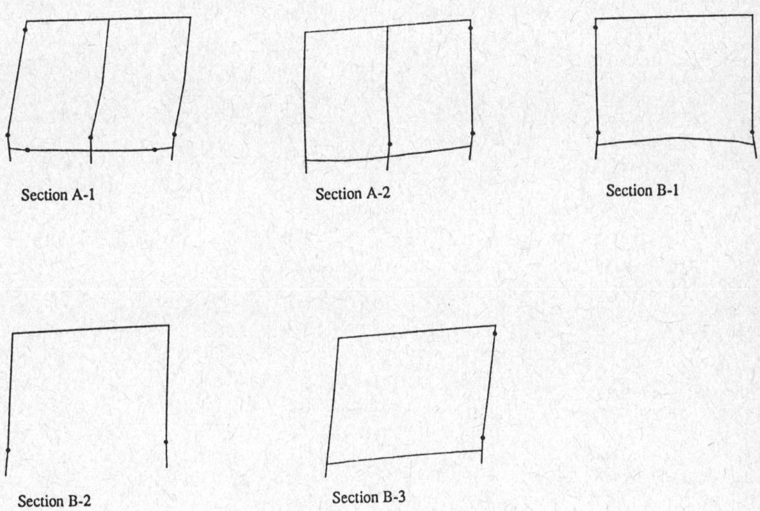

(2) $\chi = 45$ deg.

Fig. 5 Deformed Shapes of the Mode having the Lowest Path Reliability Index (to be continued)

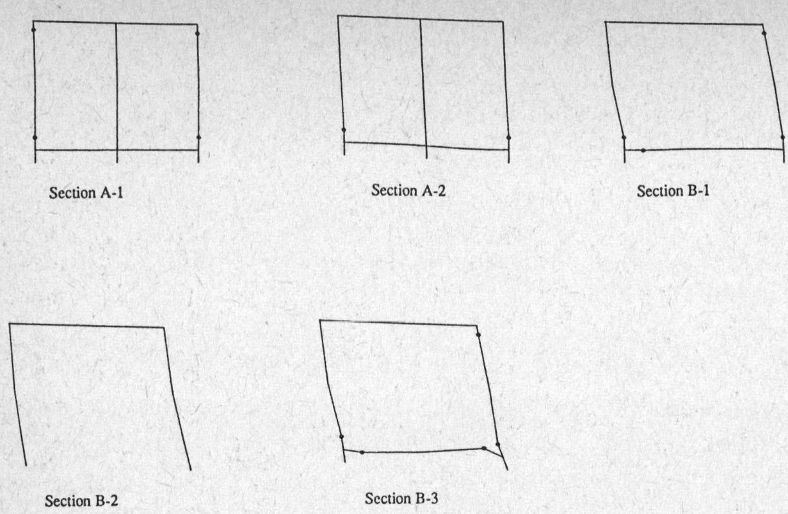

(3) χ =90 deg.

Fig. 5 Deformed Shapes of the Mode having the Lowest Path Reliability Index (continued)

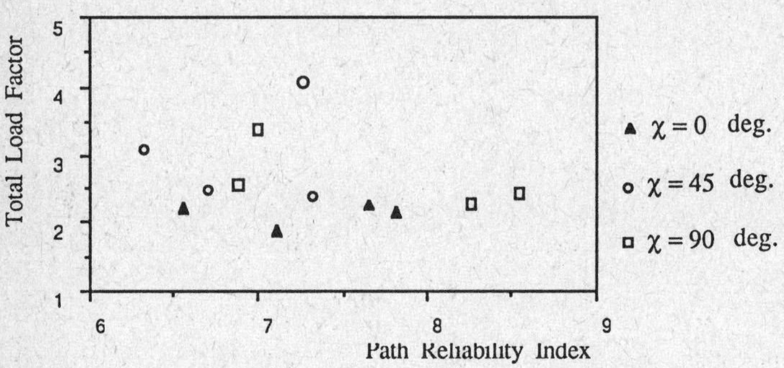

Fig. 6 Path Reliability Index (β_{path}) vs Total Load Factor (λ_T)

DOUBLE SECANT METHOD FOR NONLINEAR MATHEMATICAL PROGRAMMING

Tsung-Wu Lin
Department of Civil Engineering
National Taiwan University, Taipei, Taiwan 10764, R.O.C.

Abstract

This papaer presented a method to solve the general nonlinear programming. The nonlinear constraints and the objective function are linearized and solved by the simplex method with consideration of upper bounds for the move limits in each iteration. For each design variable, we use two variables in linearized programming problem. One variable is for the increased value, the other variable is for the decreased value. For the increased variable, x_i^+, its coefficient in the constraint equation is the divided difference of the values of constraints at $(x_1, ..., x_i + d_i^+, ..., x_n)$ and $(x_1, ..., x_i, ..., x_n)$. Similarly, for the decreased variable, x_i^-, its coefficient is the divided difference of the values of constraints at $(x_1, ..., x_i - d_i^-, ..., x_n)$ and $(x_1, ..., x_i, ..., x_n)$. Where the d_i^+ and d_i^- can be different for two directions, and are chosen to be the move limits for the next step. This situation is much different from the so called cutting plane method. In the proposed method, each constraint surface is represented by a surface formed by many plane surfaces parallel to the planes cut through the initial point. Therefore, the linearized constraint surface is not a plane in the original n-dimensional design space as in the cutting plane method but is a multiple plane surface in the original n-dimensional design space. Although the linearized constraint surface is not a plane in the original n-d space, it is indeed a plane in the 2n-dimensional enlarged super-space. The advantage is the error of the linearization from the proposed method is quite smaller than that from the cutting plane method.

Introduction

Almost all the structural optimum design problems are nonlinear , the direct use of a nonlinear programming technique is appropriate to its solution. Because of the efficiency of the linear programming algorithm, it is attractive to convert the nonlinear problem to a linear programming formulation. Due to the nonlinearities, the optimization procedure will necessarily be an iterate one, resulting in a sequence of linear programming problems. There are many different schemes for linearization of the nonlinear programming problem.

A simple algorithm for converting the nonlinear programming problem into a sequence of LP problems was first presented by Moses[1]. A complete description of this approach, coupled with a stiffness analysis procedure, was presented by Karihaloo et al.[2]

Later, Moses[3] also illustrates the pitfalls inherent in this approach. The approach will lead to an oscillation of the solution between widely separated values.

A method called cutting plane method was presented by Kelly[4]. The method is based on the property that linearized constraints in a convex problem necessary lie entirely outside the feasible region. Therefore, an envelope of such constraints may be used to represent the original nonlinear constraint to any degree of accuracy. The retention of constraints leads progressively towards the solution of larger linear programming problem at each stage and does not appear to be economical.

Another method called move limits was presented by Griffith and Stewart[5]. It can eliminate the oscillatory tendency of the solution and avoid the difficulties of the cutting plane method. The method restricts the moves to the neighborhood of the previous optimum by imposing artificial limits on the variation of the design variables. The method of move limits has been extensively applied to structural optimization [6-11].

To further increase the efficiency of move limits and the cutting plane method, a double secant method is proposed in this paper. The nonlinear constraints and the objective function are linearized and solved by the simplex method with consideration of upper bounds for the move limits in each iteration. For each design variable, we use two variables in linearized programming problem. One variable is for the increased value, the other variable is for the decreased value. For the increased variable, Δx_i^+, its coefficient in the constraint equation is the divided difference of the values of constraints at $(x_1, ..., x_i + d_i^+, ..., x_n)$ and $(x_1, ..., x_i, ..., x_n)$. Similarly, for the decreased variable, Δx_i^-, its coefficient is the divided difference of the values of constraints at $(x_1, ..., x_i - d_i^-, ..., x_n)$ and $(x_1, ..., x_i, ..., x_n)$. Where the d_i^+ and d_i^- are positive constants for the two directions, and are chosen to be the move limits for the next step.

The advantage of the proposed method is that the error is quite smaller than that from the cutting plane method. If the move amount is within d_i, the maximum error from the cutting plane method is about four times the maximum error of the proposed method for a quadratical constraint. Subroutines according to the proposed method are available and can be used directly.

Theory

1. Iterative Procedure of Linearized LP Programming

The general constrained optimization problem can be stated as follows:

$$\text{Minimize} \quad W(x_1, x_2, \ldots, x_n) = W(\underset{\sim}{x}) \tag{1a}$$

$$\text{Subject to} \quad G_j(x_1, x_2, \ldots, x_n) = G_j(\underset{\sim}{x}) = 0, \quad j \in E \tag{1b}$$

$$G_j(x_1, x_2, \ldots, x_n) = G_j(\underset{\sim}{x}) \geq 0, \quad j \in I \tag{1c}$$

where $\{\underset{\sim}{x}\}^T$ is the vector of design variables to be determined and $W(\underset{\sim}{x})$ is the objective function. $G_j(\underset{\sim}{x})$ are the equality constraints ($j \in E$) and the inequality constraints ($j \in I$). $W(\underset{\sim}{x})$ and $G_j(\underset{\sim}{x})$ are, in general, nonlinear function of $\underset{\sim}{x}$.

The constraints $G_j(\underset{\sim}{x})$ define the acceptable region, and the optimum design lies at the point of tangency of the objective function to the boundary of the acceptable region. In contrast to the optimum design in LP, the optimum needs not lie at a vertex or intersection point of the constraints.

Equation (1) can be linearized using the first term of a Taylor series expansion to produce

$$\text{Minimize} \quad W(\underset{\sim}{x}) = W(\underset{\sim}{x}^0) + \sum_{i=1}^{n} \left.\frac{\partial W}{\partial x_i}\right|_{\underset{\sim}{x}=\underset{\sim}{x}^0} (x_i - x_i^0) \tag{2a}$$

$$\text{Subject to} \quad G_j(\underset{\sim}{x}) = G_j(\underset{\sim}{x}^0) + \sum_{i=1}^{n} \left.\frac{\partial G_j}{\partial x_i}\right|_{\underset{\sim}{x}=\underset{\sim}{x}^0} (x_i - x_i^0) = 0 \quad or \quad \geq 0 \tag{2b}$$

Collecting the known terms involving x^0, eq.(2) can be written as:

$$\text{Minimize} \quad \bar{W}(\Delta \underset{\sim}{x}) = W(\underset{\sim}{x}) - W^0 = \langle c \rangle \{\Delta x\} \tag{3a}$$

$$\text{Subject to} \quad \bar{G}_j(\Delta \underset{\sim}{x}) = \langle a_j \rangle \{\Delta x\} + b_j = 0 \quad or \quad \geq 0 \tag{3b}$$

where

$$W^0 = W(\underset{\sim}{x}^0)$$

$$c_i = \left.\frac{\partial W}{\partial x_i}\right|_{\underset{\sim}{x}=\underset{\sim}{x}^0}$$

$$a_{ji} = \left.\frac{\partial G_j}{\partial x_i}\right|_{\underset{\sim}{x}=\underset{\sim}{x}^0}$$

$$b_j = G_j(\underset{\sim}{x}^0)$$

$$\Delta \underset{\sim}{x} = \underset{\sim}{x} - \underset{\sim}{x}^0$$

The problem given by eq.(3) is a standard LP problem. Note that Δx_i is unrestricted in sign, be the change in design variables.

Based on eq.(3), a simple algorithm for converting the nonlinear design problem into a sequence of LP problem is as follows[1]:

(1) Linearize the objective function and the constraints in the neighborhood of a starting point $\underset{\sim}{x}^0$ and solve the resulting LP problem given by eq.(3) for the optimal $\Delta \underset{\sim}{x}$.

(2) Redefine the point $\underset{\sim}{x} = \underset{\sim}{x}^0 + \Delta \underset{\sim}{x}$ as the new starting point and repeat the process until the $\Delta \underset{\sim}{x}$ obtained in a cycle is close to zero.

2. Move Limits

The solution obtained by this procedure may oscillate indefinitely between adjacent vertices, if the optimum is at the point of tangency of the objective function to one of the constraints. To eliminate the oscillatory tendency of the solution, the LP problem given by eq.(3) are subject to the following additional constraints[5]:

$$-d_i^- \leq \Delta x_i \leq d_i^+ \tag{4}$$

where d_i^-, d_i^+ are suitably chosen positive constants, called move limits, and may be changed at each cycle. The process is repeated until either no significant change occurs in the solution or successive solutions starts to oscillate between the move limits. In the latter case, use reduced values of d_i^+, d_i^- in the next cycle.

3. Double Secant Method

If $W(\underset{\sim}{x})$ and $G_j(\underset{\sim}{x})$ are linearized as follows:

$$\begin{aligned} W(\underset{\sim}{x}) &= W(\underset{\sim}{x}^0) + \sum_{i=1}^{n} \nabla_i^+ W \cdot \Delta x_i^+ + \sum_{i=1}^{n} \nabla_i^- W \cdot \Delta x_i^- \\ &= W^0 + \langle \nabla^+ W \rangle \{\Delta x^+\} + \langle \nabla^- W \rangle \{\Delta x^-\} \end{aligned} \tag{5a}$$

$$\begin{aligned} G_j(\underset{\sim}{x}) &= G_j(\underset{\sim}{x}^0) + \sum_{i=1}^{n} \nabla_i^+ G_j \cdot \Delta x_i^+ + \sum_{i=1}^{n} \nabla_i^- G_j \cdot \Delta x_i^- \\ &= G_j^0 + \langle \nabla^+ G_j \rangle \{\Delta x^+\} + \langle \nabla^- G_j \rangle \{\Delta x^-\} \end{aligned} \tag{5b}$$

where

$$
\begin{aligned}
W^0 &= W(\underset{\sim}{x}^0) = W(x_1^0, ..., x_i^0, ..., x_n^0) \\
G_j^0 &= G_j(\underset{\sim}{x}^0) = G_j(x_1^0, ..., x_i^0, ..., x_n^0) \\
\nabla_i^+ W &= [W(x_1^0, ..., x_i^0 + d_i^+, ..., x_n^0) - W^0]/d_i^+ \\
\nabla_i^- W &= [W(x_1^0, ..., x_i^0 - d_i^-, ..., x_n^0) - W^0]/d_i^- \\
\nabla_i^+ G_j &= [G_j(x_1^0, ..., x_i^0 + d_i^+, ..., x_n^0) - G_j^0]/d_i^+ \\
\nabla_i^- G_j &= [G_j(x_1^0, ..., x_i^0 - d_i^-, ..., x_n^0) - G_j^0]/d_i^- \\
x_i &= x_i^0 + \Delta x_i^+ - \Delta x_i^-
\end{aligned}
\tag{6}
$$

then the nonlinear programming problem, eq.(1), can be linearized as follows:

$$
\textit{Minimize} \quad \bar{W}(\Delta \underset{\sim}{x}^+, \Delta \underset{\sim}{x}^-) = W(\underset{\sim}{x}) - W^0 = \langle \nabla^+ W \quad \nabla^- W \rangle \begin{Bmatrix} \Delta x^+ \\ \Delta x^- \end{Bmatrix} \tag{7a}
$$

$$
\textit{Subject to} \quad \bar{G}_j(\Delta \underset{\sim}{x}^+, \Delta \underset{\sim}{x}^-) = G_j(\underset{\sim}{x}) = G_j^0 + \langle \nabla^+ G_j \quad \nabla^- G_j \rangle \begin{Bmatrix} \Delta x^+ \\ \Delta x^- \end{Bmatrix}
$$

$$
= 0 \quad or \quad \geq 0 \tag{7b}
$$

$$
\textit{and the move limits} \qquad d_i^+ \geq \Delta x_i^+ \geq 0, \qquad d_i^- \geq \Delta x_i^- \geq 0 \tag{7c}
$$

The problem is now changed to a standard LP with nonnegative and upper bound constraints on each variables.

4. Compare the Linearized Equations

The linearized curves are quite different between eq.(2) and eq.(5). To illustrate this difference, we examine a nonlinear equation $x^2 + y^2 = 1$ for some conditions as shown in Fig.1, where the linearized equations by eq.(2) and eq.(5) are as follows:

(a) For $\underset{\sim}{x}^0 = (x^0, y^0) = (0,0)$ and $d^+ = d^- = 0.71$,

$$
0\Delta x + 0\Delta y - 2 = 0
$$

$$
\Delta x^+ + \Delta x^- + \Delta y^+ + \Delta y^- - 2 = 0 \tag{8a}
$$

(b) For $\underset{\sim}{x}^0 = (x^0, y^0) = (0.5, 0.5)$ and $d^+ = d^- = 0.5$,

$$
4\Delta x + 4\Delta y - 2 = 0
$$

$$
3\Delta x^+ - \Delta x^- + 3\Delta y^+ - \Delta y^- - 2 = 0 \tag{8b}
$$

(c) For $\underset{\sim}{x}^0 = (x^0, y^0) = (0.71, 0.71)$ and $d^+ = d^- = 0.28$,

$$\Delta x + \Delta y = 0$$

$$\Delta x^+ - 0.66\Delta x^- + \Delta y^+ - 0.66\Delta y^- = 0 \tag{8c}$$

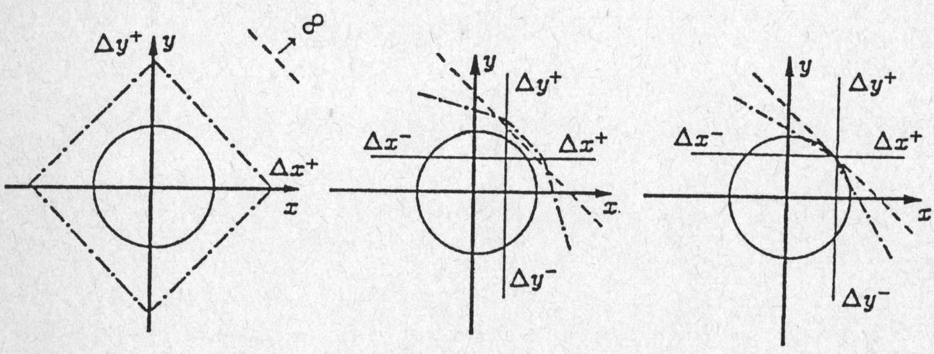

$(x^0, y^0, d) = (0, 0, .71)$ (a) $(x^0, y^0, d) = (.5, .5, .5)$ (b) $(x^0, y^0, d) = (.71, .71, .28)$ (c)

Fig.1 Linearized equations of double secant method and cutting plane method
—— nonlinear eq. —·—·— double secant - - - - cutting plane

From fig.1, it is obvious that the error from the double secant linearized equation is quite smaller than that from the tangent linearized equation. Therefore, the double secant linearized equations can represent the original curve more closely.

To reduce the computational effort, the divided differences can be calculated by the reanalysis technique as stated in the following paragraphs, instead of carrying out the whole computation for each variation of design variables.

5. Computation of Divided Difference

For a large structural system analyzed by stiffness method, the equations of equilibrium can be written as:

$$[K^0]\{D^0\} = \{P\} \tag{9}$$

where $[K^0]$ is the original stiffness matrix, $\{D^0\}$ is the correspondent displacement vector and $\{P\}$ is the loading vector.

If the changes to the original stiffness $[K^0]$, due to the change of the design variable, x_i, are denoted as

$$[\Delta K] = [K(x_1, ..., x_i + d_i^+, ..., x_n) - K^0)]/d_i^+ \tag{10}$$

then

$$[K^0 + \Delta K]\{D^0 + \Delta D\} = \{P\} \tag{11}$$

where $\{\Delta D\}$ is the changes to the original displacements $\{D^0\}$.

Subtract eq.(9) from eq.(11) to obtain:

$$[K^0 + \Delta K]\{D^0 + \Delta D\} = [K^0]\{D^0\} \tag{12a}$$

or

$$[K^0]\{\Delta D\} = -[\Delta K]\{D^0\} - [\Delta K]\{\Delta D\} \tag{12b}$$

If the changes in stiffness are collected in the upper left portion of $[\Delta K]$ and denoted as $[\Delta K_{11}]$, and all degrees associated with stiffness changes are designated as $\{D_1^0\}$, and the remainder as $\{D_2^0\}$, then the solution of $\{\Delta D\}$ from eq.(12a) will be[12]

$$\{\Delta D\} = \begin{Bmatrix} \Delta D_1 \\ \Delta D_2 \end{Bmatrix} = \begin{bmatrix} [q]^{-1} - [I] \\ -[F_{21}^0][\Delta K_{11}][q]^{-1} \end{bmatrix} \{D_1^0\} \tag{13a}$$

where

$$\begin{bmatrix} F_{11}^0 \\ F_{21}^0 \end{bmatrix} = \begin{bmatrix} K^0 \end{bmatrix}^{-1} \begin{bmatrix} [I] \\ [0] \end{bmatrix} \tag{13b}$$

$$[q] = [I] + [F_{11}^0][\Delta K_{11}] \tag{13c}$$

Eq.(13b) should be calculated by forward substitutions using lower triangular matrix $[L^0]$, then by backward substitutions using upper triangular matrix $[U^0]$, where $[L^0]$ and $[U^0]$ are the decomposed matrices of $[K^0]$ such that $[K^0] = [L^0][U^0]$. And because the order of the matrix $[q]$ is very small, $\{\Delta D\}$ can be calculated quickly from eq.(13) without recomputing the decomposed matrices of $[K^0 + \Delta K]$.

The resulting vector $\{\Delta D\}$ can then be used to calculate the divided differences required in eq.(5).

To calculate the partial derivative required in eq.(2), let $[\Delta K]$ represents $\partial[K^0]/\partial x_i$, then $\{\Delta D\}$ will represent $\partial\{D^0\}/\partial x_i$ and it can be calculated exactly by eq.(12b) by setting $[\Delta K][\Delta D] = 0$, or

$$\{\Delta D\} = -[K^0]^{-1}[\Delta K]\{D^0\} \tag{14}$$

Example

The following nonlinear programming problem has been solved by the proposed method:

$$\text{Minimize}: \quad W = 37.293229x_1 + 0.8356891x_1x_5 + 5.3578547x_3^2 - 40792.141$$

Subject to:

$$0 \leq 85.33441 + 0.0056858x_2x_5 + 0.0006262x_1x_4 - 0.0022053x_3x_5 \leq 92$$

$$90 \leq 80.51249 + 0.0071317x_2x_5 + 0.0029955x_1x_2 + 0.0021813x_3^2 \leq 110$$

$$20 \leq 9.300961 + 0.0047026x_3x_5 + 0.0012547x_1x_3 + 0.0019085x_3x_4 \leq 25$$

$$78 \leq x_1 \leq 102$$

$$33 \leq x_2 \leq 45$$

$$27 \leq x_3 \leq 45$$

$$27 \leq x_4 \leq 45$$

$$27 \leq x_5 \leq 45$$

If the starting point is $\underset{\sim}{x}^0 = (78.62, 33.44, 31.07, 44.18, 35.22)$, and $\underset{\sim}{d}$ is initial set to $0.1\,\underset{\sim}{x}$, only two LP problems are required, and the results are $W = -30665.4$ and $\underset{\sim}{x} = (78.0, 33.0, 29.995, 45.0, 36.776)$. If the starting point is changed to $\underset{\sim}{x}^0 = (50., 20., 20., 20., 20.)$, and $\underset{\sim}{d}$ is initial set to 10, then the same results are obtained.

Conclusion

Because two design variables are used in linearized programming problem, the situation is much different from the so called cutting plane method. In the proposed method, each constraint surface is represented by a surface formed by many plane surfaces parallel to the planes which cutting through the initial point. Therefore, the linearized constraint surface is not a plane in the original n-dimensional design space as in the cutting plane method but is a multiple plane surface in the original n-dimensional design space. Although the linearized constraint surface is not a plane in the original n-d space, it is indeed a plane in the 2n-dimensional enlarged super-space. The advantage is the error of the linearization from the proposed method is quite smaller than that from the cutting plane method. The disadvantage is the resulting LP problem is larger, and requires more computations to get the divided differences. But if the reanalysis technique is used, the increased computation time for the additional two-sides divided differences will be only a small fraction of the total computation time.

Reference

1. Moses, F., "Optimum Structural Design Using Linear Programming", Proc. ASCE J. of the Struct. Div., Vol.90, No.ST6, Dec. 1964, pp.89-104.
2. Karihaloo, B. L., Pathare, P. R. and Ramesh, C. K., "The Optimum Design of Space Structures by Linear Programming Using the Stiffness Matrix Method of Analysis", in Space Structures, R. M. Davies (Ed.), J. Wiley, New York, 1967.
3. Moses, F., "Some Notes and Ideas on Mathematical Programming Methods for Structural Optimization", Norwegian Technical University, Institute for Ship Design, Report SKB 11/118, 1967.
4. Griffith, R. E. and Stewart, R. A., "A Nonlinear Programming Technique for the Optimization of Continuous Processing Systems", Management Science, Vol.7, 1961, pp.251-263.
5. Romstad, K. M. and Wang, C. K., "Optimum Design of Framed Structures", Proc. ASCE, J. of the Struct. Div., Vol.94, No.ST12, Dec. 1968, pp.2817-2845.
6. Pope, G. "Optimum Design of Stressed Skin Structures", AIAA J., Vol.11, 1973, pp.1545-1552.
7. Reinschmidt, K. F., Cornell, C. A. and Brotchie, J. F., "Iterative Design and Structural Optimization", Proc. ASCE, J. of the Struct. Div., Vol.92, No.ST6, Dec. 1966, pp.281-318.
8. Bhavihatti, S. and Ramakrishnan, C., "Computational Efficiency of Improved Move Limit Method of Sequential Linear Programming for Structural Optimization", Comp. and Struct., Vol.11, 1980, pp.191-196.
9. Saka, M. P., "Optimum Design of Rigidly Jointed Frames", Comp. and Struct., Vol.11, 1980, pp.411-419.
10. Zienkiewicz, O. C. and Campbell, J. S., "Shape Optimization and Sequential Linear Programming", Ch.7 in Optimum Structural Design, R. H. Gallagher and O. C. Zienkiewicz, Editors, J. Wiley Book Co., 1973.
11. Kelly, J. A., "The Cutting Plane Method for Solving Convex Programs", J. of Soc. for Industrial and Appl. Math., Vol.8, No.4, Dec. 1960, pp.703-712.
12. McGuire, W. and Gallagher, R., Matrix Structural Analysis, Ch.10, J. Wiley Book Co., New York, 1979.

IMPROVING THE ACCEPTABILITY OF RELIABILITY-BASED SAFETY FACTORS FOR THE DESIGN OF STRUCTURAL SYSTEMS

W. Manners
Department of Engineering, University of Leicester
Leicester, LE1 7RH, United Kingdom

ABSTRACT

There has been a trend in recent years, with the introduction of partial safety factors into structural design codes, to reduce the magnitude of the safety factors applied to dead loads, compared to those applied to other loads. Concern has been expressed about the size of the resulting factors. This paper reviews possible reasons for this concern and discusses the extent to which they are valid. The use of reliability analysis techniques to evaluate safety factors is outlined. It is shown that some assumptions normally thought of as being conservative are not conservative when these techniques are used. Suggestions are made for improving the acceptability of reliability-based safety factor evaluation by taking account of the matters previously discussed.

INTRODUCTION

Recent revisions of codes of practice for structural design in the UK and elsewhere have brought partial safety factors into widespread use. In general, the effect has been that safety factors applied to dead loads are smaller than those used in previous allowable stress codes, while safety factors on live loads have tended to increase. There is, however, a lot of variation between codes in respect of both the sizes of the factors and the definition of different types of loading. A particularly marked example of the trend is BS 5400 where for structural steel dead loads the partial factor is only 1.05 (ref 1).

Small factors are justified on the grounds that dead loads, especially those due to structural weight, are known quite accurately compared with other loads. This is particularly true of items that are lifted or floated into position (as with offshore platforms, for example), where it is essential to know the weight and centre of gravity to a high degree of accuracy, and where the success of the installation validates the weight calculation.

The trend towards smaller dead load factors has been assisted by the use of reliability analysis techniques to quantify uncertainty and reliability, and to enable rational procedures to be created for the evaluation of safety factors (ref 2), although it is worth noting that the load factors in BS 5400 were not derived using such a procedure.

Concern has been expressed from time to time about the use of small safety factors, and in some cases this has limited the reduction that might otherwise have taken place. For example, in the revision of the AISC rules a dead load factor of 1.2 was proposed as this was felt to be "the lowest the profession would accept", and an additional load case of 1.4 times the dead load acting alone was introduced which "prevents the factored load effect from becoming too small" (ref 3).

The purpose of this paper is to discuss possible reasons for this concern, to investigate to what extent they are valid and to determine how they should be taken into consideration in any rational system for determining safety factors.

It is important to realise, of course, that the absolute magnitude of any safety factor is strictly irrelevant. The most important fact is the size of the overall combined factor that separates the load and the strength. In more detailed discussions the relative sizes of the factors used for different types of loading are of concern. The definitions used to determine the nominal design values of loads and strengths are part of these considerations; in this paper it will be assumed that conventional definitions of nominal loads and strengths are used, and hence that conventional ideas of what would be 'large' or 'small' safety factors may be referred to without further discussion.

The discussion in this paper will be conducted mainly in the context of steel structures, because it is in this area that the greatest reductions in load factors have been proposed.

CAUSES FOR CONCERN

(1) Uncertainty in dead load stress

The first reason for exercising caution about the size of the dead load safety factor is that in the many code calculations where design is based on elastic assumptions, it is the dead load stress, not the dead load itself that is required. The term 'dead load stress' is taken to mean the stress in the structure when only dead load is acting, and hence it includes the effects of lack-of-fit, welding, construction sequences, etc., which may be present whenever a structure as a whole (or a stress distribution within a basically determinate structure) is statically indeterminate.

Now, of course, there are situations such as residual welding stresses in steel compression members, where these effects have a major impact on the safety of a design, and here they have to be specifically allowed for in the design rules.

On the other hand, there are situations such as plastically designed steel frames where the design method calculates the ultimate strength of the structure after redistribution of stresses, and so self-equilibrating internal stresses are irrelevant.

Large parts of most codes of practice are based on elastic assumptions about load distribution in the structure and, in these cases, for the reasons above, the relationship between the structural dead load and the dead load stress at a point in a structure contains an additional uncertainty when compared with the relationship between other applied loads and the resulting stresses. Consideration of this would therefore lead to the factor on structural dead load being increased relative to that on other dead loads, such as finishes and services, and on other loads such as live loads and wind loads, at least for elastic analysis calculations.

Codes vary in how they handle the uncertainty in the structural analysis process itself. Logically it should be associated with the loads and load factors, since together they calculate the load effect, such as stress, which is then checked against some limiting value in the design calculation. However, in a code like BS 5400, dealing with steel, concrete and composite bridges, it makes more sense to identify it with the strength calculations because these are in the material-specific parts of the code, and the uncertainties in the structural analysis process are likely to depend on the material. The effect discussed above is concerned with the difference between different types of load and therefore has to be allowed for in the values of the load factors. This argument therefore leads to the conclusion that load factors cannot be fully determined by loading considerations alone; they may have to depend on other aspects of the design situation.

(2) Time-Dependence in Ductile Failure

The second reason for exercising caution about the size of the dead load factors originates in the fact that ductile failure mechanisms take a finite time to operate. Thus live loads that exceed the static strength of a ductile structure, as determined by plastic design theory, may not actually cause an observable failure of the structure if they last for only a short time. This is not however normally taken into consideration in design rules, and hence structures whose loading changes rapidly with time have an additional margin of safety which is not enjoyed by structures whose loading is predominantly static. A uniform level of safety will therefore be achieved only if safety factors for static loads are higher relative to factors for moving or changing loads than would be indicated by consideration of the statistics of the loads themselves.

It will be noted that, unlike the previous argument which applied to structural dead load specifically, this one applies to all static loads and thus includes long term live loads such as floor loads in buildings.

There seems to be little information available on the magnitude of this effect. Most structural testing is done at very low strain rates and most structures and material texts only refer to creep in the context of concrete, and metals at high temperatures. In these cases it is impossible to avoid consideration of time-dependent effects in design, whereas for ductile structures at normal temperatures, making the apparently conservative assumption that failure occurs when the applied loads reach the static failure load still leaves plenty of scope for the design of economic structures. In fact, apparently conservative assumptions are not always so when one attempts to quantify safety and calculate optimum values for safety factors; this is discussed later. It seems to be only in the study of metal-forming that the strain-rate dependence of steel at yield is normally considered, and it was using such data that Horne made an attempt to determine design values for wind loading for use in plastic design, taking account of the time-dependence of ductile failure (ref 4). Considering one particular simple portal frame, he obtained the graph reproduced in Fig 1 which relates the amount by which the applied load has to exceed the static load to the time taken for failure to occur. Failure was defined as occurring when permanent plastic defor-

Figure 1:

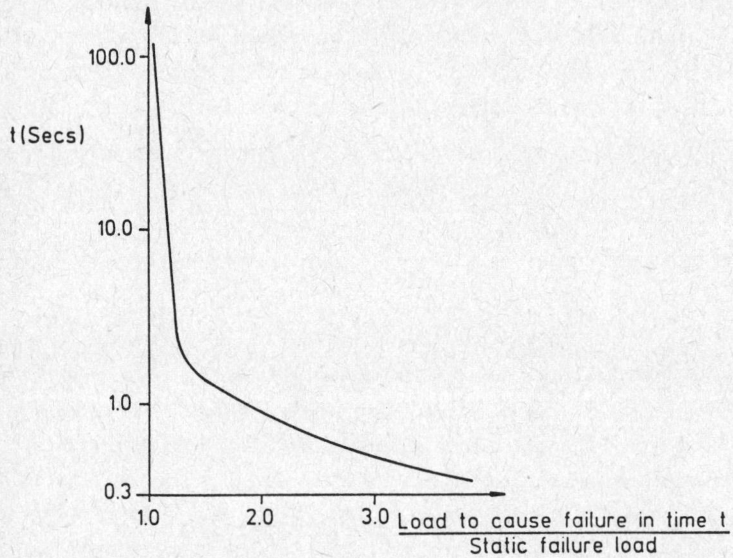

Load increase required to produce collapse in finite time.

mations were ten times the deformation at which yield first occurred. It must be emphasised that this result is calculated, not experimental, and applies strictly only to the frame considered. Also, the time-dependence is due not only to material effects, but also to the inertia of the structure (ref 5). Nevertheless it gives some idea of the possible order of magnitude of this effect for a single loading. Where loads are repeated and varied through the life of the structure, the development of residual stresses is important. Phenomena such as shakedown can occur and detailed analysis is extremely complex.

In view of these complications it remains sensible to ignore this effect in formulating design rules. However, some consideration of it should be taken when assessing safety factors, and once again the situation arises where to take account of all matters relevant to the safety of a class of structures, load factors cannot be determined by considering only the nature of the loads.

(3) Consequences of Failure

Another reason for keeping dead load factors higher, relative to other load factors, than would seem to be justified by the variability of the loads themselves was suggested by Wex and Brown (ref 6). They pointed out that a member containing a high proportion of dead load is likely to be a major component of a large structure. Since the consequences of failure in such a component are likely to be exceptionally severe, it would be reasonable for these designs to have a higher degree of safety, and this can be achieved, for example, by using the same safety factor for dead and live loads, as was the case in the older allowable stress codes.

This argument clearly has some validity, when one considers structural elements such as the towers and cables of suspension bridges. However, when one considers also structures such as masts and towers for radio communications systems, it is not clear that there is a good correlation between the proportion of the design strength that is required to carry dead load and the consequences of failure, unless the assessment is limited to a single type of structure.

It has been proposed that the consequences of failure should be explicitly allowed for in design. The ISO recommendations for safety factor format include a factor that is to be varied according to the consequences of failure. Public opinion seems reluctant to accept that safety is a relative not an absolute concept when human lives are at risk. Consequently it is only for structures such as agricultural buildings (ref 7) and masts and towers (ref 8) that British codes of practice take explicit account of the consequences of failure.

(4) Structural Stiffness

Another consequence of the trend to reduce dead load safety factors is the reduction in the stiffness of structures. It is not the reduction of member sizes as a direct result of less onerous design requirements that is the problem to be addressed here;

any undesirable consequences of this can be avoided by including appropriate deflection limits in the code, and deflection due to dead load is rarely important since the structure can be pre-cambered to compensate. Deflection limits for live loads are in fact becoming more important as improvements in material strength lead to lighter structures; any reduction in safety factors will increase this tendency.

The real cause for concern is that deflections, and natural frequencies, are normally calculated using linear elastic procedures which do not allow for the reduction in stiffness of compression members due to the presence of axial compressive forces. The full theory is given by Horne and Merchant (ref 9), but to a first order approximation, in the presence of axial compressive load P, the bending stiffness of a column should be factored by $(1 - P/Pcr)$ where Pcr is the critical elastic buckling load of the column. In the extreme case of a slender column Pcr is also the design strength and so if the column is designed to carry static load P with a total safety factor γ, the factor on stiffness becomes $(1 - 1/\gamma)$ which varies from 0.41 for $\gamma = 1.7$ to 0.17 for $\gamma = 1.2$. Columns of such extreme slenderness are rare, and if the bending stiffness of the column is important it is likely that the column will carry significant moments as well as axial load, hence in practice stiffness reductions of this magnitude are unlikely, but not impossible. Slender beam and column sway frames are likely to be particularly critical types of structure.

If the load P is not a static load, but varies with time, then the stiffness of the structure also varies with time, and the equations for the dynamic behaviour of the structure are non-linear. In many cases however the deflections will be less if the minimum stiffness is attained only momentarily than if this minimum stiffness were maintained continuously.

As long as the structure remains elastic, this aspect of behaviour can be determined exactly by calculations. It would, however, be extremely onerous to insist on non-linear deflection calculations in routine design work. A more realistic alternative is that code drafting committees should set deflection limits sufficiently conservatively to allow for the effects of ignored non-linearities, and these in turn will depend on the minimum values which the safety factors can take.

(5) Small Errors

All engineering calculations are to some degree approximate, and there is always a chance of errors creeping in. The existence of a major error should be obvious to an experienced designer and procedures for checking and approving designs aim at finding such errors. Safety factors cannot be used as a safeguard against major errors, whose magnitude (e.g. a factor of 10) may be considerably in excess of any practicable safety factor.

Small errors, and the cumulative effects of rounding errors are more difficult to avoid, and most designers would feel that, for this reason, there is a minimum acceptable value for the total safety factor present in any calculation.

This factor should perhaps depend on the nature of the structure. For a large costly structure, where the cost depends strongly on the weight of the structural elements, a small minimum factor combined with a high level of checking and supervision is sensible. Conversely, where the structural weight is a fairly small part of the total cost of the structure, a higher minimum factor and a lower level of checking and supervision may be reasonable. The difference between the minimum total factors that are possible in the steel building code (ref 10) and the bridge code (ref 1) may partially reflect this argument.

Some research is in hand (e.g. ref 11) which attempts to quantify errors in structural design. This aims to incorporate consideration of errors into the use of reliability analysis to assess structural safety.

PRELIMINARY DISCUSSION

The previous sections of this paper outlined five reasons for caution about either allowing the total safety factor to fall too low, or about making the factor applied to dead load small compared with those applied to other loads. Quantifying these effects was difficult because good data or models were generally absent. It follows that they should be taken into account in determining safety factors because allowing for them explicitly in design rules is not practicable.

One general conclusion is that it is not possible to determine factors for loads by considering only the nature of the loads themselves. Because different types of loads have different effects on different types of structure, it is necessary to consider the whole design situation when choosing factors. It may be generally practicable to decide on nominal design values of loads by considering only the statistics of the load in question, and then using different partial safety factors for different types of structure to take account of the variation in the effect of that particular type of loading on different types of structure.

The simplest approach, conceptually, is one in which values for the complete set of safety factors required for a given design code are chosen so that the resulting designs satisfy some specification for 'optimum' structures. It is clear from previous discussions that the optimality criterion must include consideration of structural safety and the risk of failure. This is most readily done by incorporating into the criterion the probability of failure of designs made in accordance with the code.

CODE CALIBRATION

The ideal way to use these probabilities of failure when determining safety factors for a code would be by a process of economic optimisation - finding the factors which give the optimum balance between the cost of initial construction, which will increase as safety factors are increased, and the cost of failures, which will decrease as increasing safety factors make failure less likely.

Attempts to implement such a programme have not so far been successful because of the difficulties in determining the true costs that follow from failures. Instead a process of code calibration has been devised (ref 2). This is based on the assumption that within the range of structures covered by the code whose factors are to be determined, structures can be found which were designed to existing design rules and whose level of safety is considered acceptable. The average probability of failure of these structures, weighted according to usage, is therefore an acceptable value for the type of structure in question, and the factors for the new code are then determined so that the probabilities of failure of future structures conform as closely as possible to this target value. Because there is usually a highly non-linear relationship between safety factors and probability of failure, it is found that maintaining the same weighted average probability of failure between an existing code and a new one, but reducing the scatter of the probabilities of failure usually produces a modest overall economy in the weight of structures when considering a representative range of designs. In the case of BS 5400 Part 3 this was estimated to be about 6% (ref 12).

The principal advantage of this calibration procedure over the optimisation concept is that since the assessments that go into the probability calculations affect both the "input" (old code) and the "output" (new code) the resulting factors are comparatively insensitive to many of these assessments. Thus the procedure is capable of producing modest improvements in code factors, based on available statistical data, but is strongly influenced by the acceptance of existing good practice.

This robust insensitivity cannot be absolutely guaranteed. Since the calibration procedure effectively combines all the designs to both old and new codes into a single "system", any input to the calculations may significantly affect any output. Some examples of unexpected consequences of this "system" emerge in the next section.

CALIBRATION EXAMPLES

The operation of the calibration process, and the effect on it of the five points outlined earlier can be illustrated by simple examples. In normal use, where a whole code is under review, some scores or hundreds of designs would be considered.

For the purpose of illustration, consider a single type of element loaded with a combination of dead load and some type of live load in a variety of ratios. The nominal design values of dead load, live load and structural resistance are D_n , L_n and R_n respectively. These are expressed in terms of the same stress or stress resultant; that is they are all values of the same stress at the same point in the structure, or, say, the loads acting on, and the strength of, a particular member. If partial factors of γ_D , γ_L , γ_R respectively are used, then the design equation is

$$\gamma_D D_n + \gamma_L L_n \lesssim R_n/\gamma_R \tag{1}$$

In what follows, as in all code reliability work, it is assumed that this equation is satisfied exactly; member sizes are not increased above what is actually required, to the nearest convenient size.

The actual dead load stress, or stress resultant, is defined as $D_n x_D$ where x_D is a probabilistic variable whose distribution represents the uncertainty about the true dead load stress in any particular case. Similarly $L_n x_L$ is the actual maximum live load during the structure's design life and $R_n x_R$ the actual resistance (assumed constant during the life). Hence the structure will fail if $M < 0$ where M, the margin of safety is given by

$$M = R_n x_R - D_n x_D - L_n x_L \tag{2}$$

$$= \gamma_R(\gamma_D D_n + \gamma_L L_n)x_R - D_n x_D - L_n x_L$$

Hence we can define a non-dimensional margin of safety M^1 by

$$M^1 = \gamma_R x_R\big((1 - \eta)\gamma_D + \eta\gamma_L\big) - (1 - \eta)x_D - \eta x_L \tag{3}$$

where $\eta = L_n/(D_n + L_n)$

The probability of failure is the probability that $M^1 < 0$, and is therefore a function of the safety factors, the live load ratio η, and the probability distributions of x_D , x_L and x_R .

Basic Data Case

The first example considered used the data set P1 in table 1 for the probability functions. This is based on the data used in ref. 12 for highway bridges. The dead load was taken to be normally distributed, the live load used the Extreme Type 1 or Gumbel distribution, and the resistance was lognormally distributed. The mean dead load is 5% above the nominal value and the standard deviation is 5% of the nominal value; this suggests that these data already include some allowance for possible residual stresses, lack-of-fit, etc. Assuming that the code to be calibrated against is a traditional allowable stress code with load factors 1.0 and resistance factor 1.67, Fig. 2 shows how the probability of failure varies with the percentage live load, η. It will be seen that, as expected, the cases with a high proportion of dead load are safer than those with a high proportion of live load.

In practice, structures with very high or very low ratios of live load to dead load are rare, so initially only cases with 50% and 75% live load were considered, and these were given equal weighting. This gives a target probability of failure of 1.07×10^{-5} .

In setting the partial factors for the new code it was assumed that the resistance factor is always 1.0. This is because, from a reliability point of view it is only the total safety factor that is important; the distribution of this factor between load and resistance is irrelevant. The best fit to the target at 50% and 75% live load is obtained by using dead load factor 1.24 and live load factor 1.87, and this is also shown on Fig. 2.

Figure 2:

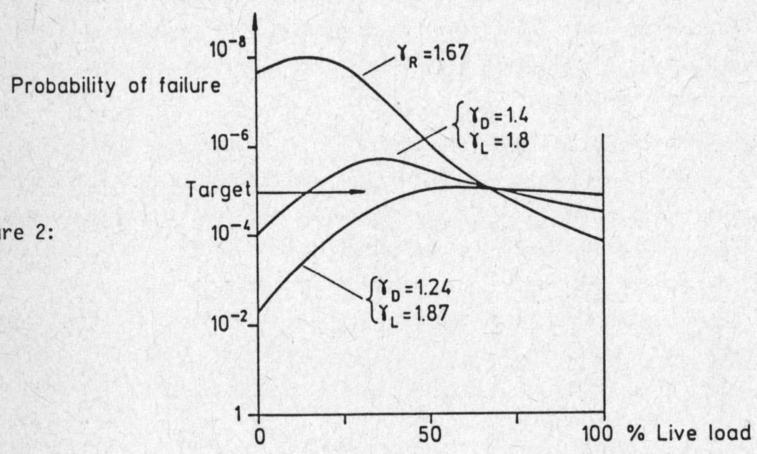

Example of typical calibration results

This immediately shows one hazard of the calibration process; concentrating on this limited range of live load ratios gives designs with a reliability well below the target for high proportions of dead load. Now of course the choice of weighting factors implies that no such structures will be built, but clearly if they are permitted by the code it would be prudent to give them some weighting.

Weighting Effect

One clear feature of the curves in Fig. 2 is that they are indeed curves. This is an inevitable result of the probability functions applied to the loads and means that a single design rule with two independent factors cannot give a uniform reliability over the whole range of load ratios. Factors therefore have to be selected to give the "best fit" to the target; refs. 2 and 12 discuss formal procedures for doing this. Fig. 2 also shows the variation of reliability due to load factors of 1.4 and 1.8 for dead and live loads respectively. This is typical of the types of result that might be obtained from a formal optimisation procedure. The exact result would depend on the procedure adopted and the weighting factors used, but since in a practical code factors would be quoted only to the nearest 0.1 or 0.05 the same factors could result from several different procedures or sets of weighting factors. As can be seen, the chosen factors have considerably reduced the range of probabilities compared with the original 'allowable stress' situation.

The wholly dead load case has the lowest reliability after calibration for both sets of factors shown in Fig. 2. The dead load factor would have to be 1.48 to give a probability of failure equal to the target for this case. This suggests that the scope for reduction in dead load factors when using a probabilistic approach is perhaps less than frequently imagined.

The only way to get a more uniform reliability would be to use more than one design rule, for example, specifying two pairs of factors of which the more critical case would govern in any particular instance. This would give greater uniformity of reliability at the expense of complicating the design procedure. At present the concensus seems to be that, except for dealing with combinations of variable loads, such as live loads, wind loads, snow loads etc., the complication of using multiple sets of load factors is not justified.

Figure 3:

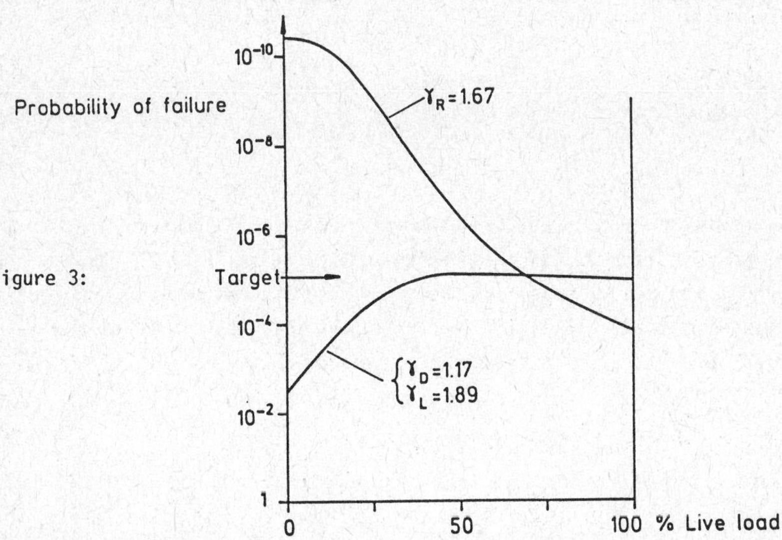

Calibration with reduced dead load variability.

Dead Load Stress Effect

Fig. 3 shows the effect of using a narrower measure of load uncertainty. The data used is in column P2 of table 1 and is more likely to be typical of the variability of the weight of structural steelwork, although still probably an overestimate for large structures whose weight control is important. As can be seen comparing Figs. 2 and 3, reducing the uncertainty in the dead load increases the spread of probabilities of failure obtained. The factors required to give the target reliability exactly at 50% and 75% live load have diverged further than before from the 'old code' value of 1.67, and hence the importance of considering weighting factors carefully is increased.

One point to note is that the target probability of failure in Figs. 2 and 3 remains virtually unchanged. This is because, in taking the average of the 50% and 75% live load probabilities of failure, it is the 75% value that dominates, and this is relatively unaffected by the change in dead load variability.

Time Dependent Failure Effect

The effect of the enhanced strength of structures when subjected to short duration loading was investigated using data set P3. The resistance to dead load alone is the same as the original data P1, but for other cases the mean and standard deviations are increased by up to 18% in proportion to the percentage live load. Fig. 4 shows the results. Because the live load resistance is enhanced, countering the greater variability of that load, the reliability is more uniform and the target probability

Figure 4:

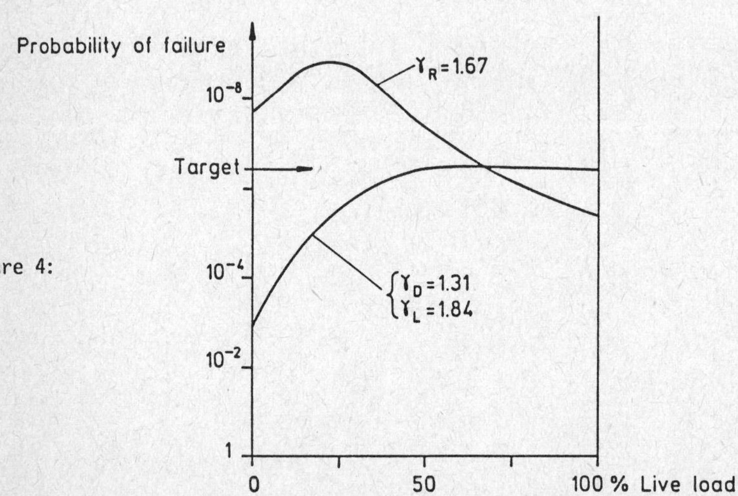

Calibration with enhanced resistance to live loads.

of failure is lower than before. Factors of 1.31 and 1.84 now give the best fit to the target at 50% and 75% live load. Comparing these values with those for data set P1 in Fig. 2 shows that the supposedly conservative assumption of ignoring the extra reserve of strength of live loads does not significantly affect the final live load factor, but has produced an unconservative value for the dead load factor of 1.24 in place of the more accurate 1.31. The effect of this on the final reliability can be seen in Fig. 5 which gives the reliabilities of designs to both sets of factors, evaluated using the more correct data P3.

This effect also occurs if more uniform weighting is used. For example, the dead load factor required to attain the target reliability of Fig. 4 exactly for a wholly dead load case is 1.59, compared with 1.48 obtained before for the target in Fig. 2. The cause of this effect is that the target is dominated by the supposedly conservative calculation. The 100% dead load case is so much safer than the live load dominated cases that it will have little effect on the value of the target, whatever weighting is used. The calibration process will then reduce the dead load factor, increasing the probability of failure of dead load dominated cases until the target is (approximately) reached. If the target has been made, by a "conservative" assumption, higher than it should have been, the reduction in dead load factor will be too great.

The lesson is that great care must be taken over assumptions of this sort, and in general 'best' estimates rather than 'conservative' estimates should be used.

Consequences of Failure

There are two ways of handling the consequences of failure, depending on whether or not a separate consequences factor is being used in the code format.

If it is not, then the weighting factors used in calculating targets and deriving factors should reflect not just the numbers of structures, but also their economic importance. If it is found that, on average, structural elements with a high proportion of dead load will have greater consequences if they fail, then they will have enhanced weighting. From the illustrations above it is clear that this is likely to mean higher dead load factors after calibration.

If an explicit consequences factor is used, then the designs considered in the calibration process should be described not just by the actual details of the elements under consideration but also by their use. For example, the designs would be distinguished not only by the percentage live load, as above, but also by whether they were in bridges, low-rise buildings, high-rise buildings, etc., and each of these groups would have its own target reliability. It is not possible to use the principle of calibration directly in such an exercise because there has been no consequences factor in past codes and so there is no directly usable past experience on which to base target values. Comparative studies of the reliability implied by a range of past codes of practice could be instructive, however.

Figure 5:

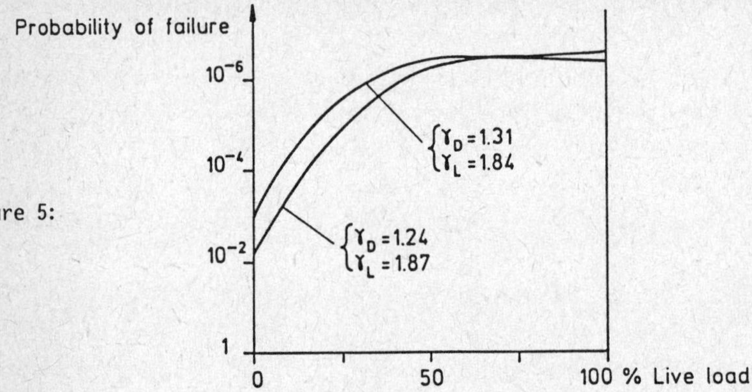

Effect of supposedly "conservative" assumption on calibration process.

Stiffness Effect

It was concluded above that the effect on stiffness of changing dead load factors was best covered by making sure that deflection limits were set sufficiently conservatively to avoid the need for non-linear calculations. There is therefore no need to consider it in the determination of safety factors for strength checks. In determining the degree of conservatism required, reliability techniques could be employed, but since the non-linear behaviour can be calculated exactly (for linear elastic materials), deterministic calculations may suffice. It would however be necessary to set the deflection limit after the load factors have been chosen.

Cumulative Small Errors

Allowing for cumulative small errors introduces another uncertainty into all the cases, thereby increasing all the probabilities of failure, but reducing their sensitivity to the percentage live load. This effect applies to the calibrated factors as well; there will be less difference between the dead load and live load factors produced by any given procedure if an allowance is made for errors.

Development of models for errors is still at an early stage. For present purposes it will suffice to demonstrate the possible effects. A uniform error effect applied identically to all load ratios will have the same effect as an increase in the standard deviation of the resistance. If the errors are allowed for by multiplying the resistance by a factor with mean of 1.0 and standard deviation of say 0.05, then because this is independent of the uncertainty in the resistance already allowed for, the combining of these two has the effect of only increasing the standard deviation of the resistance from 0.10 (data set P1) to about 0.11. If however the errors have a standard deviation of 0.1, the combined effect has a standard deviation of about 0.14. The effect of changes such as these is shown in Fig. 6. It is clear that a relatively modest increase in the standard deviation of the resistance has a marked effect on the sensitivity of the reliability to the percentage live load, and hence on the difference between dead load and live load factors that would result from an optimisation procedure.

Figure 6:

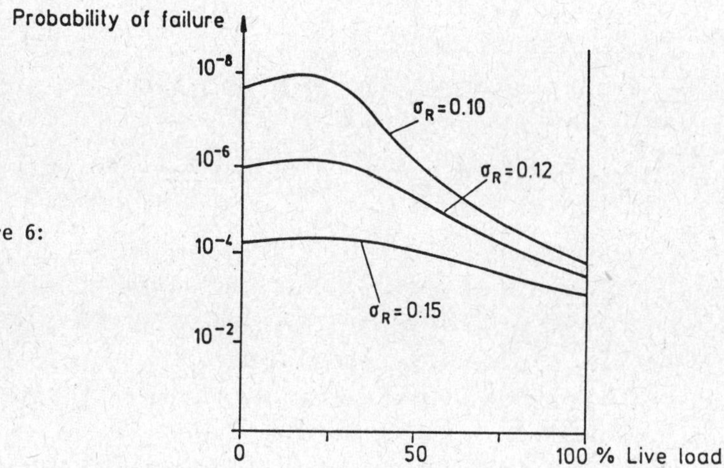

Effect of small errors or increases in variability of resistance.

CONCLUSIONS

Discussion about the values that safety factors should take inevitably involves treating safety as a quantitative concept; it will be argued, for example, in the course of such discussions that certain structures are, or should be, more, or less, safe than others.

Probability provides the best way to quantify safety, although it is clear that the techniques are still under development, and many of the data and models that would be useful are not readily available. The techniques of reliability analysis and code calibration are best seen as a support for engineering judgement, not as a substitute for it.

It has been shown that there is some legitimate cause for concern about the reduction in dead load factors that has occurred and that these matters are, in most cases, best evaluated using probabilistic reasoning.

Load factors should not be determined by considering the statistics of the loads alone; it is necessary to consider their differing effects on different types of structure. In any one code-writing exercise it may well be possible finally to fix single values for each load factor that apply to all structures covered by the code, but this should not be assumed from the outset. The possibility of using different factors for structures made from different materials, for example, should not be ruled out.

The following effects should be carefully considered in determining dead load factors (some are very obvious, others are more subtle results of the calibration process):

(1) the relative variability of the different types of load. In particular it is necessary to be clear whether it is the load itself, or the corresponding stress that is being factored in the design calculation.

(2) the weighting given to different load ratios. For future designs this must include consideration of what is possible under the code as well as what is probable.

(3) apparently conservative assumptions, such as ignoring enhanced structural strength under short term loading, should be avoided, particularly when they apply to the least reliable cases under consideration. This tends to give unconservative values for some safety factors.

(4) uncertainties that apply equally to all load ratios, such as the uncertainty in resistance or the effect of errors in calculations have a significant effect in determining the sensitivity of structural safety to the differences between different types of load, and hence on the resulting optimal factors.

(5) the effect of changed factors on the stiffness of structures (in particular on the magnitude of the non-linearities ignored in normal deflection calculations) should be considered in setting deflection limits.

REFERENCES

1. BS 5400: Steel, Concrete and Composite Bridges, Part 2: Specification for loads, BSI, London, 1978.

2. Rationalisation of safety and serviceability factors in structural codes, Report 63, CIRIA, London, 1977.

3. Ellingwood, B. et al., Probability based load criteria: load factors and load combinations, J. Struct. Div., ASCE, May 1982, Vol. 108, No. ST5, 978-997.

4. Horne, M.R., Wind loads on structures, J. Inst. Civ. Engrs., Jan. 1950, Vol. 33, 155-178.

5. Discussion on ref. 4, J. Inst. Civ. Engrs., Oct. 1950, Vol. 34, 357-360

6. Wex, B.P. and Brown, C.W., Limit states - real or imaginary in box girder bridges?, Metal Construction, Oct. 1976, 434-438.

7. BS 5502: British Standard code of practice for design of buildings and structures for agriculture: Section 1.2 Design, construction and loading, BSI, London, 1980.

8. BS 8100: Lattice towers and masts, Part 1: Code of practice for loading, BSI, London, 1986.

9. Horne, M.R. and Merchant, W., The Stability of Frames, Pergamon, Oxford, 1965.

10. BS 5950: Structural use of steelwork in building, Part 1: Code of practice for design in simple and continuous construction: hot rolled sections, BSI, London, 1985.

11. Melchers, R.E. and Stewart, M.G., Data-based models for human error in design, Proc. 4th Int. Conf. on Structural Safety and Reliability (ICOSSAR'85), IASSAR, New York, 1985.

12. Flint, A.R. et al., The Derivation of Safety Factors for Design, in Rockey, K.C. and Evans H.R. (eds), The Design of Steel Bridges, Granada Publishing, London, 1981.

Table 1 Data used for probability functions

Variable	Parameter	Data set P1	P2	P3
x_D	Mean	1.05	1.00	1.05
	Standard Deviation	0.05	0.02	0.05
x_L	Mean	1.05	1.05	1.05
	Standard Deviation	0.10	0.10	0.10
x_R	Mean	1.10	1.10	1.10 - 1.30
	Standard Deviation	0.10	0.10	0.10 - 0.118

RELIABILITY CONCEPT APPLIED TO OPTIMUM CONFIGURATION OF FIBROUS COMPOSITES

M. Miki, Y. Murotsu, T. Tanaka & S. Shao
Department of Aeronautical Engineering
University of Osaka Prefecture, Sakai, Osaka 591, Japan

Introduction

The strength and stiffness of composite materials change remarkably by changing the kinds, volume contents, and orientations of the reinforcing fibers and stacking sequences. Therefore optimum material design can be performed under a given loading condition. From this standpoint optimum material design methods have been developed under the criteria on the maximum bending stiffness[1], the maximum in-plane strength[2], the maximum bending strength[3], the maximum buckling strength[4] and the maximum fundamental frequency[5]. It is found from these studies that the maximum performance can be obtained with an optimum laminate configuration, but sometimes the optimum fiber orientations change uncontinuously when the loading condition changes and the performances have large sensitivities to the design variables and loading conditions. Therefore, the best configurations studied so far can be optimum under deterministic conditions, but there remain some problems for probabilistic conditions.

A method for evaluating the reliability for the failure of composites under probabilistic conditions is proposed in this paper. This study aims to establish an optimum material design method based on reliability.

Reliability Analysis

AFOSM-method

A mathematical expression for the failure of unidirectional composites is assumed to be given as follows:

$$M = f(X_1, X_2, \cdots, X_n) \leq 0 \tag{1}$$

where M is a safety margin, X means n basic variables which affect the strength. $M \leq 0$ means failure and $M > 0$ means non-failure. The failure probability P_f can be calculated by using the joint probability density function $f_X(X)$:

$$P_f = \iint \cdots \int_D f_{x1 \cdot x2 \cdots xn}(X_1, X_2, \cdots ,) dX_1 \cdot dX_2 \cdots dX_n \tag{2}$$

where D is the region where $M \leq 0$.

Figure 1
Coordinate systems for
Uni-directional composites.

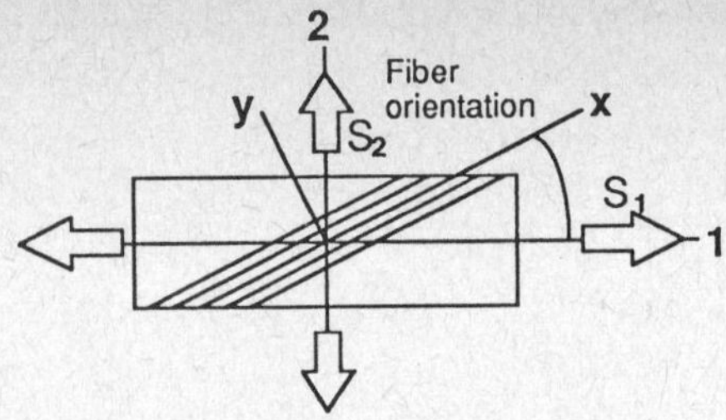

Thin unidirectional composite plates under in-plane loading condition are considered here, as shown in Fig. 1. Then applied stresses S_1, S_2 and S_6 become basic variables instead of applied loads. Another basic variables are strengths along the fiber direction R_x, R_x', R_y, R_y' and R_s, and fiber orientation angle θ. If multi-directional laminates are considered then the applied loads and each orientation angles of the laminated plies should be considered.

The function f is an appropriate failure criterion. Here Tsai-Wu criterion[6] which is most popular for unidirectional lamina is used. Then the safety margin M becomes as follows:

$$M = 1 - \left[\sum_{i=x}^{y} \left\{ \left(\frac{1}{R_i} - \frac{1}{R_i'} \right) S_i + \frac{S_i^2}{R_i R_i'} \right\} + \frac{S_s^2}{R_s^2} + \frac{2F_{xy}{}^* S_x S_y}{\sqrt{R_x R_x' R_y R_y'}} \right] \tag{3}$$

where subscrips x and y denote directions and subscript s denotes shear, and the prime denotes compressive strength. Factor $F_{xy}{}^*$ is assumed to be -1/2.

The fiber orientation angle is assumed to be deterministic in this paper. Then the number of basic random variables is eight. But the integration of Eq.(2) in such a high dimensional space is difficult and function $f_{\mathbf{x}}(\mathbf{X})$ is not clear when sufficient statistical data are not obtained. Consequently, reliability analysis is carried out by using the linear approximation of the safety margin, M, the first and second moments (mean and variance), and a safety index, β.

When the safety margin, M, is linear and the basic variables are normally distributed, safety index β is defined as follows[7]:

$$\beta = \mu_M / \sigma_M \tag{4}$$

where μ_M and σ_M is the mean and the standard deviation of safety margin M. When the safety margin is nonlinear, the approximate values of μ_M and σ_M can be obtained by linealizing the safety margin, that is, the safety margin can be expressed by Taylor series and the linear term is retained.
For the linearization of the safety margin, the linearlization point where Taylor series are evaluated affects the mean and the variance of the safety margin. If the mean values of the basic variables are

adopted as the linearization point, then the safety index, β, can be obtained analytically, but the safety index seems to be overestimated. This method is called FOSM (First Order Second Moment) method. The definition for the safety index, Eq.(4), introduces a difficulty in evaluating the safety index because the safety index varies according to the forms of functions of the safety margin even if they are equivalent. Consequently, a new safety index which is independent of the failure functions has been proposed by Hasofer and Lind[8]. In case of independent basic variables, the basic variables X_i are transformed into standard normal variables U_i, as follows:

$$U_i=\Phi^{-1}[F_{X_i}(X_i)] \tag{5}$$

where $F_{X_i}(X_i)$ and Φ are the probability distribution function of X_i and standard normal distribution function, respectively. Then the safety margin in X-space is transformed to the safety margin in U-space:

$$h(\mathbf{U}) = M(\mathbf{X}) \tag{6}$$

The linearization of the limit state function is carried out at such point $\mathbf{u}^*$ that the distance between the point on the failure surface and the origin in U-space is minimized. Point $\mathbf{u}^*$ is called β-point and the safety index, β, is given as the distance between the origin and point $\mathbf{u}^*$:

$$\beta = \min(\mathbf{u}^T \cdot \mathbf{u})^{1/2} \tag{7}$$

β-point is obtained usually by using an iterative method[9),10] or a nonlinear mathematical programming method[11]. This method is called AFOSM (Advanced FOSM) method.
The search for β-point is reduced to the following constrained optimization problem:

$$\text{Minimize} \quad f = (\mathbf{u}^T \cdot \mathbf{u})^{1/2} \qquad \text{subject to } h(\mathbf{u}) = 0 \tag{8}$$

Proposed method

To solve this problem efficiently the following method has been proposed[12]. An extended Lagrangean function is introduced as follows:

$$L_r(\mathbf{u},\mu) = (\mathbf{u}^T \cdot \mathbf{u})^{1/2} + \mu h(\mathbf{u}) + 0.5\, r\, [h(\mathbf{u})]^2 \tag{9}$$

where μ and r are constants (μ, $r > 0$).

Equation (7) can be solved easily by making use of an unconstrained optimization technique. An algorithmic procedure is as follows.

Step 1 : Specify the initial values of μ and r (for example, $r^0=5$, $\mu^0=0$), and set $k=0$, $r^k=r^0$, and $\mu^k=\mu^0$.

Step 2 : Input the initial value of $\mathbf{u}$, i.e., $\mathbf{u}^k=\mathbf{u}^0$ (for example, $u_i^0=0$, $i=1,2,\ldots,n$).

Step 3 : Solve the unconstrained optimization problem of minimizing $L_r(u,\mu)$ by a conjugate gradient method. Then, obtained the solution $\mathbf{u}^{k+1}$.

Step 4 : If the convergence condition $|h(\mathbf{u}^{k+1})| < \varepsilon$ is satisfied for a sufficiently small value of $\varepsilon(>0)$, stop the calculation. Otherwise, go to step 5.

Step 5 : Set $r^{k+1}=\omega \cdot r^k$, $\mu^{k+1}=\mu^k+r^k \cdot h(\mathbf{u}^k)$, where ω is a constant ($\omega>0$)(for example, $\omega=5$).

Step 6 : Set k=k+1, then go to Step 3.

By substituting the optimum solution u^* thus obtained into Eq. (7), the probability of failure is estimated as

$$P_f = \Phi(-\beta) \tag{10}$$

β-Point Searching Method for Multi-modal Limit State Function

When a limit state function is not uni-modal, that is, there exist more-than-one local minima. $\mathbf{u}^*$ is dependent on the initial value $\mathbf{u}^0$. Consequently, it is not sure that the global minimum is obtained. The limit state function discussed here is multi-modal, and so an extension is performed.

Consider that vector $\mathbf{u}^*$ is rotated around the origin with arbitrary angle. If vector $\mathbf{u}^*$ has the global minimum length to the limit state function, the loci of the vector always exist in the safety region after any rotational movements. Otherwise, that is, vector u^* is a local minimum, the terminus of the rotated vector may exist in the failure region after rotation as seen from a multi-modal limit state function of Fig. 2. Then another local minimum point is searched which has shorter length between the point and the origin than the previous point. β-point for a multi-modal limit state function can be obtained by repeating the procedure.

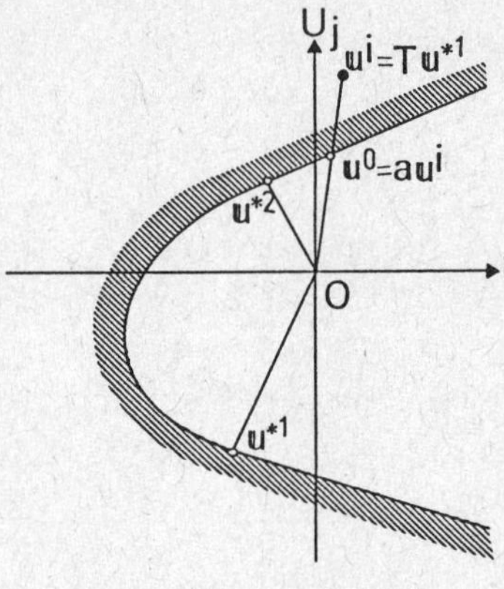

Figure 2
Method for finding the global minimum for multi-modal limit state function.

The algorithm is as follows:

Step 1 : Input $\mathbf{u}^0$, i.e., the initial value of $\mathbf{u}$.

Step 2 : Obtain a local minimum point $\mathbf{u}^*$ by using the extended Lagrangean function.

Step 3 : Set i=0 , $\mathbf{u}^i=\mathbf{u}^*$, and generate uniform random numbers.

Step 4 : Determine the rotation angle from the random numbers, and rotate $\mathbf{u}^i$ into $\mathbf{u}^{i+1}$, i.e., $\mathbf{u}^{i+1}=T\mathbf{u}^i$

Step 5 : If $h(\mathbf{u}^{i+1}) > 0$ then go to Step 6, otherwise go to Step 7.

Step 6 : Set i=i+1. If $i \geq$ Limit then set $\mathbf{u}^*=\mathbf{u}^i$ and terminate the calculation. Otherwise go to Step 4.

Step 7 : Find a such that $h(a\mathbf{u}^{i+1})=0$, and set $\mathbf{u}^0=a\mathbf{u}^{i+1}$, and go to Step 1.

Results and Discussions

Strength parameters used for the calculations

The material used for the calculations is a typical Graphite/Epoxy(T300/5208) and their strengths are assumed to be normally distributed. The mean values of the strengths along the fiber direction are as follows:

$$R_x=1.5,\ R_x'=1.5,\ R_y=0.04,\ R_y'=0.246,\ R_s=0.068\ \text{(GPa)} \tag{11}$$

The coefficients of variation are all assumed to be 0.1, for simplicity. The applied stresses, S_1, S_2 and S_6, are also assumed to be normally distributed and the respective mean values and the coefficients of variation are given appropriately.

Reliability under in-plane stress

(1) Effect of the mean value of the stress

Figure 3 shows the relation between the safety index, β, and the orientation angle, θ, under 1-axis tensile and shear stresses. The heavy solid line shows the reference stress condition and the maximum reliability is obtained at about θ=30°. The light solid lines show the results for the mean

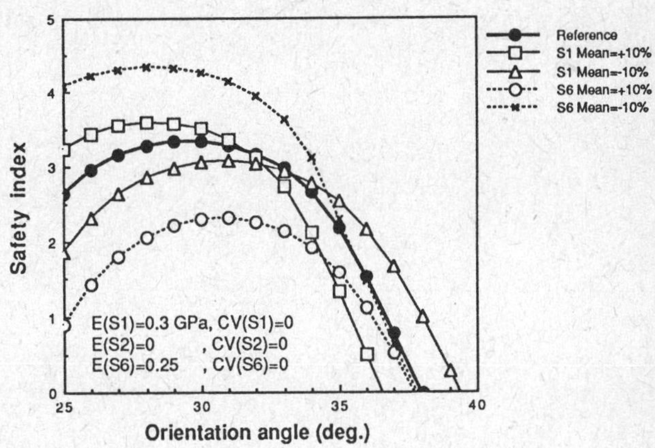

Figure 3
Effect of the mean values of S_1 and S_6.

value of 1-axis tensile stress, μ_{S1}, being increased and decreased by 10%, respectively. The safety index increases as μ_{S1} increases for small angle (θ<32°), and it decreases as μ_{S1} increases for large angle (θ>32°). And the angle which gives the maximum reliability increases with decrease in μ_{S1}.

The broken lines show the the results for the mean value of the shear stress, μ_{S6}, being changed by 10%. The safety index decreases as μ_{S6} increases and the angle which gives the maximum reliability decreases with decrease in μ_{S6}.

The results are summarized.

i) There is an angle which gives the maximum reliability under a certain stress condition.

ii) The angle which gives the maximum reliability approaches zero as the ratio of the tensile stress to the shear stress increases. On the other hand, it approaches 45° as the ratio decreases.

iii) The reliability increases even when the stress increases for a certain orientation angle. In contrast with this the reliability decreases when the stress decreases for a certain angle.

Item (iii) is very important for optimum material designs. Fig. 3 shows that the reliability can be increased by increasing the tensile stress for $\theta=30°$, for example. The direction as well as the magnitude of the principal stress will cause a critical result.

(2) Effect of CV (Coefficient of Variation) of the stress

Figure 4 shows the relation between safety index β and orientation angle θ under 1-axis tensile and shear stresses, and the effect of CV of the tensile stress is shown. The safety index decreases as CV increases and the angle which gives the maximum reliability changes a little. Figure 5 shows the effect of CV of the shear stress. The safety index decreases as CV increases and the angle which gives the maximum reliability changes considerably.

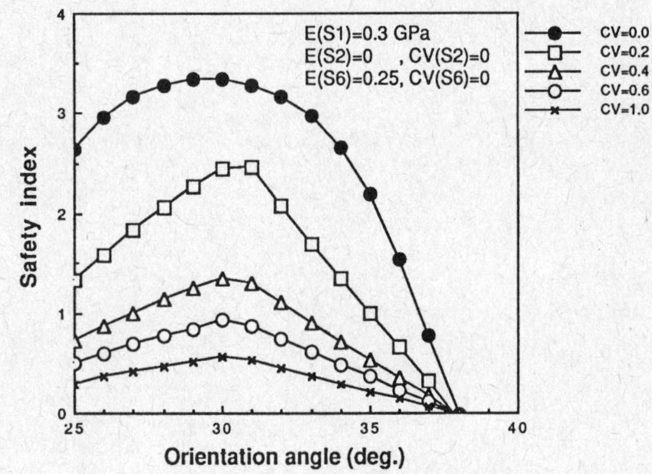

Figure 4

Effect of the coefficient of variation of S_1.

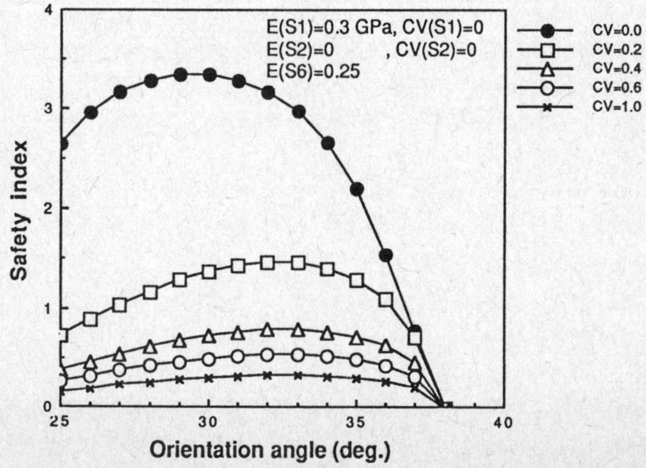

Figure 5

Effect of the coefficient of variation of S_6.

Comparison between AFOSM and FOSM methods

For nonlinear limit state functions the value of the safety index depends on the linearlization point of the function. FOSM method adopts the mean value point as a linealization point while AFOSM method adopts the β-point described above. The FOSM method is easy to calculate, but it sometimes overestimates the reliability. The results shown up to here are all obtained by AFOSM method, but the following is the comparison between AFOSM and FOSM methods.

Figures 6 and 7 show the relations between the safety index and the orientation angle. Fig. 6 shows the effect of CV of S_1 and Fig. 7 shows the effect of CV of S_6. From these figures the following observations are taken.

i) The value of β is overestimated where β>0, and this tendency becomes remarkable with increasing β.

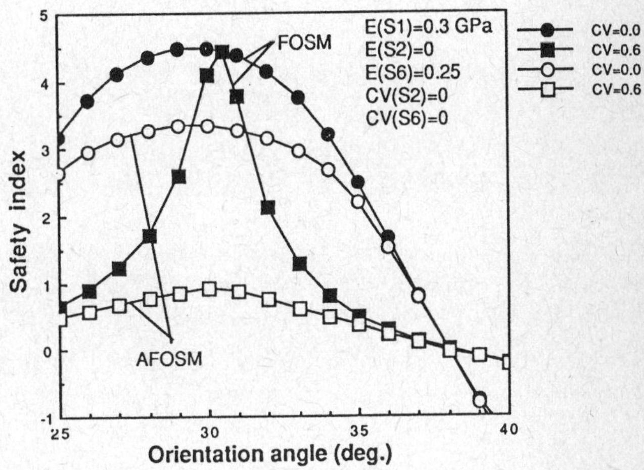

Figure 6
Comparison between AFOSM and FOSM method on the effect of CV of S_1.

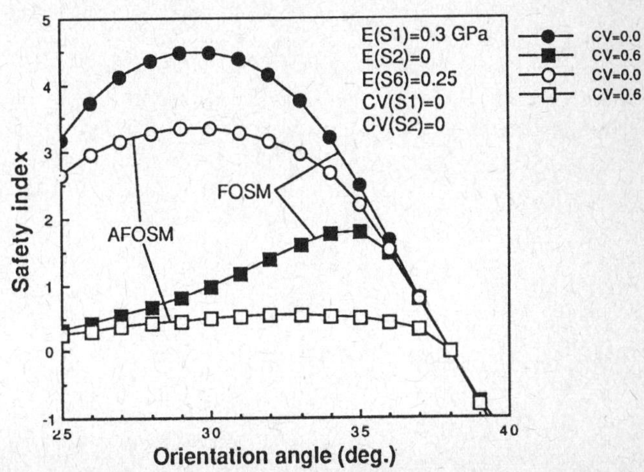

Figure 7
Comparison between AFOSM and FOSM methods on the effect of CV of S_6.

ii) FOSM method sometimes introduces erroneous results as shown in Fig. 6 where the safety index shows a peak for CV=0.6.

iii) The fiber orientation angles which give the maximum β are almost the same between these methods.

When the basic variables are normally distributed and the limit state function is linear, then the probability of failure is 50% for β=0 and 0.14% for β=3. Consequentry the value of around 3 for the safety index is important. FOSM method is not valid for the practical value of safety index.

The Effect of Strength Distributions

Normal distributions are assumed for the strength and the load variables up to here. It is considered reasonable in carrying out reliability analysis based on statistical data of composite materials. However, there exists negative values for strength variables when they are normally distributed, and a problem occurs in searching β-point by using Eq. (3) because it contains square root. Therefore the computations are performed here by imposing an additional constraint that the strength variables are positive.

The probability distributions which do not take negative values and are often used as strength distributions are log-normal and Weibull distributions. The followings are the discussions about the effect of strength distributions. Figures 8 and 9 show the difference among various strength distributions. Fig. 8 shows the case of changing the CV of S_1 and Fig. 9 shows the case of changing the CV of S_6. The followings are summarized from these figures.

i) In comparison with normal distribution, log-normal distribution gives larger safety index and Weibull distribution gives smaller. This tendency becomes remarkable when β is greater than about two.

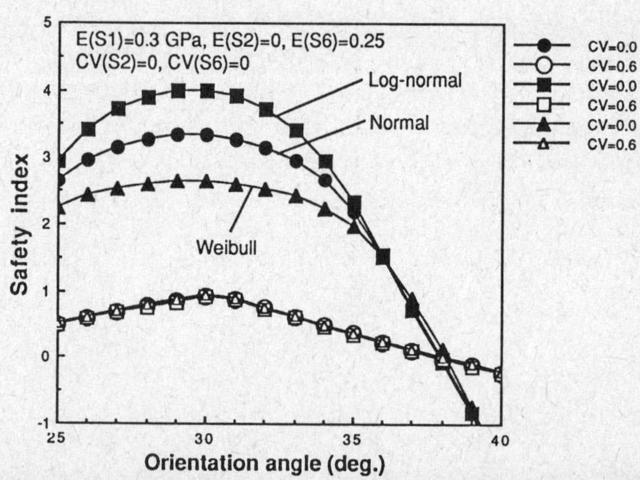

Figure 8
Difference due to strength distributions. The effect of CV of S_1 also is shown.

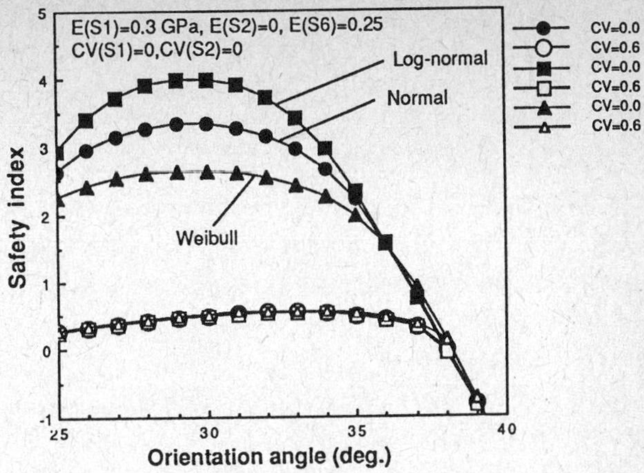

Figure 9

Difference due to strength distributions. The effect of CV of S_6 also is shown.

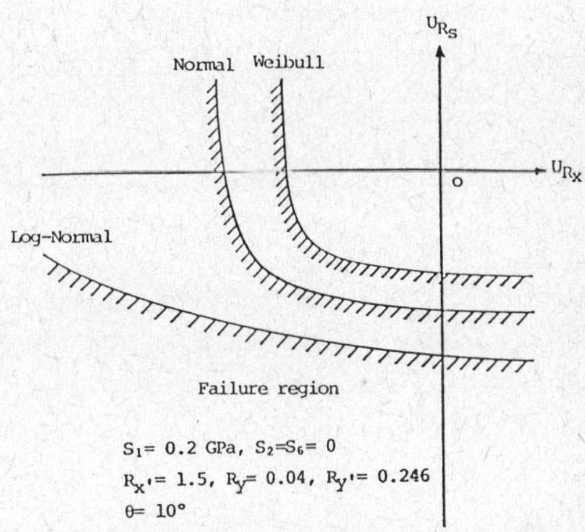

Figure 10

Comparison among various strength distributions.

ii) The fiber orientation angles which gives the maximum reliability are almost the same among those distributions.

iii) There becomes no difference when the variation of the stress is large.

Figure 10 shows the difference among various strength distributions on the limit state function in standard normal space. It can be seen that the limit state surface using Weibull distribution is the nearest to the origin and the one using log-normal distribution is farthest from the origin.

Conclusions

The following conclusions are drawn from this study:

(1) The reliability analysis of unidirectional fibrous composites can be performed by using AFOSM-method.

(2) The proposed search method for multi-modal limit state functions is found to be valid.

(3) There exists the angle which gives the maximum reliability under in-plane stress condition.

(4) The optimum orientation angle which gives the maximum reliability changes when the variation of the stress is considered.

(5) FOSM-method gives overestimated value of the safety index, and it sometimes yields erroneous result in evaluating the safety index. But it can be used for only obtaining the optimum qrientation angle since the optimum angles by AFOSM and FOSM methods are approximately same.

(6) The type of strength distribution affects the reliability. Weibull distribution gives smaller safety index while log-normal gives larger in comparison with normal distribution. But, the differences become small when the variance of stress increases.

References

[1] Tauchert, T.R. ; Adibhatla, S., J. Compos. Mater., Vol. 16, 1982, p.58.
[2] Park, W.J., J. Compos. Mater., Vol. 16, 1982, p.341.
[3] Tauchert, T.R. ; Adibhatla, S., Engineering Optimization, Vol. 8, 1985, p.253.
[4] Miki, M., Composites '86: Recent Advances in Japan and the United States, K. Kawata et al eds., Japan Society for Composite Materials, 1986, p.673.
[5] Reiss, R. ; Ramachandran, S., Composite Structures 4, Elsevier Applied Science, 1987, p.1.476.
[6] Tsai, S.W. ; Hahn, H.T., Introduction to Composite Materials, Technomic, 1980, p.280.
[7] Cornell, C.A., ACI-Journal, Vol. 66, 1969, p.974.
[8] Hasofer, A.M. ; Lind, N.C., Journ of the Eng. Mech. Div., ASCE, 1974, p.111.
[9] Rackwitz, R. ; Fiessler, B., Structural Reliability under Combined Random Sequences, Comput. Struct., Vol. 9, 1978, p.489.
[10] Thoft-Christensen, P. ; Murotsu, Y., Application of Structural Systems Reliability Theory, Springer Verlag , 1986.
[11] Shinozuka, M., Basic Analysis of Structural Safety, Journ. of the Eng. Div., ASCE, Vol. 109, No. 3, 1983, p.721.
[12] Murotsu, et.al., A Study on First-Order-Second-Moment Method in Structural Reliability, Bulletin of University of Osaka Pref., Vol. A-33, 1984, p.23.

ON A RELIABILITY-BASED SHAPE OPTIMIZATION OF TRUSS STRUCTURES

Y. Murotsu, S. Shao, M. Miki & H. Okada
College of Engineering
University of Osaka Prefecture, Sakai, Osaka 591, Japan

ABSTRACT. This paper is concerned with a shape optimization of truss structures to minimize the structural weight under the constraints on reliabilities of the members. Nodal coordinates are taken as the shape design variables together with the sizing design variables such as cross-sectional areas of the members. A two-level method is applied to facilitate the optimization with respect to the two different types of the design variables, where two separate design spaces are introduced, i.e., one for the shape variables and the other for the sizing variables. Numerical examples are provided to illustrate the validity of the proposed methods.

KEY WORDS. Reliability-based design, Shape optimization, Shape design variable, Sizing design variable, Two-level method, Failure probability, Multiplier method, Optimum configuration, Truss structure.

1. INTRODUCTION

The theories of reliability and optimization have been applied to the optimum designs of structures. Many of these have dealt with the optimization problems in which geometric cofigurations of structures are specified and only the dimensions of the members such as member areas are determined to attain the minimum structural weight or cost under the constraint on structural reliability [1]. Since the minimum structural weight is largely dependent on the geometric configuration of the structure, it is needed to treat the geometry of the structure as a design parameter. Some studies have been performed in this direction [2,3]. A method for the reliability-based shape optimization of truss structures was presented by Murotsu, et.al [4], in which nodal coordinates are specified and the optimized structure is obtained by sequentially deleting the unnecessary elements from the initial configuration which is formed by connecting the nodes with the candidate elements.

In this paper, nodal coordinates of truss structures are taken as the shape design variables in addition to the sizing design variables such as cross-sectional areas of elements. Beginning with an initial configuration, the nodes are moved until an optimum configuration under the given loading conditions is found. The applied loads and the member strengths of the truss are considered to be random variables. Changing the configuration of the structure may result in different loading conditions of the members during the design process. Consequently, the calculation of structural reliability and optimization becomes more complex than those in the case of pure sizing optimization. A new method is presented here for calculating the probability of failure of members under various loading conditions. Considering the different nature of the shape design variables and the sizing design variables, a two-level method [5] is applied to facilitate the optimization where two separate design spaces are introduced, i.e., one for the shape variables and the other for the sizing variables. Numerical examples are provided to illustrate the validity of the proposed methods.

2. FAILURE PROBABILITY OF MEMBERS

Let R_i and S_i be the strength and the applied load of member i. These are random variables with known distributions, as shown in Fig. 1. Failure of member i is defined as the event $(R_i \le S_i)$. So the failure probability of member i is given by

$$P_{fi} = \int_0^\infty F_{Ri}(s) f_{Si}(s) ds \qquad (1)$$

where $f_{Si}(s)$ and $f_{Ri}(r)$ are respectively the density functions for S_i and R_i, and $F_{Ri}(s)$ the distribution function of R_i.

It must be noted that there are two different resistances for a member of a truss, i.e., one for the tensile load and the other for the compressive load. Because the loading condition changes largely in the shape optimization, a method must be given for the calculation of failure probability in any loading condition, whether it is tensile or compressive.

Two strength variables R_{pi} and R_{ni} are defined for member i to represent resistances to positive (tensile) and negative (compressive) loads. The failure of member i can be defined as the event:

$$(R_{pi} \leq S_i,\ S_i \geq 0) \cup (R_{ni} \geq S_i,\ S_i < 0)$$

The distributions of R_{pi}, R_{ni} and S_i are assumed to be known, as shown in Fig. 2. Then the failure probability of member i is calculated as follows:

$$P_{fi} = P_{fi}^{-} + P_{fi}^{+}$$

$$P_{fi}^{-} = \int_{-\infty}^{0} [1 - F_{Rni}(s)]\, f_{Si}(s)\, ds$$

$$P_{fi}^{+} = \int_{0}^{\infty} F_{Rpi}(s)\, f_{Si}(s)\, ds \qquad (2)$$

Compared with Eq.(1), Eq.(2) should be used especially when the load S_i spans from the positive region to the negative.

3. SHAPE OPTIMIZATION OF TRUSS STRUCTURE

Consider a truss structure which consists of n members and m variable nodal coordinates. Let $A=(A_1,A_2,...,A_n)$ and $X=(X_1,X_2,...,X_m)$ be the vectors of the cross-sectional areas and the nodal coordinates, which are taken as the design variables in shape optimization problems. The optimum design is defined as attaining the structure of least volume subject to a given set of constraints on the failure probabilities of the members and on the design variables, where the strengths of the members and the loads applied to the structure are treated as Gaussian random variables with their parameters given.

In this study, the following two design problems are considered to investigate the reliability-based shape optimization of the truss structure.

Problem 1

Find X and A

$$\text{such that} \quad H_c(X,A) = \sum_{i=1}^{n} A_i \cdot l_i(X) \rightarrow \text{minimize}$$
$$\text{subject to} \quad P_{fi}(X,A) \leq P_{fai}$$
$$A_{imin} \leq A_i \leq A_{imax} \qquad (i=1,2,...,n)$$
$$X_{jmin} \leq X_j \leq X_{jmax} \qquad (j=1,2,...,m) \qquad (3)$$

where $H_c(X,A)$ is the structural volume.

$l_i(X)$ is the length of member i.

P_{fai} is the allowable failure probability of member i.

A_{imin} and A_{imax} are the minimum and maximum allowable cross-sectional areas of member i.

X_{jmin} and X_{jmax} are the minimum and maximum allowable values of nodal coordinate X_j.

Failure probability $P_{fi}(X,A)$ is determined from Eq.(2) in which the load acting on member i (random variable S_i) and its probabilistic nature can be calculated from the loading condition applied to the structure and the nodal coordinates X.

As shown in Eq.(3), the objective function H_c is described as a function of the sizing variables A and the shape variables X. This means that the optimization proceeds simultaneously with regard to two different types of design variables A and X. If the magnitude of these two types of the design variables or their gradients differs very much, convergence problems result especially when the total number of design variables is large. To overcome these problems, it is an effective method to scale the design variables.

Problem 2

In this design problem, the optimization is divided into two levels [5]. At first, a minimum volume design is performed for a given geometry with respect to the sizing variables A. That is

Level 1:

Find A

$$\text{such that} \quad H_c(X_0,A) = \sum_{i=1}^{n} A_i \cdot l_i(X_0) \rightarrow \text{minimize}$$
$$\text{subject to} \quad P_{fi}(X_0,A) \leq P_{fai}$$
$$A_{imin} \leq A_i \leq A_{imax} \qquad (i=1,2,...,n) \qquad (4)$$

where the nodal coordinate variables are set to be constant, i.e., $X=X_0$.

Then the second level of optimization is to minimize the objective function H_c with respect to the coordinate variables X, using the optimum cross-sectional areas A^* for the minimum volume design of level 1.

Level 2:

Find X

such that $H_c(X,A^*) = \sum_{i=1}^{n} A_i^* \cdot l_i(X) \rightarrow$ minimize

subject to $X_{jmin} \leq X_j \leq X_{jmax}$ $\quad (j=1,2,...,m)$ (5)

In the above problem, A^* is determined for each X, i.e., for a given geometry of the structure such that the minimum volume design of level 1 is maintained.

Therefore, the optimization proceeds in the two separate design spaces, one for the sizing design variables and the other for the shape design variables. This can avoid the trouble caused by dealing with the two different types of variables in one space. However, the total computation time of this method is very large because the minimum volume design of level 1 is repeated very many times in the optimization processes. Consequently, some approximation to the optimum design is necessary for a large scale structure as shown in the following section.

The problems 1 and 2 described above are nonlinear programming problems. They are effectively solved by using a multiplier method [6].

4. Numerical Examples

(1) Two-member Truss

Consider a two-member truss shown in Fig. 3(a) where node 3 is fixed, and nodes 1 and 2 can be moved horizontally to find their optimum positions.

First, the optimization is carried out by a deterministic method. Here the constraint on member stress is used instead of that on member failure probability. The optimum design is shown in Table 1 where the allowable stresses and the loads are taken as σ_{yi}=18 kN/cm^2 (i=1,2), L_1=50 kN and L_2=259.8 kN. Fig. 3(b) illustrates the optimum configuration thus obtained. It is observed that member 2 is arranged in the same direction as the resultant force of L_1 and L_2. This means that all the loads are supported only by member 2, and member 1 can be omitted, i.e., A_1 is negligibly small.

Now consider the reliability-based shape optimization of the two-member truss. Table 2 shows the numerical data in which the mean values of random variables are equal to the corresponding data of the deterministic design. The allowable failure probabilities of members P_{fai} are specified as $1.0*10^{-4}$. The allowable compressive and tensile stresses are equal to σ_{yi} and buckling stress is not taken into account for the moment.

The results are given in Table 3 with the correlation coefficient of the loads being -1.0, 0.0, and 1.0. The total number of design variables in this case is only 4, and then the optimization can be performed by treating all the variables in one space as described in Problem 1 of section 3, with reasonable scaling of the design variables introduced. The result obtained by the two-level method is also given in Table 3, which is almost identical to that by the other method but takes about 20 times computation time. The failure probabilities of the members are calculated with Eq.(2). As shown in Table 3 ($\rho_{L1,L2} \neq 1.0$), the failure probability of member 1 in the compressive (negative) region is considerablly large although the mean value of the applied load is positive.

Comparing the results of Table 3 with that of the deterministic optimum design, it is observed that due to the random nature of the loads and the strengths of the members, the reliability-based design gives the different shape of the structure and the minimum volume from those obtained by the deterministic design. However, the optimum shape is identical to the deterministic design when $\rho_{L1,L2}=1.0$. This is very interesting. In fact, when $CV_{L1}=CV_{L2}$ and $\rho_{L1,L2}=1.0$, the resultant force of random variables L_1 and L_2 keeps the same direction although their magnitudes vary. Just as illustrated in Fig. 3(b), L_1 and L_2 are supported only by member 2 with the area of member 1 nearly equal to 0, as shown in Table 3. This indicates an important conclusion that the geometric configuration of a structure is determined by the directions of the applied loads.

Table 4 shows the optimum designs with the buckling stress considered. The buckling stress of member i is assumed to be given by [7]

$$\sigma_{cri} = -[(\pi/20)^{1/2}(E_i/l_i)]A_i^{1/2} \tag{6}$$

where E_i, l_i and A_i are Young's modulus, the length and the cross-sectional area of member i. Comparing Table 4 with Table 3, the shape of the structure and the minimum volume are not changed in this example with the buckling stress taken into account.

Table 5 shows the effects of coefficients of variation of the loads. As the coefficients of variation become large, the minimum volume is increased. Table 6 lists the optimum designs for three different allowable failure probabilities of the members. The minimum volume becomes large when the allowable failure probability of the members is reduced.

(2) Nine-member Truss

Next consider another truss structure shown in Fig. 4(a), which consists of nine members and five nodal coordinate variables $(X_1, X_2, \ldots, X_5)$. As the total number of the design variables is very large in this case, the two-level approach is applied to the optimization. For the optimization at level 1, the following optimality criteria method is used to save the computation time. That is, the optimum solutions are assumed to be on the boundaries of the probability constraints. Consequently, the cross-sectional areas of members A are determined such that the failure probabilities of the members have the specified values P_{fai}, i.e.

$$P_{fi}(X_0, A) = P_{fai} \qquad (i=1,2,\ldots,9)$$

Following the approach by Murotsu, et.al [1], the cross-sectional areas A obtained by this way gives almost the identical volume of the structure with the minimum volume determined by the optimum design such as level 1 of Problem 2 in section 3.

Table 7 shows the optimum solutions for the various allowable failure probabilities of the members and correlation coefficients of loads L_1 and L_2 when buckling failure is not considered. The deterministic optimum design shown in Table 7 means the optimum design which is carried out under the constraints on member stresses, assuming that the strengths and the applied loads are equal to the mean values of the corresponding random variables in the reliability-based design. It is observed that, as in the case of the two-member truss, the reliability-based optimization design gives the same configuration of the structure as the deterministic design when $\rho_{L1,L2}=1.0$ and $CV_{L1}=CV_{L2}$. This is also due to the fact that the resultant force applied to the structure keeps the same direction. Fig. 4(b) compares the optimum geometry based on different design conditions. On the other hand, Table 8 corresponds to the case when buckling is taken into account. Comparing the results in Tables 7 and 8, the effect of buckling stress is very large on the optimum design. That is, the minimum volumes in Table 8 are larger than those in Table 7, and the height of the structure becomes considerably lower in Table 8 than that in Table 7. It should be also noted here that the reliability-based optimum configuration with buckling failure included is not so much differrent from the deterministic design, as illustrated in Fig. 4(c).

5. CONCLUSIONS

The nodal coordinates were taken as the shape design variables to study the shape optimization of the truss structure based on reliability concept. As a result, the following conclusions were drawn.

(1) The convergence problems owing to the different nature of the shape design variables and sizing design variables can be effectively overcome by scaling the design variables or by introducing two separate design spaces, i.e., two-level method.

(2) The method for calculating the failure probabilities of the member both in the tensile and compressive regions is necessary for the shape optimization.

(3) The optimum shape and minimum volume of truss structure are dependent on the probabilistic nature of the random variables, e,g, the member strengths and the loads applied to the structure.

(4) The buckling stress has large effect on the shape optimization of truss structures.

REFERENCES

1. P. Thoft-Christensen and Y. Murotsu, Application of Structural Systems Reliability Theory, Springer Verlag (1986).
2. N. Shiraishi and H. Furuta, 'On Geometry of Truss,' Memoirs of the Faculty of Engineering, Kyoto University, XLI, 4 (1979), pp. 498-517.

3. P. Thoft-Christensen, 'Application of Optimization Methods in Structural Systems Reliability Theory,' Structural Reliability Theory Paper, 53, Institute of Building Technology and Structural Engineering, Aalborg University (1987).
4. Y. Murotsu, M. Kishi and M. Yonezawa, 'On the Shape Optimization of Truss Structures Based on Reliability Concept,' in G.I.N.Rozvany and B.L.Karihaloo(eds.), Structural Optimization, pp.193-200(1988), Kluwer Academic Publishers.
5. G. N. Vanderplaats and F. Moses, 'Automated Design of Trusses for Optimum Geometry,' J. Struct. Div., ASCE, 98, ST3 (1972), pp. 671-690.
6. H. Konno and H. Yamashita, 'Nonlinear Programming' (in Japanese), Nikkagiren Press (1978).
7. M. W. Dobbs and L. P. Felton: 'Optimization of Truss Geometry,' J. Struct. Div., ASCE, 95, ST10, October 1969, pp. 2105-2118.

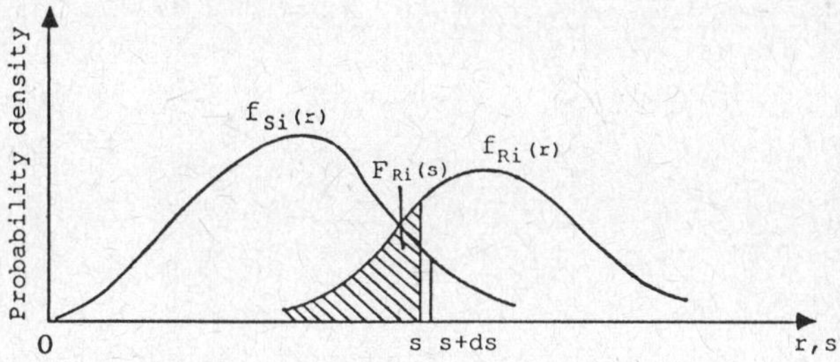

Fig. 1 Strength and load density functions for member i.

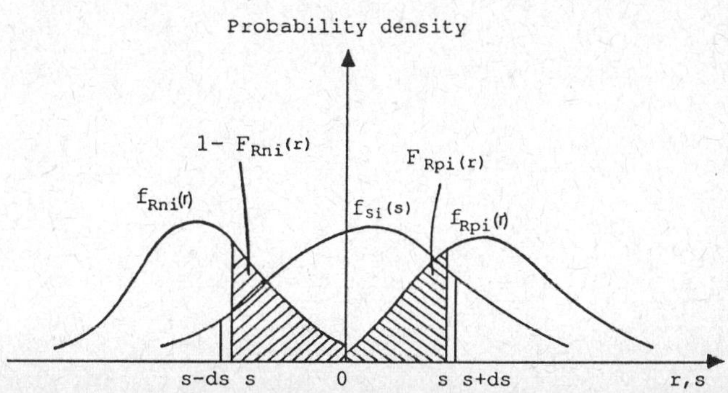

Fig. 2 Density functions when load spans positive and negative regions

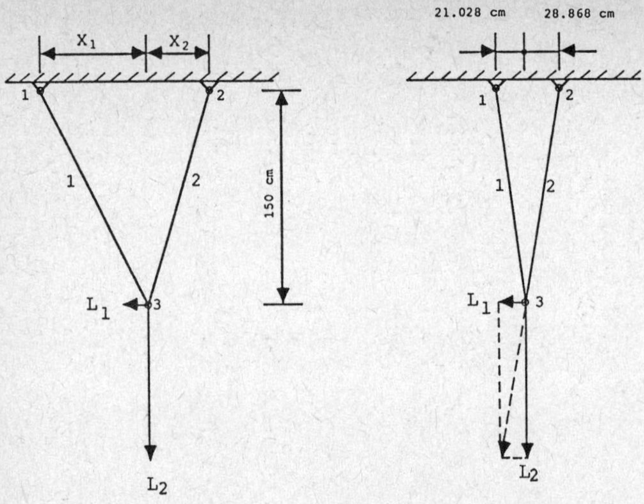

(a) General configuration

(b) Deterministic optimal geometry

Fig. 3 Two-member truss

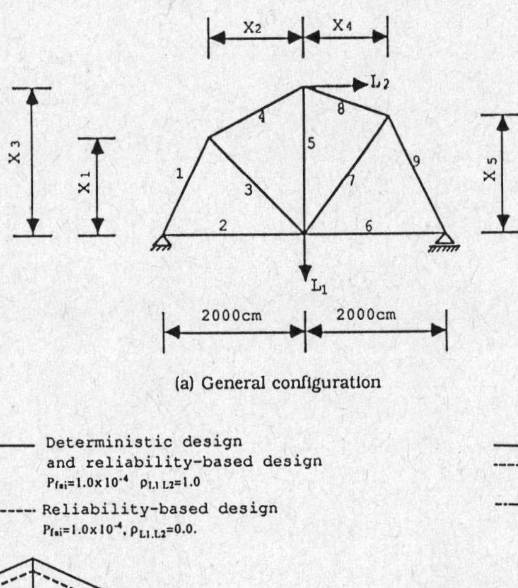

(a) General configuration

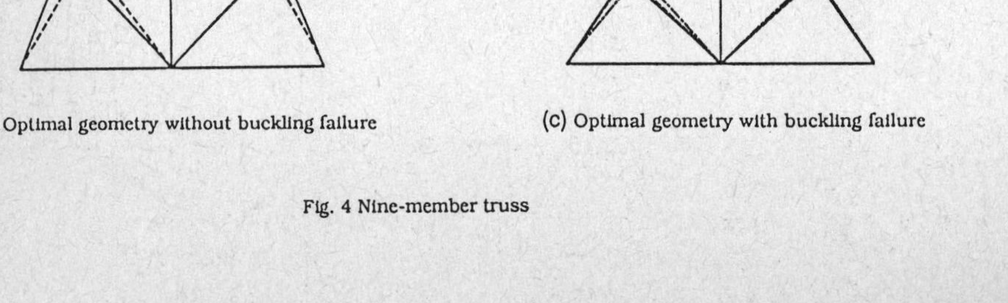

(b) Optimal geometry without buckling failure

(c) Optimal geometry with buckling failure

Fig. 4 Nine-member truss

Table 1 Deterministic optimum design of two-member truss

X_1(cm)	X_2(cm)	A_1(cm^2)	A_2(cm^2)	H_c(cm^3)
21.028	28.868	1.3463×10^{-6}	14.698	2245.2

Table 2 Numerical data of two-member truss

$\overline{\sigma}_{yi}$ (kN/cm^2)	$CV_{\sigma yi}$	$\overline{E}_i$ (kN/cm^2)	CV_{Ei}	$\overline{L}_1$ (kN)	$\overline{L}_2$ (kN)	CV_{L1}	CV_{L2}
18	0.05	7103	o.o5	50	259.8	0.2	0.2

Table 3 Reliability-based optimum designs of two-member truss (without buckling failure). $P_{fai}=1.0\times10^{-4}$

$\rho_{L1,L2}$	Nodal coordinate X_1(cm) X_2(cm)	Member i	Cross-sectional area A_i(cm^2)	Load μ_{Si}(kN) σ_{Si}	Failure probability P_{fi}^- P_{fi}^+	Volume H_c(cm^3)	Shape
-1.0	48.569	1	9.3252	10.036	1.1807×10^{-5}	4321.2	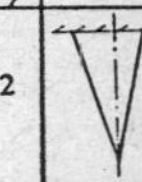
				41.232	8.8193×10^{-5}		
	31.823	2	18.593	255.82	0		
				13.016	1.0000×10^{-4}		
0.0	80.901	1	4.9951	4.4272	1.6354×10^{-5}	4535.2	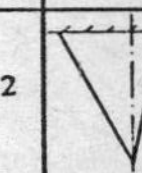
				22.262	8.3647×10^{-5}		
	30.540	2	24.066	261.15	0		
				40.872	1.0000×10^{-4}		
1.0	21.044	1	5.9053×10^{-4}	-5.8496×10^{-3}	9.9701×10^{-5}	4079.8	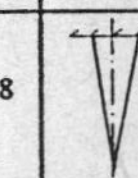
				1.1699×10^{-3}	0		
	28.867	2	26.708	264.57	0		
				52.915	9.9808×10^{-5}		
1.0*	21.843	1	4.3160×10^{-2}	4.2758×10^{-1}	0	4082.7	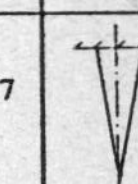
				8.5515×10^{-2}	1.0000×10^{-4}		
	28.951	2	26.682	264.16	0		
				52.833	9.8081×10^{-5}		

*Two-level optimization

Table 4 Reliability-based optimum designs of two-member truss (with buckling failure). $P_{fai}=1.0\times10^{-4}$

$\rho_{L1,L2}$	Nodal coordinate X_1(cm) X_2(cm)	Member i	Cross-sectional area A_i(cm^2)	Load μ_{Si}(kN) σ_{Si}	Failure probability P_{fi}^- P_{fi}^+	Volume H_c(cm^3)
-1.0	48.559	1	9.3263	10.040	1.1800×10^{-5}	4321.2
				41.236	8.8200×10^{-5}	
	31.824	2	18.591	255.82	0	
				13.012	1.0000×10^{-4}	
0.0	82.299	1	4.9454	4.2540	1.7063×10^{-5}	4535.2
				22.063	8.2937×10^{-5}	
	30.487	2	24.101	261.31	0	
				41.001	9.9999×10^{-5}	
1.0	21.147	1	5.7066×10^{-3}	-4.4089×10^{-3}	9.9796×10^{-5}	4089.5
				8.8179×10^{-4}	0	
	28.868	2	26.766	264.57	0	
				52.914	9.3432×10^{-5}	

Table 5 Effect of coefficients of variation of the loads. $P_{fai}=1.0\times10^{-4}$, $\rho_{L1,L2}=0.0$.

$CV_{L1}=CV_{L2}$	X_1(cm)	X_2(cm)	A_1(cm^2)	A_2(cm^2)	H_c(cm^3)
0.1	86.146	29.615	2.3945	20.339	3524.0
0.2	82.299	30.487	4.9454	24.101	4535.2
0.3	79.771	31.497	7.6134	27.981	5582.2

Table 6 Optimum designs for three values of allowable failure probability. $\rho_{L1,L2}=0.0$

P_{fai}	X_1(cm)	X_2(cm)	A_1(cm^2)	A_2(cm^2)	H_c(cm^3)
1.0×10^{-2}	78.551	31.209	3.3950	20.160	3663.6
1.0×10^{-4}	82.299	30.487	4.9454	24.101	4535.2
1.0×10^{-6}	83.261	30.224	6.2083	27.193	5226.0

Table 7 Optimum designs of nine-member truss (without buckling failure)

Reliability-based optimum design $\overline{\sigma}_{yi}$=12 kN/cm², $CV_{\sigma yi}$=0.05 $\overline{L}_1$=200 kN, $\overline{L}_2$=100 kN, $CV_{L1}=CV_{L2}$=0.2	ρ_{L1L2}=-1.0		ρ_{L1L2}=0.0		ρ_{L1L2}=1.0	
	Pfai=1.0x10⁻²					
X_1(cm) A_1(cm²)	1463.8	12.209	1407.5	10.280	1434.2	6.7496
X_2(cm) A_2(cm²)	1049.4	12.687	1168.9	14.985	1398.3	15.084
X_3(cm) A_3(cm²)	1880.2	7.2807	1862.6	6.6965	2004.0	5.1575
X_4(cm) A_4(cm²)	1373.8	11.718	1373.1	10.200	1420.3	6.7075
X_5(cm) A_5(cm²)	1364.4	10.194	1353.3	10.102	1413.7	10.297
A_6(cm²)		6.8579		7.8387		7.6769
A_7(cm²)		12.735		14.506		15.531
A_8(cm²)		16.977		19.380		20.234
A_9(cm²)		16.440		18.649		20.235
H_c(cm³)	180160		190477		190278	
	P_{fai}=1.0x10⁻⁴					
X_1(cm) A_1(cm²)	1454.1	16.774	1405.1	13.778	1433.7	8.1955
X_2(cm) A_2(cm²)	1018.5	13.946	1119.9	17.932	1398.7	18.311
X_3(cm) A_3(cm²)	1853.7	9.7898	1828.3	8.7598	2003.4	6.2651
X_4(cm) A_4(cm²)	1366.0	16.114	1361.1	13.655	1419.8	8.1467
X_5(cm) A_5(cm²)	1353.0	12.359	1334.9	12.195	1413.7	12.496
A_6(cm²)		7.8493		9.5230		9.3263
A_7(cm²)		14.391		17.147		18.853
A(cm²)		19.250		23.152		24.565
A_9(cm²)		18.500		22.058		24.564
H_c(cm³)	214258		230990		230979	
	P_{fai}=1.0x10⁻⁶					
X_1(cm) A_1(cm²)	1447.6	20.301	1397.8	16.473	1434.0	9.3371
X_2(cm) A_2(cm²)	1005.4	14.995	1102.3	20.241	1399.0	20.863
X_3(cm) A_3(cm²)	1835.9	11.801	1810.6	10.349	2003.6	7.1411
X_4(cm) A_4(cm²)	1360.0	19.540	1355.8	16.349	1420.2	9.2809
X_5(cm) A_5(cm²)	1344.3	14.035	1326.3	13.867	1413.7	14.238
A_6(cm²)		8.6919		10.826		10.621
A_7(cm²)		15.745		19.295		21.487
A_8(cm²)		21.149		26.143		27.992
A_9(cm²)		20.221		24.778		27.990
H_c(cm³)	241453		263040		263212	

Deterministic optimum design σ_{yi}=12 kN/cm² L_1=200 kN L_2=100 kN	
1434.1	X_1(cm)
1398.3	X_2(cm)
2003.4	X_3(cm)
1419.8	X_4(cm)
1413.7	X_5(cm)
4.5110	A_1(cm²)
10.079	A_2(cm²)
3.4482	A_3(cm²)
4.4836	A_4(cm²)
6.8761	A_5(cm²)
5.1326	A_6(cm²)
10.377	A_7(cm²)
13.520	A_8(cm²)
13.519	A_9(cm²)
127125	H_c(cm³)

Table 8 Optimum designs of nine-member truss (with buckling failure)

Reliability-based optimum designs $\overline{\sigma}_{yi}$=12 kN/cm², $CV_{\sigma yi}$=0.05, $\overline{E}_i$=7103 kN/cm², CV_{Ei}=0.05 $\overline{L}_1$=200 kN, $\overline{L}_2$=100 kN, $CV_{L1}=CV_{L2}$=0.2	ρ_{L1L2}=-1.0		ρ_{L1L2}=0.0		ρ_{L1L2}=1.0	
	P_{fai}=1.0x10⁻²					
X_1(cm) A_1(cm²)	1049.4	18.790	1047.2	17.260	1095.7	14.515
X_2(cm) A_2(cm²)	1124.0	14.526	1137.7	17.348	1199.6	18.113
X_3(cm) A_3(cm²)	1459.1	7.8688	1456.5	7.1515	1524.1	6.0833
X_4(cm) A_4(cm²)	1188.8	18.540	1187.7	17.307	1206.5	15.055
X_5(cm) A_5(cm²)	1075.3	11.648	1076.1	11.478	1105.3	11.464
A_6(cm²)		10.823		11.995		12.366
A_7(cm²)		11.925		13.256		13.885
A_8(cm²)		22.965		24.570		25.697
A_9(cm²)		22.001		23.568		24.731
H_c(cm³)	205099		214462		216374	
	P_{fai}=1.0x10⁻⁴					
X_1(cm) A_1(cm²)	1048.4	22.819	1040.3	20.517	1106.2	16.463
X_2(cm) A_2(cm²)	1108.5	15.973	1126.9	20.564	1202.7	21.861
X_3(cm) A_3(cm²)	1454.5	10.223	1447.0	9.0833	1537.1	7.3517
X_4(cm) A_4(cm²)	1186.4	22.161	1185.0	20.419	1206.5	17.013
X_5(cm) A_5(cm²)	1075.3	14.215	1071.6	13.986	1118.7	13.802
A_6(cm²)		12.365		14.306		14.866
A_7(cm²)		13.640		15.765		16.918
A_8(cm²)		25.034		27.537		29.118
A_9(cm²)		24.030		26.397		28.260
H_c(cm³)	235939		251189		254646	
	P_{fai}=1.0x10⁻⁶					
X_1(cm) A_1(cm²)	1049.7	25.667	1045.4	23.139	1140.3	18.219
X_2(cm) A_2(cm²)	1104.6	17.122	1103.5	23.173	1179.1	24.840
X_3(cm) A_3(cm²)	1458.1	12.036	1459.9	10.158	1557.1	8.2760
X_4(cm) A_4(cm²)	1188.6	24.798	1231.2	22.375	1263.4	18.025
X_5(cm) A_5(cm²)	1079.8	16.244	1048.2	16.591	1107.0	15.753
A_6(cm²)		13.506		15.548		15.950
A_7(cm²)		15.062		17.722		19.730
A_8(cm²)		26.648		30.622		32.685
A_9(cm²)		25.589		27.686		29.862
H_c(cm³)	259889		279742		284509	

Deterministic optimum design σ_{yi}=12 kN/cm² E_i=7103 kN/cm² L1=200 kN L2=100 kN	
1081.3	X_1(cm)
1191.2	X_2(cm)
1501.4	X_3(cm)
1195.7	X_4(cm)
1093.6	X_5(cm)
11.187	A_1(cm²)
12.227	A_2(cm²)
4.1063	A_3(cm²)
11.616	A_4(cm²)
7.6523	A_5(cm²)
8.4295	A_6(cm²)
9.2669	A_7(cm²)
19.614	A_8(cm²)
18.925	A_9(cm²)
154672	H_c(cm³)

PREDICTION OF EXTREME VALUES OF LINEAR COMBINATION OF LOAD EFFECTS

Arvid Naess
Department of Civil Engineering, Norwegian Institute of Technology
N-7034 Trondheim, Norway

Abstract The accuracy of the SRSS (square root of sum of squares) combination formula for estimating extremes of linear combination of load effects is investigated by means of specific examples. It is pointed out that the SRSS formula is nearly optimal in the case of independent Gaussian load effects. However, when the load effects are non-Gaussian, it is shown that the use of the SRSS formula may lead to significant errors.

Introduction

In structural dynamics one frequently meets the problem of determining a design value for a load effect combination. This could e.g. be the total stress in one of the tendons of a tension leg platform designed for offshore oil production. This would typically represent a linear load effect combination problem, given as follows:

$$X(t) = X_1(t) + \ldots + X_n(t) \tag{1}$$

where $X_j(t)$, $j = 1,\ldots,n$, are stochastic processes.

In this situation the designer might be faced with the problem of determining a response level which the total load effect $X(t)$ does not exceed by a preset probability throughout some specified time interval. Let $\hat{X}(T)$ denote the largest value of $X(t)$ over the time interval $(0,T)$, i.e. $\hat{X}(T) = \max(X(t);\ 0 \leq t \leq T)$. Ideally, it would be desirable to know the probability distribution function $F_{\hat{X}(T)}(\xi) = \mathrm{Prob}\{\hat{X}(T) \leq \xi\}$. In the case of jointly Gaussian component processes $X_j(t)$, then $X(t)$ is also a Gaussian process and good estimates of $F_{\hat{X}(T)}$ can be provided. However, in the general case this is only rarely obtainable. Hence various approximate procedures have been developed. In the present paper we shall investigate the case of stationary and mutually independent component processes. One particular procedure for estimating design values in this case, which is often used in design of offshore structures, is the so-called SRSS (square root of sum of squares) combination formula [1]. An investigation of the accuracy of this formula will be the main theme of this paper. In addition, the point-crossing formula of Cornell and Larrabee [2] will be considered. For a review of other combination procedures the reader may consult Pierce and Wen [1].

The SRSS Combination Formula

If the desired statistical descriptor of $\hat{X}(T)$ is its expected value, then the SRSS formula associated with the load effect formula of equation (1) for mutually independent, zero-mean and stationary processes is as follows:

$$E[\hat{X}(T)] \approx \{ \sum_{j=1}^{n} E[\hat{X}_j(T)]^2 \}^{1/2} \tag{2}$$

where $E[\cdot]$ denotes the expectation operator. It should be clear that no loss in generality is suffered by assuming zero mean values. If instead of the expected largest value one seeks the level $\hat{\xi}_\alpha$ so that $\hat{X}(T)$ is less than $\hat{\xi}_\alpha$ with probability $1-\alpha$, then the corresponding SRSS formula takes the form

$$\hat{\xi}_\alpha \approx \{ \sum_{j=1}^{n} \hat{\xi}^2_{\alpha,j} \}^{1/2} \tag{3}$$

where $\hat{\xi}_{\alpha,j}$ satisfies $F_{\hat{X}_j(T)}(\hat{\xi}_{\alpha,j}) = 1 - \alpha$.

In the subsequent discussions, we shall deal with equation (2). However, the conclusions reached apply equally well to equation (3). It is convenient to consider the Gaussian and non-Gaussian cases separately.

Gaussian case

Assuming that the $X_j(t)$ are mutually independent, zero-mean and stationary Gaussian processes with spectral densities $S_j(\omega)$, respectively, then $X(t)$ is also a zero-mean and stationary Gaussian process with spectral density $S(\omega) = \Sigma_j \; S_j(\omega)$. Let $m_k(m_{k,j})$ denote the k'th spectral moment of $X(t)$ $(X_j(t))$, then clearly $m_k = \Sigma_j \; m_{k,j}$. The mean zero-upcrossing rate, $v^+_X(0)$ say, of $X(t)$ is then given by

$$v^+_X(0) = \frac{1}{2\pi} \{\frac{m_2}{m_0}\}^{1/2} = \frac{1}{2\pi} \{ \sum_{j=1}^{n} m_{2,j} / \sum_{j=1}^{n} m_{0,j} \}^{1/2} \tag{4}$$

For any stationary, zero-mean Gaussian process satisfying certain regularity conditions and with mean zero upcrossing rate $v^+_X(0)$, it can be shown that

$$E[\hat{X}(T)] = \sqrt{m_o}\{\sqrt{Q} + \gamma/\sqrt{Q}\}, \quad T \to \infty \tag{5}$$

where $Q = 2 \ln(\nu^+_X(0)T)$ and $\gamma = 0.5772 \ldots$ is Euler's constant [3,4]. Based on equation (5), it can be shown [5] that for Gaussian load effect components

$$E[\hat{X}(T)] \geq \{ \sum_{j=1}^{n} E[\hat{X}_j(T)]^2 \}^{1/2} \tag{6}$$

Equality holds in equation (6) if and only if $\nu^+_{X_1}(0) = \ldots = \nu^+_{X_n}(0)$ where $\nu^+_{X_j}(0)$ denotes the mean zero upcrossing rate of $X_j(t)$.

Since this conclusion is based on equation (5), it is dependent on the accuracy of that equation. It is known that the right hand side of equation (5) may significantly overestimate $E[\hat{X}(T)]$ when $X(t)$ is narrow-banded. However, from the structure of the problem, one would expect the conclusion to be valid, at least approximately, also for narrow-band processes.

Even if the inequality in relation (6) is strict in the case of different crossing frequencies, the effect is not very strong in most cases of practical interest. As indicated in the section Numerical Examples, the discrepancy will only rarely exceed a few per cent.

Non-Gaussian case

To investigate the performance of the SRSS formula for non-Gaussian load effect components, the following processes are introduced:

$$Z_j(t) = |X_j(t)|^{\alpha-1} X_j(t), \quad 0 < \alpha < 2 \tag{7}$$

Here $X_j(t)$, $j = 1,\ldots,n$ denote stationary independent Gaussian $N(0,1)$-processes. The Gaussian case discussed above corresponds to the special case $\alpha = 1$.

The probability distribution function of the extreme value of the combined load effect $Z(t) = Z_1(t) + \ldots + Z_n(t)$, as well as for each component $Z_j(t)$, will be calculated by using the approximation

$$F_{\hat{Z}(T)}(\zeta) = \exp\{-\nu_Z^+(\zeta)T\}, \quad T \to \infty \tag{8}$$

where $\nu_Z^+(\zeta)$ denotes the mean ζ-upcrossing rate of $Z(t)$. Equation (8) is based on the assumption that upcrossing of high levels are statistically independent events. This is usually a good approximation except for narrow-band processes, cf. previous discussion. The parameter $v_Z{}^+(\zeta)$ is calculated by using the Rice formula

$$v_Z^+(\zeta) = \int_0^\infty s f_{Z\dot{Z}}(\zeta, s) ds \tag{9}$$

where the overdot denotes a time derivative.

For simplicity, in the subsequent developments, only the special case n = 2 will be considered, i.e.

$$Z(t) = Z_1(t) + Z_2(t) \tag{10}$$

By using the laws of marginal and total probability together with transformation of variables, equation (9) can be rewritten as follows

$$v_Z^+(\zeta) = \int_{-\infty}^{\infty}\int_0^\infty s\, f_{X_1 Z\dot{Z}}(x_1,\ \zeta,\ s) ds\, dx_1$$

$$= \int_{-\infty}^{\infty}\int_0^\infty s\, f_{X_1 X_2 \dot{Z}}\,(x_1,\ x_2,\ s)\alpha^{-1}|x_2|^{1-\alpha}\, ds\, dx_1$$

$$= \alpha^{-1}\int_{-\infty}^{\infty}\{\int_0^\infty s\, f_{\dot{Z}|X_1 X_2}(s|x_1, x_2) ds\}|x_2|^{1-\alpha} f_{X_1 X_2}(x_1, x_2) dx_1 \tag{11}$$

Here x_2 is determined by the equation

$$\zeta = |x_1|^{\alpha-1}x_1 + |x_2|^{\alpha-1}x_2 \tag{12}$$

From the equations (7) and (10), it follows that

$$\dot{Z}|_{X_j = x_j} = \sum_{j=1}^{2} \alpha|x_j|^{\alpha-1}\dot{X}_j \tag{13}$$

which implies that the conditional variable $\{\dot{Z}|X_i = x_i\}$ is Gaussian. Its mean value and variance are found to be

$$E[\dot{Z}|X_i = x_i] = 0 \tag{14}$$

and

$$\mathrm{Var}[\dot{Z}|X_i = x_i] = \ddot{\sigma}^2 = \sum_{j=1}^{2} \alpha^2 |x_j|^{2(\alpha-1)} \dot{\sigma}_j^2 \tag{15}$$

where $\dot{\sigma}_j^2 = \mathrm{Var}[\dot{X}_j]$. Taking account of this, it can be shown that

$$\nu_Z^+(\zeta) = \frac{\dot{\sigma}_1 \zeta^{\beta/2}}{(2\pi)^{3/2}\alpha} [\int_0^\infty \{|1-u|^{\beta(1-\alpha)} + \lambda^2 u^{\beta(1-\alpha)}\}^{1/2} \exp\{-\frac{\zeta^\beta}{2}(u^\beta + |1-u|^\beta)\}du$$

$$+ \int_0^\infty \{(1+u)^{\beta(1-\alpha)} + \lambda^2 u^{\beta(1-\alpha)}\}^{1/2} \exp\{-\frac{\zeta^\beta}{2}(u^\beta + (1+u)^\beta)\}du] \tag{16}$$

where $\beta = 2/\alpha$ and $\lambda = \dot{\sigma}_2/\dot{\sigma}_1$. Since $0 < \alpha < 2$, $\beta > 1$. Note that λ will also be the ratio between the mean zero upcrossing rates.

The evaluation of the integrals in equation (16) cannot be achieved in closed form for general values of β (>1). However, Laplace's method for asymptptic evaluation of integrals [6] can be used to provide accurate estimates for levels ζ of interest for the calculation of extreme values. Application of this method gives the following result

$$\nu_Z^+(\zeta) \underset{\zeta\to\infty}{=} \frac{\dot{\sigma}_1\sqrt{1+\lambda^2}}{2\pi\sqrt{2(2-\alpha)}} \exp\{-(\frac{\zeta}{2})^\beta\} \tag{17}$$

It is interesting to note that this asymptotic estimate is exact in the Gaussian case, i.e. for $\alpha = 1$ ($\beta = 2$).

The point crossing formula of Larrabee and Cornell [2] provides an upper bound for $\nu^+_Z(\zeta)$. Let $\tilde{\nu}^+_Z(\zeta)$ denote the mean upcrossing rate given by the point crossing formula, then [2]:

$$\nu_Z^+(\zeta) \leq \tilde{\nu}_Z^+(\zeta) = \nu_{Z_1}^+ * f_{Z_2}(\zeta) + f_{Z_1} * \nu_{Z_2}^+(\zeta) \tag{18}$$

where * signifies convolution. Again invoking Laplace's method for the asymptotic evaluation of integrals, it can be shown that

$$\tilde{\nu}_Z^+(\zeta) \underset{\zeta\to\infty}{=} \frac{\dot{\sigma}_1(1+\lambda)}{2\pi\sqrt{2(2-\alpha)}} \exp\{-(\frac{\zeta}{2})^\beta\} \tag{19}$$

Hence it follows that $\tilde{\nu}_Z^+(\zeta)/\nu_Z^+(\zeta) = (1+\lambda)/\sqrt{1+\lambda^2}$, $\zeta\to\infty$, the asymptotic equality being an exact equality in the Gaussian case. Note that the ratio is (asymptotically) independent of α (and ζ). The maximum (asymptotic) error is incurred when $\lambda=1$, in which case the ratio assumes the value $\sqrt{2}$. For most applications in reliability and design, such an error is usually insignificant. It should also be added that for such applications, usually asymptotic crossing rates are very accurate, specifically so for the cases studied here. This has been verified by numerical calculation of the rhs of equation (16).

Having established the asymptotic crossing rate for the combined process Z(t), equation (8) is used to estimate the distribution function of the extremes. This can be written

$$F_{\hat{Z}(T)}(\zeta) = \exp\{-\exp[-\theta(\zeta)]\} \tag{20}$$

where

$$\theta(\zeta) = (\frac{\zeta}{2})^{\beta} - \ln(MT) \tag{21}$$

and

$$M = M(\alpha,\lambda) = \dot{\sigma}_1\sqrt{1+\lambda^2}/(2\pi\sqrt{2(2-\alpha)}) \tag{22}$$

$E[\hat{Z}(T)]$ can now be calculated by using the formula $E[\hat{Z}(T)] = \zeta_o + \gamma/\theta'(\zeta_o)$, where ζ_o is the appropriate solution of the equation $\theta(\zeta) = 0$ [7,8]. It is found that

$$E[Z(T)] = 2[\ln(MT)]^{\alpha/2} + \alpha\gamma[\ln(MT)]^{\alpha/2 - 1},\ T \to \infty\ (\gamma = 0.5772\ldots) \tag{23}$$

To determine the corresponding expressions for $Z_1(t)$ and $Z_2(t)$, the following simple and useful result is applied [9, 10]: Assume that $Y(t) = h[X(t)]$ where h is a deterministic function and X(t) is a stationary stochastic process. Provided that the necessary regularity conditions are satisfied, then

$$\nu_Y^+(\eta) = \sum_{j=1}^{n} \nu_X^+(\xi_j) \tag{24}$$

where $\{\xi_1,\ldots,\xi_n\} = h^{-1}(\eta)$, that is $\xi_1,\ldots,\xi_n$ are all possible x-solutions of the equation $\eta = h(x)$, observing the domain of definition of h. Furthermore, $\nu_X^+(\cdot)$ and $\nu_Y^+(\cdot)$ denote the mean upcrossing rates of X(t) and Y(t).

Applying this result to the processes $Z_j(t)$, $j = 1,2$, the mean upcrossing rate of $Z_j(t)$ is found to be

$$\nu_{Z_j}^+(\zeta) = \frac{\dot{\sigma}_j}{2\pi}\exp\{-\frac{\zeta^{\beta}}{2}\} \quad (\zeta > 0) \tag{25}$$

Following the same procedure as above, it is obtained that ($M_j = \dot{\sigma}_j/2\pi$, j=1,2)

$$E[\hat{Z}_j(T)] = [2\ln(M_jT)]^{\alpha/2} + \alpha\gamma[2\ln(M_jT)]^{\alpha/2-1}, \quad T \to \infty \tag{26}$$

From equations (23) and (26), the accuracy of the SRSS combination formula for the class of load effect component processes $Z_j(t)$ can now be investigated. This is done in the next section.

Numerical Examples

Gaussian case

Example 1 Consider the case of two independent Gaussian load effects: $X(t) = X_1(t) + X_2(t)$. Using the previous notation, two parameters a and b are defined by $m_{0,2} = a\, m_{0,1}$ and $m_{2,2} = b\, m_{2,1}$. The underestimation incurred by the SRSS formula is given in % as a function of a and b in Table 1. Except for the two rather extreme situations giving rise to the deviations of 20 and 21%, it is seen that the accuracy of the SRSS formula is quite good for the more likely values of a and b.

Underestimation in %

Table 1

a \ b	1	10	100	1000
1	0	2	6	21
10	3	0	1	3
100	20	1	0	0

Non-Gaussian case

The special case $Z(t) = Z_1(t) + Z_2(t)$ is considered. Here $Z_j(t)$, j = 1,2, are defined as above.

Example 2 It is assumed that $\dot{\sigma}_1 = 2\pi$ and $\lambda = 1$, cf. equation (22). In Fig. 1

it is plotted the deviation D in % as a function of the transformation parameter α for a one hour time interval. A positive D indicates that the SRSS formula overestimates the true value, while a negative D indicates underestimation. It is seen that the accuracy of equation (2) deteriorates rather rapidly as the transformation parameter α deviates from 1, which corresponds to the Gaussian case. The deviation D is positive (negative) when the tails of the probability density functions of the load effect components lie above (below) the corresponding Gaussian tails. Entirely analogous conclusions are reached for other values of the parameters $\dot{\sigma}_1$ and T. The effect of an increasing λ is similar to that in the Gaussian case in the sense that the curve in Fig. 1 will be shifted downwards.

<u>Example 3</u> In this example it is assumed that $\dot{\sigma}_1 = 2\pi$, $\lambda = 1$ and $\alpha = 1.5$. The two component processes then have the same mean level crossing rates. The numerics of the example are given in Table 2. It is seen that the SRSS term overestimates the expected extremes of the total load effect by 16-17%. $E[\hat{Z}(T)]$ has also been estimated by using the crossing rate given by the point crossing formula, cf. equations (18) and (19). This estimate, which is denoted by $E_p[\hat{Z}(T)]$, deviates less than 3% from $E[\hat{Z}(T)]$ over the tested range of time intervals.

Table 2

T (hours)	1	10	100	1000
$E[\hat{Z}(T)]$	10.48	12.42	14.25	16.01
$E_p[\hat{Z}(T)]$	10.79	12.71	14.53	16.28
$E[\hat{Z}_j(T)]$	8.57	10.21	11.76	13.25
SRSS	12.12	14.44	16.63	18.74
D[%]	16	16	17	17

<u>Example 4</u> Let $\dot{\sigma}_1 = 2\pi$, $\lambda = 10$ and $\alpha = 1.5$. This example differs from the previous one in that the two load effect components have different mean level crossing rates by a factor 10. The numerics of this example are given in Table 3. It is seen that the overestimation incurred by the SRSS formula is reduced to 10-13%. This is in accordance with comments made in Example 2: The effect of an increasing λ is to shift the deviation curve downwards, cf. Fig. 1. Whether this will lead to an increase or a decrease in the absolute error, depends on the value

of the parameter α. The estimates $E_p[\hat{Z}(T)]$ provided by the point crossing formula are seen to be very close to $E[\hat{Z}(T)]$ in this case, which was to be expected since the ratio $(1+\lambda)/\sqrt{1+\lambda^2}$ is very close to 1.0.

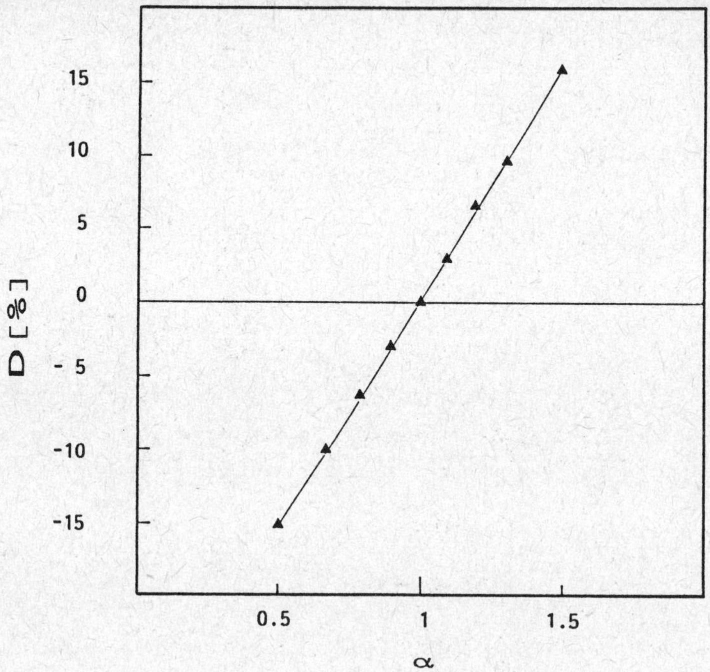

Fig. 1 Deviation in per cent between left and right hand side of equation (2) as a function of the transformation parameter α. $T = 1$ hour, $\dot{\sigma}_1 = 2\pi$ and $\lambda = 1$.

Table 3

T (hours)	1	10	100	1000
$E[\hat{Z}(T)]$	12.13	13.98	15.76	17.46
$E_p[\hat{Z}(T)]$	12.22	14.06	15.83	17.53
$E[\hat{Z}_1(T)]$	8.57	10.21	11.76	13.25
$E[\hat{Z}_2(T)]$	10.21	11.76	13.25	14.68
SRSS	13.33	15.57	17.72	19.78
D[%]	10	11	12	13

<u>Example 5</u> Assume $\dot{\sigma}_1 = 2\pi$, $\lambda = 1$ and $\alpha = 0.5$. For this value of α, the load effect components have a peak response behaviour somewhat similar to a white noise excited oscillator with nonlinear restoring force given by a third order polynomial. An example would be the Duffing oscillator [11]. The numerics of the present example are given in Table 4. It is seen that the SRSS underestimates the target value by 15-16% over the tested range of time intervals. The corresponding estimates given by the point crossing formula are accurate to within about 1% or less.

Table 4

T (hours)	1	10	100	1000
$E[\hat{Z}(T)]$	3.42	3.63	3.81	3.97
$E[\hat{Z}_p(T)]$	3.46	3.66	3.84	3.99
$E[\hat{Z}_j(T)]$	2.05	2.17	2.27	2.37
SRSS	2.90	3.07	3.21	3.35
D[%]	-15	-15	-16	-16

<u>Example 6</u> Let $\dot{\sigma}_1 = 2\pi$, $\lambda = 10$ and $\alpha = 0.5$. From previous comments one should expect that the SRSS formula underestimates the target value even more than in Example 5. From Table 5 it is seen that the underestimation is increased to about 17%. The accuracy of the estimates $E_p[\hat{Z}(T)]$ are again seen to be very good.

Table 5

T (hours)	1	10	100	1000
$E[\hat{Z}(T)]$	3.60	3.78	3.94	4.09
$E_p[\hat{Z}(T)]$	3.63	3.81	3.97	4.11
$E[\hat{Z}_1(T)]$	2.05	2.17	2.27	2.36
$E[\hat{Z}_2(T)]$	2.17	2.27	2.36	2.45
SRSS	2.98	3.14	3.27	3.40
D[%]	-17	-17	-17	-17

In the examples presented above the component processes all had the same variance. If the relative size of the two load effects become significantly different, the inaccuracy of applying equation (2) diminishes. This is clearly to be expected.

On the other hand, increasing the number of load effect components to three or more, will further increase the error. For instance, if in Example 3 the number of load effect components had been increased to three, the overestimation of the SRSS formula would increase to more than 20% over the same span of time intervals.

CONCLUSIONS

In the paper has been presented an investigation of the accuracy of the so-called SRSS combination formula for estimating the extreme values of a total load effect resulting from a linear sum of component load effects. It has been pointed out that the SRSS formula is nearly optimal in a certain sense in the case of independent Gaussian load effect components. However, in that case a combination formula for extremes is not particularly important, since the extreme value of the total load effect can be calculated directly.

The accuracy of the SRSS combination formula for non-Gaussian load effect components is investigated by studying particular examples. The degree of non-normality is controlled by a transformation parameter. The result from the cases considered can be stated briefly as follows: If the tails of the probability density functions of the component load effects lie below (above) that of a Gaussian law the SRSS formula leads to underestimation (overestimation) of the extreme values of the combined load effect. It has been shown that the error incurred by using the SRSS formula may be quite significant in the case of non-Gaussian load effects. The accuracy of estimates provided by the point crossing formula of Larrabee and Cornell has also been studied, and in all cases considered the accuracy of these estimates are very good.

ACKNOWLEDGEMENTS

This work was financially supported in part by The Royal Norwegian Council for Scientific and Industrial Research (NTNF), The Norwegian State Oil Company (Statoil) and Norsk Hydro through the NTNF Research Programme on Marine Structures.

REFERENCES

1. Wen, Y.K. and Pearce, H.T.: "Recent developments in probabilistic load combinations". Proc.Prob.Meth. Structural Engng. ASCE, St. Louis, No., 1981.

2. Larrabee, R.D. and Cornell, C.A.: "Combination of various load processes". Journal of Structural Division, ASCE, Vol. 107, pp. 223-239, 1981.

3. Leadbetter, M.R., Lindgren, G. and Rootzén, H.: Extremes and Related Properties of Random Sequences and Processes. Springer-Verlag, New York, 1983.

4. Davenport, A.G.: "Note on the Distribution of the Largest Value of a Random Function". Proceedings Inst. Civil Engineering, Vol. 28, 1964.

5. Naess, A.: Linear combination of load effects. Report 1.3.3 NTNF Research Programme on Marine Structures. STF71 A 87060, SINTEF, Trondheim, 1987.

6. Bender, C.M. and Orszag, S.A.: Advanced Mathematical Methods for Scientists and Engineers. McGraw-Hill, New York, 1978.

7. Gumbel, E.J.: Statistics of Extremes. Columbia University Press, New York, 1958.

8. Naess, A.: "The Joint Crossing Frequency of Stochastic Processes and its Application to Wave Theory". Journal of Applied Ocean Research, Vol. 7. No. 1, 1985.

9. Naess, A.: "Prediction of Extremes of Morison Type Loading - An Example of a General Method". Ocean Engineering, Vol. 10, No. 5, 1983.

10. Grigoriu, M.: "Crossings of Non-Gaussian Translation Processes". Journal of Eng.Mech., ASCE, Vol. 110, No. 4.

11. Lin, Y.K.: Probabilistic Theory of Structural Dynamics. McGraw-Hill, New York, 1967.

PARAMETRIC DECOMPOSITION FOR SYSTEM RELIABILITY EVALUATION

Avinash M. Nafday* & Ross B. Corotis**

***Project Engineer, EQE Engineering Incorporated**
3150 Bristol Avenue, Suite 350, Costa Mesa, California 92626

****Hackermann Professor and Chair, Department of Civil Engineering**
The Johns Hopkins University, Baltimore, Maryland 21218

Introduction

Structural systems are quite complex due to their high degree of redundancy and the random nature of both load and resistance variables. Rigorous analytical procedures for the solution of these systems are in their nascent stage; therefore, system reliability for practical structures is often estimated by simulation-based procedures in conjunction with a specified structural system model. Due to typically low probability of failure, these procedures usually require large computational effort. The inefficiency of a high number of simulation repetitions is compounded by two other factors: a large number of correlated variables to be simulated, and the necessity of solving the structural system for each set of loads and resistances. This latter feature is especially time-intensive.

All efforts at increasing efficiency of computation thus far have concentrated on reducing the number of simulation experiments by strategies such as directional simulation and importance sampling. Not much attention has been given to the related problem of the selection of a suitable structural system model and improvement of the efficiency in the repeated solution of such models. This is somewhat surprising since much more effort is needed to solve the structural system for a given set of simulated values compared to that required for generating these values.

As a first consideration, one measure of the relative efficiency of a scheme involves consideration of the fraction of simulation points in the failure region. Directional simulation and importance simulation increase this efficiency substantially, but linear programming (LP) models have even greater potential because every simulation trial results in a failure mode (Nafday and Corotis, 1988). For example, a system reliability obtained in just 100 simulations was comparable to that obtained by 10^7 trials of Monte Carlo simulation. In addition, it is possible to avoid the repeated solution of structural system altogether by following a procedure based on the decomposition of parametric space. Augusti et al. (1984) allude to such a general approach in a discussion of this topic, but do not actually present procedures for the parametric decomposition aspects. The aim in this paper is to elucidate strategies for parametric decomposition in single and multiparametric linear programming models for rigid plastic frames.

Description of the Problem

One popular method to evaluate the reliability of a redundant structural system (brittle or ductile) is based on determining the system resistance distribution from information about marginal distributions of member and modal resistances, and convolving it with the load distribution (Thoft-Christensen and Baker, 1982; Thoft-Christensen and Murotsu, 1986). In the space of load and resistance variables (basic variable space), let G_i be the subdomain corresponding to failure by mode i, if load and resistance values were scaled such that failure occurred. Denoting $L_{(i)}$ and $R_{(i)}$ as the load and resistance scalar measures, respectively, in this subdomain, one can write

$$1 - \mathit{Reliability} = P\left[\bigcup_{i=1}^{k} L_{(i)} > R_{(i)} \cap \underline{L}, \underline{R} \in G_i\right] \tag{1}$$

Because the G_i regions must be mutually exclusive and collectively exhaustive, Equation (1) can be written as

$$1 - \mathit{Reliability} = \sum_{i=1}^{k} P\left[L_{(i)} > R_{(i)} \cap \underline{L}, \underline{R} \in G_i\right] \tag{2}$$

Since there has heretofore been no practical method for computing all G_i, this procedure has had limited success. With LP-based modeling, however, definition and delineation of parametric domains seems relatively tractable.

LP-Based Models

LP-based models for rigid plastic structural systems offer some unique advantages (Casciati and Faravelli, 1985). Ditlevsen (1987) discusses the relevance of the rigid plastic assumption for various structures and shows that it is often a practical choice for material modeling. LP-based models are applicable to a variety of structures with a single stress resultant (Munro and Smith, 1972). There is no need to specify the entire limit state, and every solution of the model leads to a kinematically admissible mode that actually contributes to the probability of failure (Bjerager et al., 1987). This is an inherent computational advantage over methods that do not always lead to a failure mode or for which not every failure mode reached is kinematically admissible. Morever, it is more appropriate when considering failure to use models of plastic behavior than elastic theory. Complete member modeling would require the details of the elastic behavior of structural components between the active yield hinges, which is not necessary for the problem concerning plastic mechanisms (Ditlevsen and Bjerager, 1986). Further, extremely efficient algorithms are available for the solution of LP models, and the possibility of devising even better special purpose algorithms exists.

Formulations

There are two basic formulational approaches that can be followed for the modeling of the rigid plastic system in the LP format. One is the static formulation, based on the lower bound theorem of plasticity, and the other is the kinematic formulation, providing an upper bound (Nafday and Corotis, 1987; Nafday, et al., 1987b). These two formulation represent mathematical duals.

The static formulation states that if a set of moments can be found that everywhere satisfies equilibrium and violates no yield conditions, then the associated set of loads does not lead to a collapse mechanism. The static LP model, which might be termed the random variable or direct approach, leads to a specification of the limit state. This mathematical determination of the limit state is a difficult problem, but the computation of the probability that the random loads and strengths lead to an outcome beyond the safe limit state is quite straightforward. In particular, advanced simulation methods, such as directional simulation (Bjerager, 1988) and importance sampling (Schueller and Stix, 1987) appear to be very efficient. One further convenience with the static formulation is that the method can be extended to include multiple independent loads through the use of a multiobjective linear program (Nafday, et al., 1988).

The kinematic formulation states that if a kinematically admissable collapse mechanism can be found for a set of loads, then no equilibrium solution exists. Since the kinematic LP model is based on discrete collapse modes, it could be called the random event or indirect approach. The identification of all significant collapse modes is a formidable problem, but one that may be approached a number of different ways. The computation of the probability of occurrence of each mode is relatively tractable, although the appropriate combination of these probabilities into a system reliability offers some complexities.

Parametric Decomposition of the Kinematic Linear Model

The kinematic LP model has been shown to produce a polyhedral feasible region whose extreme points represent different failure modes of the structure (Nafday et al., 1987a). For a fixed range of load proportionalities, there is a range of values of $\underline{R}$ for which a given extreme point is optimal, i.e., occurrence regions for failure by a particular mode. Thus, it is theoretically possible to identify the basic variable resistance regions G_i corresponding to each of the extreme points or collapse mechanisms. This enables one to one replace the consideration of an infinite number of parameter combinations with a finite number of parametric regions, which are polyhedral and mutually exclusive (Casciati, 1978; Podrebarac and Sengupta, 1971).

The general kinematic linear programming model may be expressed as (Cohn et al., 1972):

$$\lambda^+ = Min \left[\sum_{j=1}^{J} M_{pj} \left(\theta_j^+ + \theta_j^- \right) \right]$$

$$s.t. \;\; \theta_j^+ - \theta_j^- = \sum_{k=1}^{N} t_k \theta_{kj}; j = 1, ..., J$$

$$\sum_{k=1}^{N} t_k e_k = 1$$

$$\theta_j^+, \theta_j^- \geq 0 \qquad (3)$$

in which J = number of critical hinge locations; θ_j = rotation at section j, $\theta_j = \theta_j^+$ - θ_j^-, where θ_j^+ and θ_j^- represent the positive and negative values of hinge locations, respectively; t_k = a coefficient indicating the contribution of the kth elementary mode to collapse; and λ^+ = collapse load factor for the kinematic LP. The parameters are θ_{kj} = hinge rotation of elementary mechanism k at joint j; e_k = external work associated with elementary mechanism k; N = number of equations of equilibrium/number of elementary mechanisms; and M_{pj} = member capacity at section j. The potential plastic hinge locations are at J critical sections.

There are two basic forms of parametric decomposition problems. One involves variation of the constraints by the perturbation of the right side of the constraint equations. This form has direct implication in terms of load ratio variability in the structural system problem. The other consists of variation of the objective function coefficients, which reflect moment capacity variability. The second type will be illustrated in this paper.

Procedures for linear parametric programming fundamentally involve a single (scalar) parameter (Hadley, 1963). Gal and Nedoma (1972) have also presented a computational procedure for simultaneous consideration of multiple (vector) parameters, based on geometric decomposition principles.

The concept of objective function decomposition has also been applied in multiobjective linear programming problems by the assignment of parametric weights to the objectives in order to enumerate noninferior extreme points (Yu and Zeleny, 1976). For example, in the multiparametric limit analysis problem (Nafday, et al., 1988), parametric decomposition leads to partitioning in the space of load parameters. These regions are also non-overlapping and polyhedral.

From the above discussion, it is clear that the principle of parametric decomposition is applicable for the generation of domains in either resistance or load space. The procedure does not in any way depend on probabilistic information. This information can then be subsequently utilized to compute system reliability.

Consider, as an illustration of objective function variation, a problem that has been solved for a given ensemble of proportional load vectors using mean plastic moment capacities. Interpolating between corresponding limit state performance points on the load vectors produces a response limit state surface. If the resistances are now randomized, the question arises of how much the moment capacities can vary without changing the mode of failure for a given load proportionality. This can be easily addressed with a kinematic LP model.

Utilizing single parameter decomposition, the vector of original moment capacities, $\mathbf{M}$, in the objective function is replaced by an augmented vector, $\mathbf{M}^+$,

$$\mathbf{M}^+ = \mathbf{M} + c\mathbf{f} \tag{4}$$

in which $\mathbf{f}$ is a vector of the same order as $\mathbf{M}$ but having zero components except for the particular moment capacities to be varied. The non-zero components of $\mathbf{f}$ are the proportionalities of changes of the moment capacities in $\mathbf{M}$, and c is an arbitrary scalar. For all variables in the basis corresponding to the failure mode being investigated, one computes the following statistic:

$$\frac{reduced\ cost\ for\ variable\ i}{f_B y_i - f_i} \tag{5}$$

in which $\mathbf{f}_B$ is the subvector of $\mathbf{f}$ containing those moment capacities appearing in the current basis,

y_i is the value of the ith variable appearing in the final simplex tableau, f_i is the ith component of the vector **f**, and the reduced cost is found from the simplex tableau. The statistic in Equation 5 represents the relative change in the objective function of the original linear program to changes in the particular moment capacities being varied. If the denominator in Equation 5 is negative for all values of i, then no value of c in Equation 4 will lead to a change in basis. However, if there is at least one value of i for which the denominator in Equation 5 is positive, then there is a finite maximum value of c, c_{max}, after which the basis changes. That maximum value, c_{max}, is found as the minimum value of the statistic in Equation 5, considering all values of i for which the denominator is positive.

$$c_{max} = \min_{\substack{i \\ f_B y_i - f_i > 0}} \left[\frac{reduced\ cost\ for\ variable\ i}{f_B y_i - f_i} \right] \tag{6}$$

Therefore, the new moment capacity vector that represents the maximum change without altering the basis is

$$\mathbf{M}^+ = \mathbf{M} + c_{max}\mathbf{f} \tag{7}$$

A Simple Example

Consider the simple example of the portal frame shown in Figure 1, in which hinge locations 2 and 4 are assumed to be in the columns. There are two elementary mechanisms, the beam and sidesway, and one combined mechanism. Equation 3 becomes

$$\lambda^+ = Min\left[M_c\left(\theta_1^+ + \theta_1^- + \theta_2^+ + \theta_2^- + \theta_4^+ + \theta_4^- + \theta_5^+ + \theta_5^-\right) + M_b\left(\theta_3^+ + \theta_3^-\right)\right]$$

$$s.t. \left\{ \begin{array}{ccc} \theta_1^+ & - & \theta_1^- \\ \theta_2^+ & - & \theta_2^- \\ \theta_3^+ & - & \theta_3^- \\ \theta_4^+ & - & \theta_4^- \\ \theta_5^+ & - & \theta_5^- \end{array} \right\} = \begin{bmatrix} 1 & 0 \\ -1 & 1 \\ 0 & -2 \\ 1 & 1 \\ -1 & 0 \end{bmatrix} \left\{ \begin{array}{c} t_1 \\ t_2 \end{array} \right\}$$

$$t_1 L_1 h + t_2 L_2 L/2 = 1$$

$$t_1, t_2 \geq 0 \tag{8}$$

For this example, the complete solution will be shown with variation in both the loads and the resistances. However, this simple problem can be solved without recourse to the more general solution techniques mentioned above.

Equations 8 may be rewritten as

$$\lambda^{+} = min\left[2M_c t_1 + M_c(t_1 + t_2) + M_c|t_1 - t_2| + 2M_b t_2\right]$$

$$s.t.\ \ t_1 L_1 h + t_2 L_2 L/2 = 1$$

$$t_1, t_2 \geq 0 \tag{9}$$

Solving for t_2 in terms of t_1 and substituting gives

$$\lambda^{+} = min\left[\left\{M_c\left(3 - \frac{2L_1 h}{LL_2}\right) - 4M_b\frac{L_1 h}{LL_2}\right\} t_1 + M_c\left|t_1\left(1 + \frac{2L_1 h}{LL_2}\right) - \frac{2}{LL_2}\right| + (4M_b + 2M_c)/LL_2\right] \tag{10}$$

In order to illustrate a set of results, let L = h, and without loss of generality define $M_b = R_m M_c$ and $L_2 = R_L L_1$. Then Equation 10 becomes

$$\lambda^{+} = min\left[M_c\left\{3 - \frac{2}{R_L} - 4\frac{R_m}{R_L}\right\} t_1 + M_c\left\{\frac{2 + 4R_m}{LR_L L_1}\right\} + M_c\left|t_1\left(1 + \frac{2}{R_L}\right) - \frac{2}{LR_L L_1}\right|\right] \tag{11}$$

It is clear that for small values of t_1, the absolute value term will reverse the sign of its internal terms. Then the minimum will occur at $t_1 = 0$ if the slope of the t_1 terms exceeds zero:

$$3 - \frac{2}{R_L} - 4\frac{R_m}{R_L} - \left(1 + \frac{2}{R_L}\right) > 0$$

$$R_M < \frac{1}{2}R_L - 1 \tag{12}$$

For large values of t_1, the absolute value term has its natural sign. The minimum occurs when t_1 is as large as possible $(1/L_1 L)$ if the slope of the t_1 terms is less than zero:

$$3 - \frac{2}{R_L} - 4\frac{R_m}{R_L} + \left(1 + \frac{2}{R_L}\right) < 0$$

$$R_m > R_L \tag{13}$$

The regions defined by Equations 12 and 13 are shown in Figure 2, along with the associated modes.

Conclusions

In the proposed approach, structural aspects are shown to depend on the parametric structure of the model, which delineates the sample space for random variables. The sample space is decomposed into non-overlapping convex polyhedral sets such that each of these sets represents a single "optimal" solution for all points in the set. The information from the parametric decomposition can be utilized to evaluate system reliability.

An important advantage of the proposed methodology is the separation of the structural and probabilistic aspects of the problem. This is because the parametric regions do not depend on the distribution of the random variables, and hence can be used to solve the problem with various alternative probability distributions for random variables. This facilitates investigation of different loading conditions and probabilistic assumptions since reliability evaluations can be obtained without requiring any further structural analyses.

References

Augusti, G., Baratta, A., and Casciati, F. (1984), Probabilistic Methods in Structural Engineering, Chapman and Hall, Ltd, New York, 556p.

Bjerager, P. (1988), "Probability Integration by Directional Simulation," Journal of Engineering Mechanics, ASCE. Vol. 114, No. 8, August, pp. 1285-1302.

Bjerager, P., Karamchandani, A., and Cornell, C.A. (1987), Failure Tree Analysis in Structural System Reliability, ICASP5, Institute for Risk Research, University of Waterloo, Waterloo, Canada, pp. 985-996.

Casciati, F. (1978). "Elastic-Plastic Deformation Analysis: A Parametric Linear Programming Model," Journal de Mecanique Appliquee, 2(3), 269-290.

Casciati, F. and Faravelli, L. (1985). "Structural Reliability and Structural Design Optimization," ICOSSAR 4, Kobe, Japan, May 27-29, III 61-70.

Cohn, M.Z., Ghosh, S.K., and Parimi, S.R. (1972), "Unified Approach to Theory of Plastic Structures," Journal of the Engineering Mechanics Division ASCE, Vol. 98, No. 5, October, pp. 1133-1158.

Ditlevsen, O. (1987), "Probabilistic Statics of Discretized Ideal Plastic Frames," DCAMM 358, The Technical University of Denmark, Lyngby, Denmark, 34p.

Ditlevsen, O. and Bjerager, P. (1986), "Methods of Structural System Reliability," Structural Safety, Vol. 3, Elsevier, Amsterdam, pp. 195-229.

Gal, T. and Nedoma, J. (1972), "Multiparametric Linear Programming," Management Science, Vol. 18, No. 7, March, pp. 406-422.

Hadley, G. (1963), Linear Programming, Addison-Wesley, Reading, Massachusetts, pp. 380-387.

Munro, J. and Smith, D.L. (1972), "Linear Programming Duality in Plastic Analysis and Synthesis," International Symposium on Computer-Aided Structural Design, Peter Peregrinus Publishing, Stevenage, U.K., A1.22-A1.54.

Nafday, A.M. and Corotis, R.B. (1987), "Failure Mode Enumeration for System Reliability Assessment by Optimization Algorithms," First IFIP WG 7.5 Working Conference, Aalborg, Denmark, Springer-Verlag, pp. 297-306.

Nafday, A.M. and Corotis, R.B. (1988), "A Kinematic Approach to Simulation-Based System Reliability Evaluation," ASCE Joint Specialty Conference, May, Virginia Polytechnic Institute and State University, Blacksburg.

Nafday, A.M., Corotis, R.B., and Cohon, J.L. (1987a), "Failure Mode Identification for Structural Frames," Journal of Structural Engineering, ASCE, Vol. 113, No. 7, July, 1415-1432.

Nafday, A.M., Corotis, R.B., and Cohon, J.L. (1987b), "System Reliability of Rigid Plastic Frames," ICASP5, Institute for Risk Research, University of Waterloo, Waterloo, Canada, pp. 119-126.

Nafday, A.M., Corotis, R.B., and Cohon, J.L. (1988), "Multiparametric Limit Analysis of Frames: Part I-Model and Part II-Computations, Journal of Engineering Mechanics, ASCE, Vol. 114, No. 3, March, pp. 397-403.

Podrebarac, M.L. and Sengupta, S.S. (1971), "Parametric Linear Programming: Some Extensions," INFOR, Vol. 9, No. 3, November, pp. 305-319.

Schueller, G.I. and Stix, R. (1987), "A Critical Appraisal of Methods to Determine Failure Probabilities," Structural Safety, Vol. 4, Elsevier, Amsterdam, pp. 293-309.

Thoft-Christensen, P. and Baker, M.J. (1982), Structural Reliability Theory and its Applications, Springer-Verlag, Berlin, 267p.

Thoft-Christensen, P. and Murotsu, Y. (1986), Applications of Structural Systems Reliability Theory, Springer-Verlag, Berlin, 339p.

Yu, P.L. and Zeleny M. (1975), "The Set of All Nondominated Solutions in Linear Cases and a Multicriteria Simplex Method," Journal of Mathematical Analysis and Applications, Vol. 49, pp. 430-468.

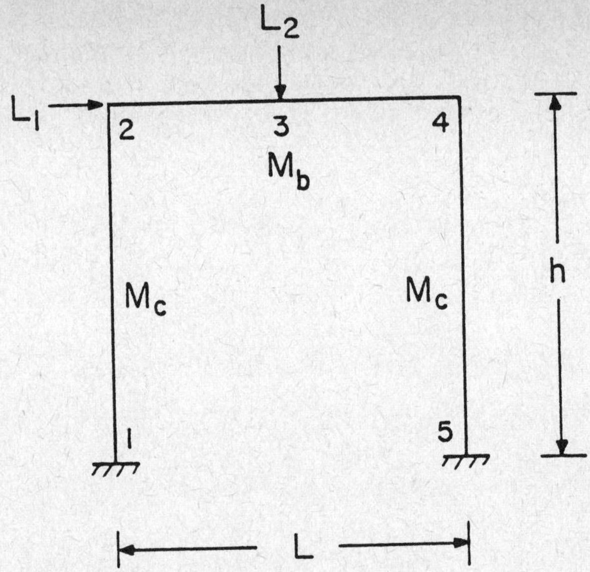

Figure 1. Simple Portal Frame

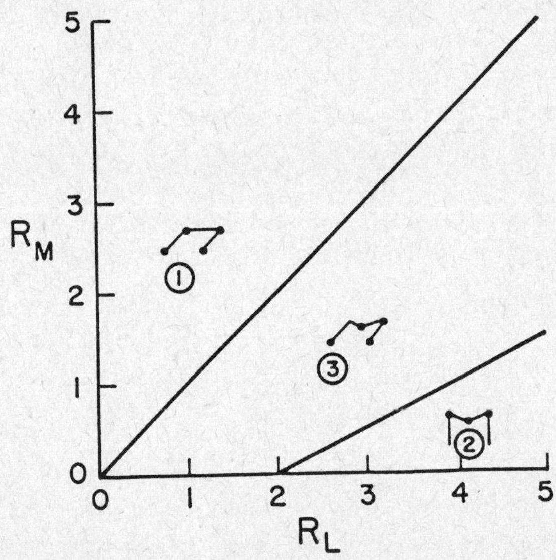

Figure 2. Load and Resistance Ratio Parametric Decomposition by Modes

INTEGRATION OF MULTINORMAL DENSITIES ON SURFACES

G. Schall*, S. Gollwitzer** & R. Rackwitz*
*Technical University of Munich, Arcisstr. 21, D-8000 Munich 2, FRG
**RCP-Consult, Munich, FRG

Introduction

Many reliability problems involve the computation of simple volume integrals of probabilities over certain domains in multidimensional spaces. This can conveniently be done by making use of first–order or asymptotic second–order concepts in the numerical analysis as described in [1,2]. Those domains must be given as inequalities.

However, the computation of outcrossing rates for Gaussian processes involves the determination of surface integrals which in some special cases involves simply an integration of probability densities over equalities [3]. In the reliability analysis of failure trees for systems the calculation of the probability of intersections of failure domains is required. But some further restrictions sometimes are given in the form of equalities, for example, if the initiating event is a discrete event. In elastic–plastic structures the analysis of successive yielding of hinges until collapse requires that the hinges except the temporary last one fulfill the yield conditions which again are given in terms of equalities. Discrete events depending on further random variables and to be introduced as equalities are also encountered in human error control as shown in [4]. Furthermore, the calculation of conditional probabilities involving equalities sometimes is required when updating structural reliability in view of special forms of inspection results [5].

In this note suitable computation schemes for integrals of the type

$$I(D) = \int_D \varphi_n(x)\, ds(x) \tag{1}$$

where

$$D = E \cap F = \bigcap_{i=1}^{\ell} \{E_i\} \cap \bigcap_{j=\ell+1}^{m} \{F_j\} = \bigcap_{i=1}^{\ell} \{e_i(X) = 0\} \cap \bigcap_{j=\ell+1}^{m} \{g_j(X) \le 0\}$$

are developed. Herein, $\mathbf{X} = (X_1,\ldots,X_n)^T$ is a n–dimensional random vector with given distribution function. ds(x) means surface integration. Since the vector $\mathbf{X}$ can always be transformed into an independent standard normal vector $\mathbf{U} = (U_1,\ldots,U_n)^T$ as shown in [6], it is sufficient to determine the integral in the so–called standard space. Of course, it is $\ell < n$. The surface integral in eq. (1) will be understood as a surface integral of the first kind. Thus, the results will be applicable precisely in the examples mentioned before. First, the available first–order results will briefly be reviewed also further highlighting the meaning of eq. (1). Then, asymptotic second–order results will be derived. Finally, an importance sampling scheme is discussed for updating volume and surface integrals over arbitrary domains.

Review of first–order results for integrals of multinormal densities over equalities

Let $\mathbf{U} = (U_1,\ldots,U_n)^T$ be an independent, standard normal vector with density $\varphi_n(\mathbf{u}) = \Pi_{i=1}^{n}\varphi(u_i)$. A first–order approximation of integrals as in eq. (1) can be deduced from the considerations in [7]. In particular, let the events in eq. (1) be given by their linearizations

$$l_{e,i}(\mathbf{u}) = \alpha_{e,i}^T\,\mathbf{u} - \alpha_{e,i}^T\,\mathbf{u}^* = 0 \text{ and } l_{g,j}(\mathbf{u}) = \alpha_{g,j}^T\,\mathbf{u} - \alpha_{g,j}^T\,\mathbf{u}^* \leq 0,$$

respectively, where $\mathbf{u}^*$ is the so–called joint β–point (see [2]) and $\alpha_{.,.}$ the normalized gradient of these functions in $\mathbf{u}^*$. Alternatively, as in "crude" first–order reliability [2], the individual β–points $\mathbf{u}_i^*$ can be chosen with usually some loss of accuracy. Then, by viewing a volume integral as the limit of an integration of a sequence of appropriately chosen surface integrals the integral in eq. (1) can be given as:

$$\begin{aligned} I(E \cap F) &= I\left(\bigcap_{i=1}^{\ell} \{e_i(\mathbf{u}) = 0\} \cap \bigcap_{j=\ell+1}^{m} \{g_j(\mathbf{u}) \leq 0\}\right) \\ &= \frac{\partial^\ell}{\partial c_1 \ldots \partial c_\ell} P\left(\bigcap_{i=1}^{\ell} \{l_{e,i}(\mathbf{u}) \leq 0\} \cap \bigcap_{j=\ell+1}^{m} \{l_{g,j}(\mathbf{u}) \leq 0\}\right) \\ &= \frac{\partial^\ell}{\partial c_1 \ldots \partial c_\ell} P\left(\bigcap_{i=1}^{\ell+m} \{Z_i \leq c_i\}\right) \end{aligned} \qquad (2)$$

with $c_i = -\alpha_{e,i}{}^T\mathbf{u}_i^*$ or $c_i = -\alpha_{e,i}{}^T\mathbf{u}_i^*$ $(i = 1,\ldots,\ell)$ and $z_i = \alpha_i{}^T\mathbf{u}$ and $z_i - c_i = 0$ the generic linear representation of the failure surfaces. It is worth noting that this interpretation does not cover all applications of surface integrals in probability theory. Evaluation of eq. (2) is best be performed by applying a recursive scheme. By conditioning on the first variable the first step of calculation is:

$$I(E_1 \cap F) = \frac{\partial}{\partial c_1}\left[\int_{-\infty}^{c_1} P\left(\bigcap_{i=2}^{\ell+m} (z_i \leq c_i \mid z_1 = s)\right) \varphi(s)ds\right] \qquad (3)$$

Introducing y_i as variables which are stochastically independent of z_1 for $i \neq 1$ with $z_i = \rho_{1i} z_1 + \sqrt{1-\rho_{1i}^2}\, y_i$ and $\rho_{1i} = \alpha_1^T \alpha_i / (\|\alpha_i\| \, \|\alpha_1\|)$ eq. (3) can be rewritten as:

$$I(E_1 \cap F) = \varphi(c_1)\, P\Big(\bigcap_{i=2}^{\ell+m} y_i \leq \frac{c_i - \rho_{1i} c_1}{\sqrt{1-\rho_{1i}^2}}\Big) \tag{4}$$

This scheme can be applied ℓ times so that the first–order result finally becomes

$$I(E \cap F) = \prod_{i=1}^{\ell} \varphi(\tilde{c}_i)\, P\Big(\bigcap_{j=\ell+1}^{m} \tilde{y}_j \leq \tilde{c}_j\Big) \tag{5}$$

where the quantities $\tilde{c}_i$ and $\tilde{y}_j$ indicate the modifications to be performed in each step. At most a $m-\ell$–dimensional normal integral needs to be evaluated. Note that explicit surface integration is avoided, i.e. by a suitable parameterization of the equalities.

Second–order results for integrals of multinormal densities over equalities

Define now the maximum point (β–point) of $\varphi_n(\mathbf{u})$ on E as usual by $\mathbf{u}^* = \min\{\|\mathbf{u}\|\}$ for $E = \{\mathbf{u}: e(\mathbf{u}) = 0\}$ with $\beta = \|\mathbf{u}^*\|$. Assume that by an orthogonal transformation the β–point lies on the negative u_n–axis, i.e. $\mathbf{u}^* = (0,0,\ldots,-\|\mathbf{u}^*\|)^T$, the equality $e(\mathbf{u}) = 0$ is at least twice differentiable near the $\mathbf{u}^*$ and the mixed second derivatives of $e(\mathbf{u}^*)$ vanish which again can be achieved by a suitable orthogonal transformation [8]. The curvatures of $e(\mathbf{u})$ in $\mathbf{u}^*$ then are given by:

$$\kappa_i = -\frac{1}{\frac{\partial e(\mathbf{u}^*)}{\partial u_n}} \frac{\partial^2 e(\mathbf{u}^*)}{\partial u_i^2} \quad \text{for } i = 1,2,\ldots,n-1 \tag{6}$$

In view of the results in [2,9] it is admissible to choose the following approximate parameterization for u_n around $\mathbf{u}^*$:

$$u_n = e^{-1}(u_1,\ldots,u_{n-1}) = \psi(u_1,\ldots,u_{n-1}) \approx -\beta + \frac{1}{2} \sum_{i=1}^{n-1} \kappa_i u_i^2 \tag{7}$$

This is a special hyperparaboloid. The gradient of $e(\mathbf{u})$ is not vanishing in $\mathbf{u}^*$. The surface integral can then be calculated from the following volume integral:

$$I(E) = \int_{\mathbb{R}^{n-1}} \varphi_n(u_1,\ldots,u_{n-1},\psi(u_1,\ldots,u_{n-1}))\, \Big(1 + \sum_{i=1}^{n-1} \Big(\frac{\partial \psi(\mathbf{u}^*)}{\partial u_i}\Big)^2\Big)^{1/2} \prod_{i=1}^{n-1} du_i \tag{8}$$

Applying the usual scaling by the factor $b > 1$ as proposed in [2,9] and making the transformation $v_i = u_i/b$ yields:

$$I(bE) = b^{n-1} \int_{\mathbb{R}^{n-1}} \varphi_n(b\mathbf{v}, \psi(b\mathbf{v}))\, (1 + \sum_{i=1}^{n-1} (\frac{\partial \psi(b\mathbf{v}^*)}{\partial v_i})^2)^{1/2} \prod_{i=1}^{n-1} dv_i \tag{9}$$

In writing

$$(2\pi)^{n/2} \varphi_n(b\mathbf{v}, \psi(b\mathbf{v})) = \exp[-\tfrac{1}{2}(\sum_{i=1}^{n-1} (bv_i)^2 + (-\beta + \tfrac{b^2}{2} \sum_{i=1}^{n-1} (\kappa_i v_i^2))^2)]$$

and expanding the second term to first order

$$(-\beta + \tfrac{b^2}{2} \sum_{i=1}^{n-1} (\kappa_i v_i^2))^2 \approx \beta^2 - b^2 \beta \sum_{i=1}^{n-1} (\kappa_i v_i^2) + \ldots$$

one can, since the last term in eq. (9) is unity, easily verify that it is:

$$I(bE) \sim b^{n-1} \varphi(\beta) \prod_{i=1}^{n-1} (1-\beta\kappa_i)^{-1/2} \prod_{i=1}^{n-1} \int_{-\infty}^{+\infty} (1-\beta\kappa_i)^{1/2} \varphi(bv_i\, (1-\beta\kappa_i)^{1/2})\, dv_i \tag{10}$$

The last product term in eq. (10) is also unity. Therefore, one has for large β ($b = 1$) the approximation:

$$I(E) \approx \varphi(\beta) \prod_{i=1}^{n-1} (1-\beta\kappa_i)^{-1/2} \tag{11}$$

This result was first given in [10] and is re–derived here on somewhat different lines.

For several equality constraints the intersection $E = \{\cap_{i=1}^{\ell} E_i\}$ with $\ell \leq n-1$ and $E_i = \{e_i(u) = 0\}$ is a $n-\ell$–dimensional surface to which the foregoing considerations apply with obvious modifications. The result is

$$I(E) \approx \prod_{i=1}^{\ell} \varphi(\alpha_i^T \mathbf{u}^*) \prod_{i=1}^{n-\ell} (1-\beta\kappa_i)^{-1/2} \tag{12}$$

where the α_i are the gradients of the equalities in $\mathbf{u}^*$.

In order to demonstrate the generalization for a mixture of equality and inequality constraints consider now the integral

$$I(E \cap F) = \int_{\mathbb{R}^n \cap E \cap F} \varphi_n(u)\, du \tag{13}$$

where $E = \{e_1(u) = 0\}$ and $F = \cap_{i=2}^{m} F_i$ with $F_i = \{g_i(u) \leq 0\}$ and $g_i(0) > 0$ for at least one i. Also, $e_1(u)$ and $g_i(u)$ are at least twice differentiable near the point u^* obtained from:

$$u^* = \min\{\|u\|\} \text{ for } \{u: e_1(u) = 0 \cap \bigcap_{i=2}^{m} g_i(u) \leq 0\} \tag{14}$$

In closely following the arguments in [11], the following result can be obtained:

$$I(E \cap F) \approx \varphi(a_1^T u^*)\, \Phi_{k-1}(c;R) \frac{1}{(\det(I-D))^{\frac{1}{2}}} \tag{15}$$

where

$$c = \{u^{*T} a_s - (u^{*T} a_1)(a_s^T a_1);\ s = 2,\dots,k\}$$

$$R = \{a_s^T a_t - (a_s^T a_1)(a_t^T a_1);\ s,t = 2,\dots,k\}$$

$$D = \{\sum_{s=1}^{k} \gamma_s \frac{\partial^2 g_s(u^*)}{\partial u_i \partial u_j};\ i,j = k+1,\dots,n\}$$

$$I = \{\delta_{ij};\ i,j = k+1,\dots,n\}$$

a_r $(r = 1,\dots,k)$ is the normalized gradient of constraint r at the point u^* and the $\gamma_s < 0$ are obtained as the solution of:

$$u^* = \sum_{s=1}^{k} \gamma_s a_s \quad \text{for } s = 1,2,\dots,k$$

k is the total number of equality constraints and active inequality constraints in u^*. The first two factors represent the first–order result which could have been derived in a more elementary manner as illustrated before.

The result eq. (15) is invariant under orthogonal transformations. Therefore, recursive application for the case of more than one equality constraint yields the general result. It is also possible to condition on all equality constraints simultaneously so that the result eq. (15) generalizes to:

$$I(\bigcap_{i=1}^{\ell} (e_i = 0) \cap \bigcap_{j=\ell+1}^{m} (g_j \leq 0)) \approx \varphi_\ell(c_\ell; \Sigma_{\ell\ell})\, \Phi_q(c_{q|\ell}; \Sigma_{q|\ell}) \frac{1}{(\det(I-D))^{\frac{1}{2}}} \tag{16}$$

with

$$q = m - \ell;\ \ \ell \leq k \leq m$$

$$c = \begin{bmatrix} c_\ell \\ c_q \end{bmatrix};\ \ R = \begin{bmatrix} R_{\ell\ell} R_{\ell q} \\ R_{q\ell} R_{qq} \end{bmatrix}$$

$$\Sigma_{\ell\ell} = R_{\ell\ell}$$

$$c_{q|\ell} = R^T_{q\ell} R^{-1}_{\ell\ell} c_\ell$$

$$\Sigma_{q|\ell} = R_{qq} - R^T_{q\ell} R^{-1}_{\ell\ell} R_{\ell q}$$

D as in eq. (15) and

$$\varphi_\ell(z\,;\Sigma_{\ell\ell}) = (2\pi)^{\ell/2} (\det(\Sigma_{\ell\ell}))^{-1/2} \exp[-\tfrac{1}{2} z^T \Sigma^{-1}_{\ell\ell} z]$$

the ℓ–dimensional standard normal density.

In order to demonstrate the validity and accuracy of the derived approximations they are compared with some exact "probabilities" on surfaces. One such exact result are the probability densities on non–central hyperspheres in multidimensional Gaussian spaces which are given by the derivative with respect to the radius r of the non–central chi–square distribution. For even n a suitable formula is

$$\frac{d}{dr} P_{\chi'^2}(r^2 | n, \lambda) =$$

$$= \sum_{j=0}^{\infty} \frac{1}{j!} e^{-\lambda/2} (\tfrac{\lambda}{2})^j \sum_{k=0}^{\frac{n}{2}+j-1} \frac{1}{k!} e^{-r^2/2} (\tfrac{1}{2})^k\, r\, (r^{2k} - 2\,k\, r^{2(k-1)}) \tag{17}$$

where r is the radius of the sphere, n the dimension of the standard space and $\lambda = \Sigma\, \delta_i^2$ the non–centrality parameter with the δ_i the offsets of the center of the sphere from the origin. In fig. 1 the ratio of the approximations according to the one dimensional versions of eq. (12), and eq. (16) and the exact result eq. (17) is presented. The curves are calculated for a constant ratio $r/\beta = 5$. For comparison, the corresponding results for the volume integral are given in fig. 2. It is seen that even in this extreme example (same curvatures in all directions) the ratios converge to unity for growing β for all n. The results for the volume integral are slightly better than for the surface integral.

The same example also allows to check the general case. In particular, if

$$E \cap F = \{(\sum_{i=1}^{n} (U_i + \delta_i)^2 - r^2 = 0) \cap \bigcap_{j=2}^{m} (U_j \geq 0)\}$$

it is easy to see that the exact probability density of this event for $\delta = (-\beta, -\epsilon, \ldots, -\epsilon)$ with $\epsilon \to 0$ is:

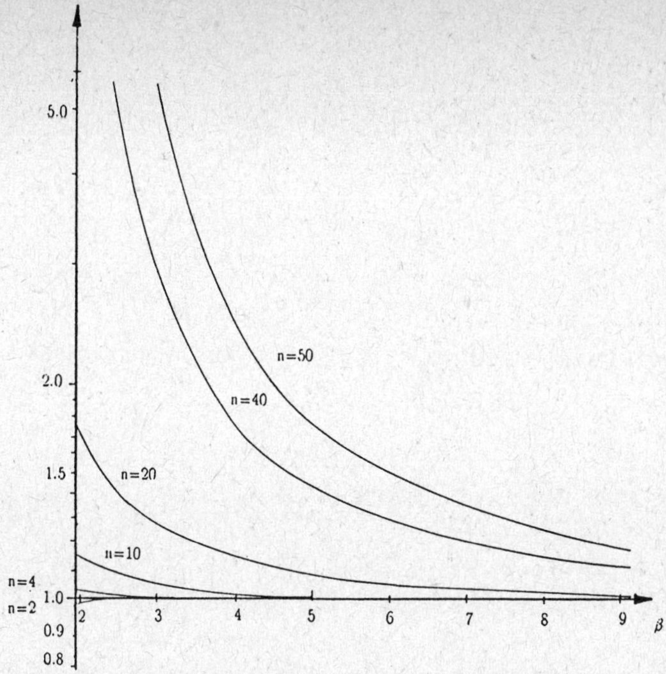

Figure 1: Ratio of density functions p(eq.11)/$p_{\chi'^2}$ with respects to β and n as parameter

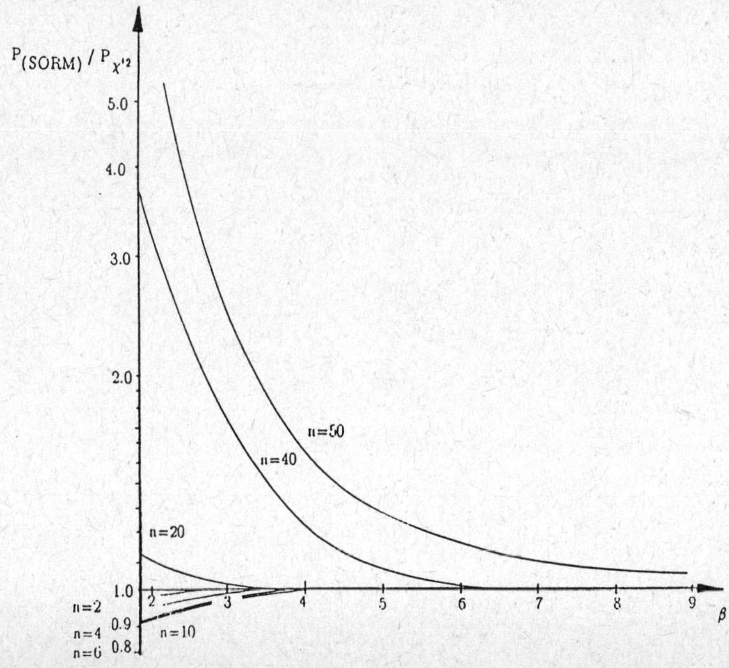

Figure 2: Ratio of CDF P(SORM)/$p_{\chi'^2}$ with respects to β and n as parameter

$$I(E \cap F) = \frac{\frac{d}{dr} P_{\chi'^2}(r^2 | n, \lambda)}{2^m}$$

On the other hand, the second order approximation must be

$$I(E \cap F) \approx \frac{\varphi(\beta) \prod_{i=1}^{n-1} (1-\beta\kappa_i)^{-\frac{1}{2}}}{2^m}$$

which can also be verified numerically by applying eq. (16). Therefore, fig. 1 is also valid in this case.

Correction of integrals of multinormal densities by importance–sampling

Convergence to the exact results with growing distance of $\|\mathbf{u}^*\|$ appears to be slightly slower for surface integrals than for volume integrals. Arbitrarily exact results can always be achieved by a correction obtained by importance sampling at the expense of some more numerical effort. We first consider a correction for volume integrals.

The studies in [12] and [13] concentrated on simple differentiable failure surfaces. In this note we generalize the arguments in [12] to arbitrary failure domains that is to unions of intersections. We shall, in following the basic idea in [12], determine a correction factor together with its error in terms of the coefficient of variation. The general form of a probability estimate then is:

$$P(F) = P(A) \frac{P(F)}{P(A)} = P(A)\, C \tag{18}$$

where P(A) is the probability of an approximation of the failure surface and the second factor C is the correction factor to be determined by importance sampling.

While the importance sampling method where only information about the location of the β–point is needed, is shown in [13] to be the least efficient among the mentioned alternatives it must be considered as a robust method because it requires only simple function calls which determine whether a sampled point is in the failure domain or not. For given or only approximate location of the β–point and smooth failure surfaces one can expect "hits" with probability around 0.5 at each sampling. An even higher hit probability can be achieved if one samples only outside a sphere with radius β. Unfortunately, the same argument no more holds if the failure domain is an intersection domain which, in extreme cases, can be a very small subdomain of $\mathbb{R}^n$. Then, the probability of "hits" in this method can reduce substantially and tends to zero in the mentioned extreme cases. The variance of the estimate then increases significantly. The method can become inefficient and, therefore, is discriminated as a suitable, general importance sampling method for cases where the failure domain is of a more complex nature. The probability of "hits" is substantially increased if a directional sampling scheme is used. For example in following [12], one can sample on a plane going through the origin with normal equal to the direction cosines of the β–point. In this case it is not only possible to integrate analytically in one dimension. It is even possible to use information about the curvature of the failure surface in the β–point. This scheme must be considered as the most efficient importance sampling method for smooth failure surfaces. It produces the smallest coefficient of variation of the probability estimate among the alternatives because the variability of the sampling density is adjusted to the actual curvatures as indicators for the "important" sampling domain. It has even been found that this adjustment is necessary in cases of extreme curvatures because not adjusted sampling densities most likely produce too many points outside the important region and there is no indication, e.g. by the magnitude of the coefficient of variation of the estimate, that the important region has not yet been sampled unless the number of sampled points is very large. The same phenomenon must be expected for arbitrary failure domains and especially for intersections with small probability content. Therefore, a sampling scheme must be designed with a large likelihood to hit the important region. Also, we require the scheme to be as robust as possible.

If D denotes a "small" intersection (at least one $g_i(\mathbf{0}) > 0$)) of inequality constraints and m denotes the number of constraints g_j of the intersection, $k \leq m$ linearizations define the linear form. As before k equals the number of active constraints.The linearizations are the linearizations at the common β–point $\mathbf{u}^*$ (i.e. where $\cap_{i=1}^{k} g_j(\mathbf{u}^*) = 0$).

The linear form first is used for the multinormal integral which evaluates P(A) as defined in eq. (18), i.e.:

$$P(F) \approx \Phi_k(-\beta\,;\,R)\,C \tag{19}$$

with

$$\beta = \{-\mathbf{u}_{(j)}^{*T}\,\alpha_j\} \quad \text{and} \quad R = \{\rho_{i,j}\} = \{\alpha_i^T\,\alpha_j\}$$

and $P(A) = \Phi_k(-\beta\,;\,R)$. The α_j are the normalized ($\|\alpha_j\|=1$) gradients of the constraints at the point(s) $\mathbf{u}^*_{(j)}$.

The information supplied by the linear form is now used to define a sampling direction and the mean values and standard deviations of the sampling densities. An orthogonal transformation is performed such that the point $\mathbf{u}^*$ lies on the n–th axis, i.e. has coordinates $(0,\dots,0,\|\mathbf{u}^*\|)^T$. Then, the integral for the correction factor is written as:

$$C = \int_F \frac{1}{P(A)} \varphi_n(\mathbf{v})\, d\mathbf{v} = \int_{\mathbb{R}^{n-1}} \frac{1}{P(A)} \int_{\mathbb{R}} 1_F(\tilde{\mathbf{v}}, v_n)\, \varphi(v_n)\, dv_n\, \varphi_{n-1}(\tilde{\mathbf{v}})\, d\tilde{\mathbf{v}} \tag{20}$$

where $1_F(\tilde{\mathbf{v}},v_n)$ is the indicator function of the failure domain and $\tilde{\mathbf{v}} = (v\ ,\ldots v_{n-1})^T$. The integration over v_n can be carried out analytically and by introducing a suitable sampling density $\psi(.)$ one obtains:

$$C = \int_{\mathbb{R}^{n-1}} \frac{1}{P(A)} \int_{\mathbb{R}} 1_F(\tilde{\mathbf{v}}, v_n)\, \varphi(v_n)\, dv_n\, \varphi_{n-1}(\tilde{\mathbf{v}})\, d\tilde{\mathbf{v}}$$

$$= \int_{\mathbb{R}^{n-1}} \frac{\Phi(-\lambda_F(\tilde{\mathbf{v}}))}{P(A)} \frac{\varphi_{n-1}(\tilde{\mathbf{v}})}{\psi_{n-1}(\tilde{\mathbf{v}})} \varphi_{n-1}(\tilde{\mathbf{v}})\, d\tilde{\mathbf{v}}$$

$$= E\left[\frac{\Phi(-\lambda_F(\tilde{\mathbf{v}}))}{P(A)} \frac{\varphi_{n-1}(\tilde{\mathbf{v}})}{\psi_{n-1}(\tilde{\mathbf{v}})}\right]$$

$$\sim \frac{1}{N} \sum_{r=1}^{N} \frac{\Phi(-\lambda_F(\tilde{\mathbf{v}}_r))}{P(A)} \frac{\varphi_{n-1}(\tilde{\mathbf{v}}_r)}{\psi_{n-1}(\tilde{\mathbf{v}}_r)} \tag{21}$$

where $\lambda_F(\tilde{\mathbf{v}})$ is the distance from the origin to the failure surface and the $\tilde{\mathbf{v}}_r$ are simulated points according to $\psi_{n-1}(.)$ (see also figure 3).

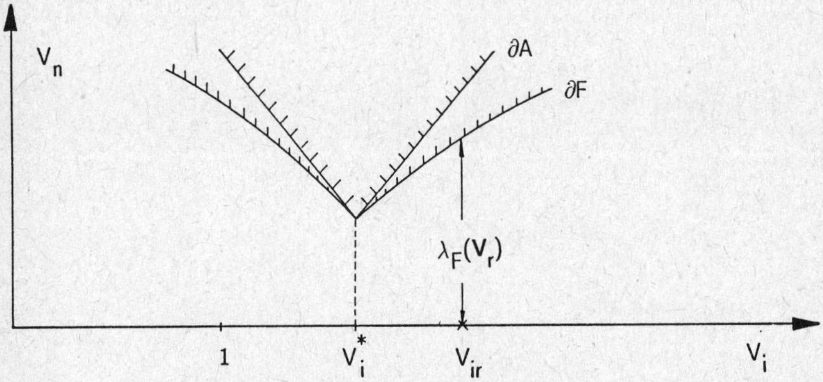

Figure 3: Illustration of sampling scheme for intersections

$\psi_{n-1}(.)$ now needs to be chosen such that the important region is sampled most. For example, points are sampled according to the normal distribution on the plane going through the point $\mathbf{u}^*$ and perpendicular to some representative value α_S of the gradients of the active constraints. More specifically, the sampling points are obtained from independent normal variables V_i with the coordinates of $\mathbf{u}^*$ as their mean values. Points on the failure surfaces are then sought in the direction of

α_s. The sampling direction α_s is chosen as the mean of the gradients of the active constraints $\alpha_{j,a}$. The simplest choice for the standard deviation of the V_i is a value of one which also is chosen as the upper limit. The angles between α_s and the $\alpha_{j,a}$ however, allow a more efficient assessment of the standard deviations of the V_i. For example, the simple linear relationship

$$\sigma(V_i) = \varphi_i/\pi \tag{22}$$

where $\sigma(V_i)$ is the standard deviation and φ_i the largest angle spanned by the i–th components of the gradients of the linear form. It concentrates sampling points around $\mathbf{u}^*$. The concentration becomes denser for decreasing angles. In this way it is assured that $0 \leq \sigma \leq 1$ for $0 \leq \varphi \leq \pi$. Note that $\varphi > \pi$ would violate the definition of $\mathbf{u}^*$ as the minimal distance of the intersection of the linearizations to the origin.

Now denote $\tilde{\mathbf{v}}_s$ a sampled point on a plane through the origin with gradient α_s and λ_A the minimal distance from $\tilde{\mathbf{v}}_s$ to the intersection of the linear form in direction α_s. This distance can be computed analytically from the α_j, $\mathbf{v}^*$ and possibly the $\mathbf{v}^*{}_j$ from $\mathbf{v}_s$ and from α_s. The point $\mathbf{v}_s + \lambda_s\ \alpha_s$ lies on one of the linearizations, l_t say, and defines the starting solution for a Newton search in direction α_s to the point $\tilde{\mathbf{v}}_t = \tilde{\mathbf{v}}_s + \lambda_s\ \alpha_s$ where $g_t(\mathbf{v}_t) = 0$. For this point it is checked whether it is in the intersection of the m nonlinear constraints. If not, the Newton search is repeated with respect to the constraint (t') for which $g_{t'}(\mathbf{v}_t) > 0$ to obtain a suitable point.

For a "large" intersection (all $g_i(0) < 0$) a slightly different sampling scheme is proposed. The linear form is given by the linearizations of all m constraints at their individual β–points $\mathbf{u}^*{}_j$ with

$$\beta = \{-\mathbf{u}_j\ \alpha_j\} = \{-\|\mathbf{u}_j\|\}\ \ (\text{all } \beta_j < 0) \tag{23}$$

The complement of F in this case is interpreted as a "small" union:

$$\overline{F} = \left\{\bigcup_{i=1}^{m} [g_i(U) > 0]\right\} = \left\{\bigcup_{i=1}^{m} [-g_i(U) \leq 0]\right\} \tag{24}$$

and P(F) is approximated by:

$$P(F) = 1 - P(\overline{F}) \approx 1 - [\Phi_m(+\beta\,;\,R)\ C] \geq 1 - \sum_{j=1}^{m} \Phi_1(+\beta_j) \tag{25}$$

Here, for each constraint the sampling density is obtained from the V_i with the coordinates of the $\mathbf{u}^*{}_j$ as mean values and standard deviations $\sigma(V_i)$ all set to one. Note that a better estimate of σ could be given according to [12] which requires the second–order derivatives of the g_j. For each constraint a number of points $\mathbf{u}_{j,s}$ are sampled on a plane through the origin with gradient α_j and, in analogy to the small intersection case, the $\lambda_{j,s}$ denote the distance from the sampled point to the nonlinear form in the direction $\alpha_s = \alpha_j$. The root of $g_j(\mathbf{u}_{j,s} + \lambda_{j,s}\,\alpha_j) = 0$ must again be found by a Newton search. As for the small intersection it is checked whether the point $\mathbf{u}_t = \mathbf{u}_{j,s} + \lambda_{j,s}\,\alpha_j$ is in the large intersection (or alternatively in the small union). If $g_i(\mathbf{u}_t) > 0$ for any $i \neq j$, the Newton search is repeated with respect to constraint number i (but still in direction α_j) which yields a point $\mathbf{u}_{t,i}$. Consequently $\lambda_{j,s}$ has to be updated. $\lambda_{j,s} = \alpha_i^T\,\mathbf{u}_{t,i}$ now is the distance of $\mathbf{u}_{t,i}$ to a plane through the origin with gradient α_i.

The right hand side of eq. (25) also allows to distribute the total computational effort according to the contribution of the constraints to the probability of the small union. For example, one could take

$$N_j = N\frac{\Phi_1(\beta_j)}{\sum\limits_{j=1}^{m}\Phi_1(\beta_j)}, \quad \sum_{j=1}^{m} N_j = N \tag{26}$$

with N the number of samples for the large intersection and N_j the number of sample points for each constraint provided that the denominator in eq. (26) is smaller than unity. Otherwise one has to normalize appropriately. The correction factor C for the case of a large intersection then is:

$$C = \frac{1}{N}\sum_{j=1}^{m}\sum_{s=1}^{N_j}\frac{\Phi(-\lambda_F(\tilde{\mathbf{v}}_{j,s}))}{P(A)}\,\frac{\varphi_{n-1}(\tilde{\mathbf{v}}_{j,s})}{\psi_{n-1}(\tilde{\mathbf{v}}_{i,s})} \tag{27}$$

If the failure set is given as a (minimal) cut set of small–probability intersection domains (cuts)

$$F = \left\{\bigcup_{t=1}^{m}\ \bigcap_{j=1}^{k(t)} F_{t,j}\right\} \tag{28}$$

sufficiently narrow probability bounds can usually be derived on the basis of the material given in [1] and [16] provided that the cut set probabilities all are relatively small. Unfortunately, those bounds cannot be improved easily (see [1]). Therefore, a check of the accuracy of the results can be especially interesting for m very large and/or $P(F_{t,j})$ not small, i.e. when these bounds can become unsatisfactory wide. The procedure is very similar to the one just described. The total number of sampling points first is distributed to the various cuts according to eq. (26) with obvious modifications. Then, the individual cuts are treated as before with the additional task to check whether projected points are also in other cuts and are closer to the plane on which the sampling points have been generated. If projected points are, in fact, also in other cuts one proceeds as described for the case of large intersections.

If now equality constraints are present the importance sampling scheme becomes even simpler. In this case a sampled point must lie on the intersection of the equality constraints and must fulfill the inequality constraints. The integral is written as:

$$I(D) = I(A)\frac{I(D)}{I(A)} = I(A)\int 1_D(\mathbf{u})\,\frac{\varphi_l(\mathbf{u})}{I(A)}\,\frac{\varphi_{n-l}(\mathbf{u})}{\psi_{n-l}(\mathbf{u})}\,\psi_{n-l}(\mathbf{u})\,ds(\mathbf{u})$$

If there are ℓ equality constraints $n-\ell$ independent sample points are generated according to some normal sampling density which, of course, is centered around the joint β–point. The rest of the coordinates is found by solving the system of equations for the equality constraints. A suitable starting point for the non–linear equation solver is the solution of the corresponding linear form. If no detailed information about the behavior of the intersection of the equalities in the neighborhood of the β–point is available the obvious choice for the standard deviation of the sampling densities is unity. This essentially is a concept according to method B. It certainly can be improved but we will make no attempt to do so herein.

Conclusions

The well–known asymptotic results for volume integrals can be generalized to certain surface integrals in a straightforward manner. The asymptotic surface integrals represent a significant improvement of the corresponding first–order results. Convergence to the exact results with growing distance of the β–point from the origin appears to be slightly slower for surface integrals than for volume integrals. However, arbitrarily exact results can always be achieved by a correction obtained by importance sampling for both volume and surface integrals at the expense of some more numerical effort.

Acknowledgements: Part of this study was financially supported by A. S. Veritas Research, Oslo. The authors also appreciate fruitful comments by P. Bjerager and L. Tvedt.

References

[1] Hohenbichler, M.; Rackwitz, R., First–Order Concepts in System Reliability, Struct. Safety, 1, 3, 1983, pp. 177–188

[2] Hohenbichler, M.; Gollwitzer, S.; Kruse, W.; Rackwitz, R., New Light on First– and Second–Order Reliability Methods, Struct. Safety, 4, 1987, pp. 267

[3] Veneziano, D.; Grigoriu, M.; Cornell, C.A., Vector–Process Models for System Reliability, Journ. of Eng. Mech. Div., ASCE, Vol. 103, EM 3, 1977, pp. 441–460

[4] Rackwitz, R., Human Error in Design and Structural Failure, Proc. of NSF–Workshop on Modeling Human Error in Structural Design and Construction, Ann Arbor, June, 1986, ASCE Publication, 1986, pp. 216–224

[5] Madsen, H.O., Model Updating in Reliability Theory, Proc. ICASP 5, Vol. 1, Vancouver, 1987, pp. 564–577

[6] Hohenbichler, M.; Rackwitz, R., Non–Normal Dependent Vectors in Structural Safety, Journ. of the Eng. Mech. Div., ASCE, Vol. 107, No. 6, 1981, pp. 1227–1249

[7] Gollwitzer, S.; Rackwitz, R., Equivalent Components in First–order System Reliability, Rel. Eng., Vol. 5, 1983, pp. 99–115

[8] Fiessler, B.; Neumann, H.–J.; Rackwitz, R., Quadratic Limit States in Structural Reliability, Journ. of the Eng. Mech. Div., ASCE, Vol. 105, EM4, 1979, pp. 661–676

[9] Breitung, K., Asymptotic Approximations for Multinormal Integrals, Journ. of the Eng. Mech. Div., Vol. 110, No.3, 1984, pp. 357–366

[10] Breitung, K., Asymptotic Approximations for Multinormal Domain and Surface Integrals, Proc. ICASP 4, I, Florence, 1983, pp. 755–768

[11] Hohenbichler, M., Rackwitz, R., Asymptotic Crossing Rate of Gaussian Vector Processes into Intersections of Failure Domains, Probabilistic Engineering Mechanics, Vol. 1, No. 3, 1986, pp. 177 – 179

[12] Hohenbichler, M.; Rackwitz, R., Improvement of Second–order Reliability Estimates by Importance Sampling, accepted for publication in Journ. of Eng. Mechanics, ASCE, 1988

[13] Fujita, M.; Rackwitz, R., Updating First– and Second– order Reliability Estimates by Importance Sampling, Structural Eng./ Earthquake Eng., JCSE, 5(1), 1988, pp. 31s–37s

[14] Hohenbichler, M.; Rackwitz, R., A Bound and an Approximation to the Multivariate Normal Distribution Function, Math. Jap., Vol. 30, 5, 1985, pp. 821–828

[15] Gollwitzer, S.; Rackwitz, R., An Efficient Numerical Solution to the Multinormal Integral, Probabilistic Engineering Mechanics, 3 (2), 1988, pp. 98–101

[16] Ditlevsen, O., Narrow Reliability Bounds for Structural Systems, J. Struct. Mech., 7(4), 1979, pp. 453–472

[17] Fujita, M.; Schall, G.; Rackwitz,R., Time–Variant Component Reliabilities by FORM/SORM and Updating by Importance Sampling, Proc. ICASP 5, Vancouver, May, 1987, I, pp. 520–527

OPTIMUM DESIGN OF DAMAGE TOLERANT MARINE STRUCTURES

W. B. Shi, P. Sen & J. B. Caldwell
Department of Naval Architecture & Shipbuilding
University of Newcastle upon Tyne, UK

Abstract: The sources of potential hazards in the sea are many, and risks of for marine structures being damaged are, sometimes, totally unpredictable. Hence a rational design philosophy incorporating the concepts of damage tolerance would be of benefit in marine structural design. In particular, the two aspects that need to be developed are, firstly, quantitative measures of structural damage tolerance and, secondly, explicit inclusion of damage tolerance considerations in design synthesis procedures. The assessment of structural redundancy in terms of reserve and residual strength factors can be combined with the ultimate behaviour of a structural system to determine damage tolerance for either skeletal or continuous structures. Such a damage tolerant design methodology is proposed based on a multi-level and multi-goal decision model. An illustrative example of design of a barge hull girder is provided to show the influence of incorporating the concept of damage tolerance in optimum marine structural design.

1. Introduction

Although considerations of damage tolerance have long existed implicitly in marine structural design by virtue of the provision of bulkheads, for example, explicit consideration and quantification of damage tolerance is of relatively recent origin. Some of the early work in this area dealt with airborne structures[1]. However, research in the domain of marine structural design is being increasingly reported[2][3][4][5]. Obviously damage tolerant design philosophy can affect all the principal aspects of structural design like load prediction, strength analysis and safety assurance as well as the design synthesis process itself. In this context, it would be rational to have structural damage tolerance concepts built into international regulations, such as those promulgated by IMO for the assessment of survivability of ships after damage[6][7]. Such an approach would require measures of damage tolerance, methods for including considerations of damage tolerance in design[1] [8] and damage statistics for marine structures[9] on the basis of which likely damage scenarios can be established.

The variety of potential hazards encountered by marine structures is large and the consequences of having to meet these hazards can be very serious. On average total losses account for about 0.5% of the would fleet annually. For minor accidents, the effects of various types of damage were examined in terms of average repair cost index and time by Burke[10]. Most of the ship casualty data used for the above investigations come from "Casualty Returns" for the world fleet published annually by Lloyds Register of Shipping. Casualty data relating to offshore structures can be found for example in Ref. [11].

Broadly speaking, damage or failure is a direct result of some kind of error or uncertainty. Errors and uncertainties can be grouped into three types[12]:-

- Random uncertainties (e.g. variations of wave bending moments, deviations of material properties from standard values),

- Systematic uncertainties (e.g. errors in idealised design models, variations in test equipment),

- Blunders (e.g. human negligence in design, construction and operation).

More often, the first two types are classified respectively as objective and subjective uncertainties. Objective uncertainties are associated with parameters which can be statistically represented and examined. Subjective uncertainties refer to those caused by lack of information and knowledge[13], and they can be treated only by using accumulated experience and sound engineering judgement. There is a voluminous body of literature on objective uncertainties, and probability-based methods are available to quantify them. A quantitative approach to subjective uncertainties by using fuzzy set theory also seems to be promising[14]. Surveys from many publications, however, indicate that most disastrous consequences are initiated by blunders. In fact, about 90% of catastrophic failures are due to human errors in the design, construction and service phases, or unpredictable events in service[15]. These blunders cannot be satisfactorily represented by statistical means, although the consequences of total failures due to blunders which may include loss of human lives, huge economic losses and environmental pollution, are far more severe than those due to other types of uncertainty.

In contrast with early practice, when redundant material used to be built into structures with consequent generous reserve and residual strengths, the current trend is towards more efficient use of scarce resources. Damage tolerance concepts can therefore be expected to become more and more important in this context. To pursue greater structural efficiency or to simplify production procedures, modern marine structures have tended to make use of higher slenderness ratios and simpler layouts. Simple structural layouts can be production-kindly and hence cheaper to build, but they may offer less residual strength in the presence of initial damage. The weaknesses of such an approach might be:-

- Less reserve strength in components, for example, components with lower plastic shape factors,

- Local and overall instability problems which may reduce post-collapse strength,

- Occurrence of interactions between failure modes,

- Excessive local deflections and yielding under service loads,

- Local fatigue failures at important joints or connections,

- Unfavourable dynamic response characteristics.

Undoubtedly, there is some conflict between the twin objectives of structural weight efficiency and damage tolerance. Some design methods can produce so-called "optimum" designs which may be highly efficient in terms of weight, cost, or other design criteria. However such structures may possess deficiencies as outlined above[16]. These "optimum" structures may be particularly sensitive to imperfections such as residual stress, eccentricity and other production-induced defects, and sudden collapses of whole structural systems may occur under extreme loading conditions or in the presence of initial damage generated by blunders if components are designed to their ultimate strength levels.

2. Assessment of Damage Tolerance in Structures

The ambiguity regarding the definitions of redundancy and damage tolerance is perhaps one of the factors delaying explicit applications in design. Several definitions of damage tolerance have been suggested[2][17][18], but so far there is still no unique and consistent definition. In this paper, reserve strength factor and residual strength factor are used as measures of structural damage tolerance.

As mentioned earlier, a damage, in the general sense, can occur in a structure during the design and construction phases, or in service. The general term "damage" can be defined micro- or macroscopically. Micro-structural damage can be viewed as any small defect arising during design, construction or operation. Examples of such damage typically are those caused by welding strains and stresses, mis-alignment of connections and fabrication flaws. The severity or frequency of occurrence of this type of damage depends on the quality and frequency of inspection during construction and service, and on maintenance. Macroscopically, damage refers to any large scale deterioration of overall strength caused progressively or suddenly by external loading and/or environmental conditions during service. Initial damages due to collision and contact are of this type. Only macroscopic types of damage are dealt with in this work.

Structural damage tolerance can be stated as the ability to resist abnormal loads and the ability to sustain initial damage. This obviously implies that the loads which a damaged structure should be capable of bearing should reflect the likely loads over the period of time the structure is expected to remain in the damaged state. In other words, a damage tolerant structure should have adequate reserve strength to avoid failure under some plausible extreme loading conditions, and sufficient residual strength to minimize the consequences of initial damage[8]. For marine structures, there can be two kinds of extreme loads[19]. The first type is those which can be predicted based on statistical calculations, such as the long term wave distribution and order statistics[20]. The second type includes rare events which cannot be statistically interpreted, e.g.

so called freak waves. From the above definitions, damage tolerance is related to the "safe-fail" of the structural system. There are several alternative measures of structural damage tolerance as discussed below, and they can be interpreted either deterministically or statistically[18]. In this paper a deterministic approach is used but comments are given later on possible incorporation of reliability considerations.

2.1. Reserve Strength Factor

In order to be consistent with commonly used design formulations, a reserve strength factor is used to define the reserve strength aspect of the damage tolerance of a structure. The reserve strength factor γ_i is defined as the ratio of the ultimate strength of an intact structural system to the characteristic load. Therefore

$$\gamma_i = \frac{C_i}{D_d} \tag{1}$$

in which C_i is the ultimate strength of the intact system and D_d is the characteristic load. It follows, therefore, that the reserve strength factor γ_i is a measure of the overall ability of the structural system to survive abnormal load conditions. Conventional measures of safety margin such as load factors, partial safety factors and reliability indices are directly related to the reserve strength factor γ_i. The damage tolerance of a structure, however, cannot be fully determined with only γ_i, since the residual strength of the structure in damage states is not reflected by γ_i. Applying a higher reserve strength factor γ_i in design will result in an increase in scantlings of the structure, but this approach may produce an irrational and uneconomic structure, particularly when the basic damage patterns are known. It is preferable to adopt a direct measure to determine the survival capability of the structure after damage.

2.2. Residual Strength Factor

The residual strength factor γ_d is defined as the ratio of the residual strength of a structure with initial damage to the reserve strength of the intact system. Therefore

$$\gamma_d = \frac{C_d}{C_i} \tag{2}$$

in which C_d is the residual strength of the structural system in a damage condition. By definition, the residual strength factor γ_d generally ranges between zero and one. However the value of γ_d can exceed one in certain cases, as shown in the example of the damage tolerant analysis of a hull girder. If all components in a system exhibit ductile and strain hardening behaviour, the residual strength factor γ_d will be less than one. However for components with brittle or partial brittle behaviour e.g. post-buckling collapse, γ_d may be greater than one depending on the loading, as the damaged structure under the applied loads can have failure paths that are different from those of the original structure. The numerical value of the product of γ_i and γ_d indicates whether the damaged structure will survive the design load, as $\gamma_i \times \gamma_d = \dfrac{C_d}{D_d} \geq 1.0$ implies survival.

After the extreme loading states and postulated damage scenarios for an existing structure have been defined, all the reserve strength factors and residual strength factors can be estimated. Based on these, the critical loading states and damage states associated with key components in the structure can be identified. Thereby, special treatments such as damage tolerant re-design, high level quality assurance and maintenance strategies can be organised. Examples are given below to demonstrate how these definitions relate to actual structures, and some relevant aspects of design are emphasised.

2.3. Examples of Damage Tolerance Analysis

a) Damage Tolerance of a Five Bar Frame

A five bar frame subject to a vertical downward point load P is considered here. The topological arrangement is shown in Fig. 1. The applied load P is kept constant. All bars are assumed to have the same cross section area of A. The material is assumed to be perfectly plastic with a yield stress of σ_y. The reserve strength factor γ_i is required to be equal to one. As shown in Table 1, one intact and fourteen damage states were considered, on the assumption that any damaged element loses all its load bearing capability. The optimum area refers to that required by each of the intact elements such that the structure as a whole can withstand the load P. The residual strength factor is merely the ratio of the residual strength and the intact strength.

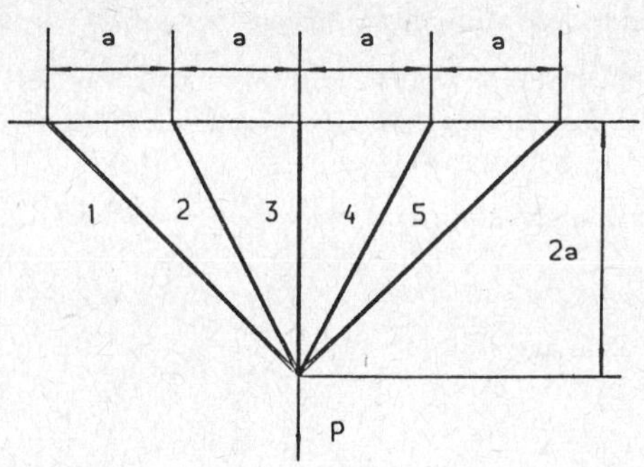

Fig. 1 A five bar frame work

It is obvious that even for this very simple bar structure, the total number of basic damage states exceeds fourteen, and some of the possible damage states are not listed in the table. As the complexity of a structure increases, it soon becomes impossible for a designer to consider all potential damage states. Therefore adequate redundancy and damage tolerance should be built in to support the design loads under the most likely damage states.

b) Damage Tolerance of a Plane Frame

A wing tank model with two struts is analysed next under four damage states, as shown in Fig. 2, by using plastic frame analysis, which involves plastic hinge insertion[21]. As the framework has distinct structural components, it may be identified as a skeletal structure. In the example,

Table 1 Damage tolerant analysis of a five bar frame

Damaged element number	Residual strength	Optimum area	Residual strength factor
(0)	$4.2030\sigma_y A$	$0.2379P/\sigma_y$	1.0000
(1) or (5)	$3.4959\sigma_y A$	$0.2861P/\sigma_y$	0.8317
(2) or (4)	$3.3086\sigma_y A$	$0.3022P/\sigma_y$	0.7872
(3)	$3.2029\sigma_y A$	$0.3122P/\sigma_y$	0.7621
(1) or (5)	$2.7888\sigma_y A$	$0.3586P/\sigma_y$	0.6635
(1,2) or (1,4)	$2.6015\sigma_y A$	$0.3844P/\sigma_y$	0.6190
(5,2) or (5,4)	$2.6015\sigma_y A$	$0.3844P/\sigma_y$	0.6190
(1,3) or (3,5)	$2.4959\sigma_y A$	$0.4007P/\sigma_y$	0.5938
(2,4)	$2.4142\sigma_y A$	$0.4142P/\sigma_y$	0.5744

four damage states were postulated. In damage state 1, point 1 is represented by a hinge. Damage state 2 represents the damage of the joint node at point 3, which is therefore represented as a hinge. Damage state 3 assumes that point 7 is damaged, and in the last damage state component 5-6-7 is supposed to lose all its load bearing capability. All the load deflection curves are shown in Fig. 3, including the curve for the intact structure. From the numerical results, the residual strength factor γ_d of each damage state is calculated as below:-

$$\text{Damage state 1}: \quad \gamma_d = \frac{20.21}{31.25} = 0.6467$$

$$\text{Damage state 2}: \quad \gamma_d = \frac{25.66}{31.25} = 0.8211$$

$$\text{Damage state 3}: \quad \gamma_d = \frac{26.30}{31.25} = 0.8416$$

$$\text{Damage state 4}: \quad \gamma_d = \frac{6.948}{31.25} = 0.2223$$

The lowest value of γ_d indicates the most severe damage to the system. Although obviously not all of the damage states were considered here, it is advisable to check the most likely damage states during the design process.

c) Damage Tolerance of a Hull Girder

A continuous structure such as a hull girder presents a different problem compared to the skeletal

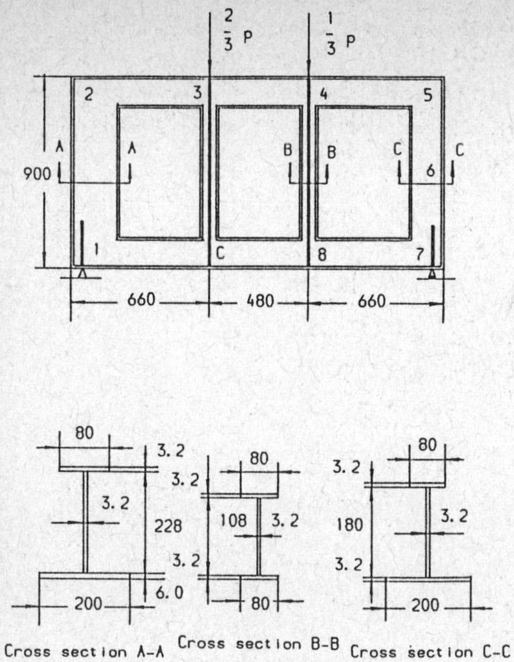

Fig. 2 Wing tank frame model with two struts

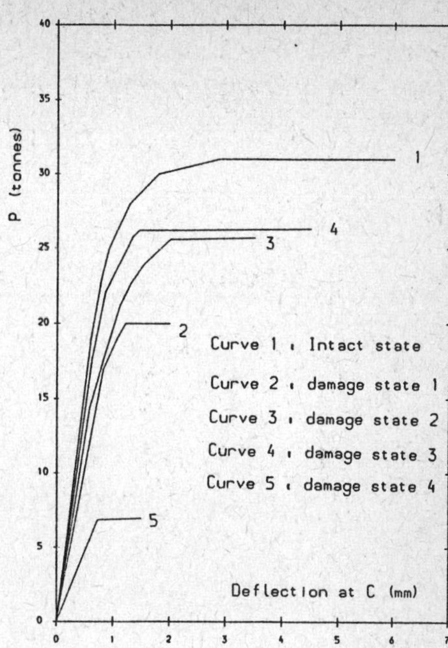

Fig. 3 Load deflection curves of the frame model

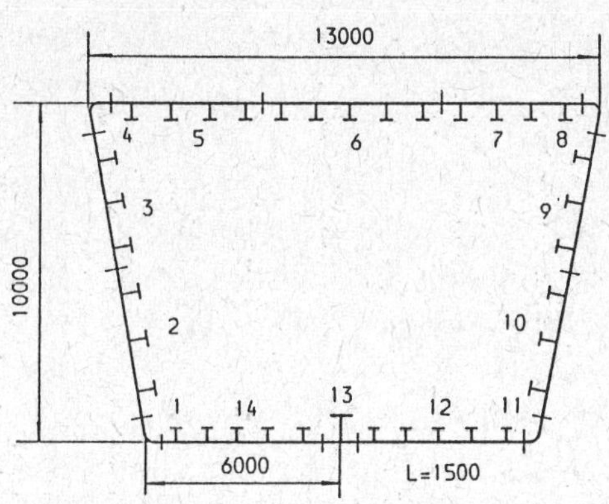

Fig. 4 Cross section of a hull girder for damage tolerant analysis

structure considered above. To examine the effects of local damage on the overall system, a hull girder was analysed for six different damage states. The overall geometry of the cross section of the hull girder is shown in Fig. 4, and the scantlings of its components and their material properties are contained in Table 2. The six damage states considered, in which various combinations of components were assumed to be rendered ineffective by damage, were as follows:-

Damage state 1 : components 7, 8

Damage state 2 : components 8, 9

Damage state 3 : components 9, 10

Damage state 4 : components 10, 11

Damage state 5 : components 11, 12

Damage state 6 : components 12, 13

Table 2 Component sizes and material properties

Component	Size (mm)	σ_y (N/mm^2)	E (N/mm^2)
Deck plate	10.0	240.0	205000
Deck stiffener	$100 \times 6 \times 150 \times 5$	240.0	205000
Bottom plate	12.0	240.0	205000
Bottom stiffener	$120 \times 7 \times 180 \times 6$	240.0	205000
Keel	$200 \times 15 \times 250 \times 15$ 1000×20	240.0	205000
Side plate	9.0	240.0	205000
Side stiffener	$100 \times 6 \times 150 \times 5$	240.0	205000

The intact structure was considered at first, with biaxial bending moments applied to the hull girder. A hull girder, depending on its function, has to withstand combinations of vertical and horizontal bending moments; hogging and sagging conditions represent the hull bending downwards at the ends and in the middle respectively. By varying the ratio of the vertical and horizontal bending moments, closed diagrams can be generated[22][23] for residual and reserve strengths, as shown in Fig. 5 in which M_{vult} and M_{hult} are the maximum bending moments in vertical and horizontal directions respectively. It can be observed that the curve of reserve strength is symmetric with respect to the horizontal bending moment axis, and the residual strength is directional, so that changing the location of damage in the structure alters the point corresponding to the maximum reduction of the ultimate strength. Using plots of this kind, the variation of reserve strength factor γ_i and residual strength factor γ_d can be conveniently studied. As pointed out before and as Fig. 5 shows, the residual strength factor γ_d might exceptionally exceed one, because the residual strength may turn out to be larger than the reserve strength in certain ranges of applied loads. This is due to the differences in the failure paths for the intact and damaged structures.

3. Integrated Safety Assurance

The performance of a marine structure depends on the conditions encountered during the stages of design, manufacture and operation, and coherent safety measures should be taken with marine structures throughout their lifetimes. The concept of "multiple defence lines" introduced for mobile offshore units[24] can also be applied to other types of marine structure. The "multiple defence lines" include several aspects of safety assurance, namely, design and manufacture, inspection and maintenance, and residual safety. In the following discussion, the main comments are related to damage tolerant structural design and hence address design and residual safety only.

3.1. Selection of Reserve Strength Factor

Relatively recent developments in structural reliability theory have provided the facilities for the calibration and determination of partial safety factors in limit state design, and therefore for the evaluation of the reserve strength factor γ_i. Structural reliability methods can be grouped into three levels. Level three full statistical approach is based on a complete probabilistic analysis for the whole structural system or its structural components. The level two approximate approach involves certain iterative algebraic procedures to obtain an approximation to the failure probability of a structure or structural element. The reliability of the structure is expressed either as a failure probability or a safety index. In level one partial safety factor methods, an approximate measure of structural reliability is provided for a structural system or structural component by the use of a number of partial safety factors related to pre-defined characteristic or nominal values of the major loading and structural variables. The level one methods consider a number of separate limit states e.g. progressive failure limit states, ultimate failure limit states, serviceability failure limit states and fatigue failure limit states. For example, a three coefficient design format may be written as

$$\gamma_m C_i \geq \gamma_c \gamma_f D_d \tag{3}$$

where γ_c is a coefficient taking account of some of the subjective uncertainties to cater for economic and social consequences and the nature of the structure. γ_m is a coefficient which is used to cover possible reductions in the nominal strengths of materials in the structure as a whole compared to the characteristic values deduced from control test specimens. In addition, the coefficient covers possible weaknesses in the structure arising from any causes other than a reduction in strength of the material, for example imperfections and residual stresses. γ_f is a coefficient which takes account of the possibility of unfavourable deviations of the loads from the characteristic external loads representing extreme values. Therefore, the reserve strength factor γ_i can be written as

$$C_i \geq \gamma_i D_d \tag{4a}$$

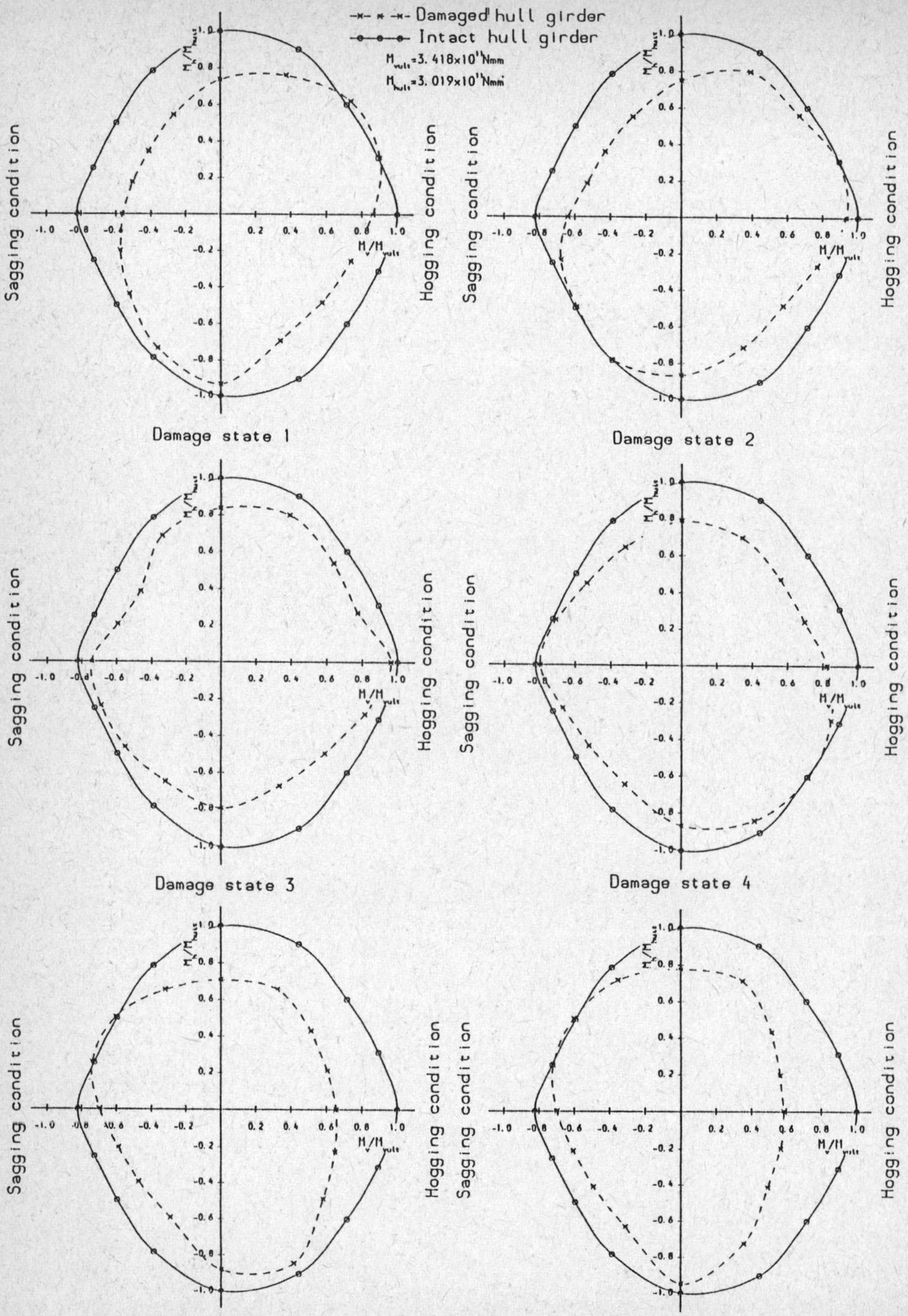

Fig. 5 Damage tolerance of ship hull grider

$$\gamma_i = \frac{\gamma_c \gamma_f}{\gamma_m} \tag{4b}$$

Here, γ_i is the reserve strength factor as defined previously; and it is called an overall partial safety factor[25].

The selection of acceptable partial safety factors is of key importance to the applications of level one methods. This procedure not only needs to use higher level reliability methods, but also requires sound engineering judgement and treatment of subjective uncertainties. At present, level two methods are a major validating means for partial safety factors, although level three methods can also be used in some simple cases. Typical values of the reserve strength factor γ_i for marine structures normally lie within the following range[12]:-

$$1.5 \leq \gamma_i \leq 2.0 \tag{5}$$

3.2. Selection of Residual Strength Factor

As indicated before, a structure with sufficient reserve strength factor γ_i may not necessarily possess an adequate level of residual strength, and may lead to rapid catastrophic failure in some damage conditions. A damage tolerant structure is likely to consume more resources, and to involve a higher initial investment as well as other unfavourable characteristics. However, for such a design inspection costs can be reduced, and lower levels of risk in damage conditions can be anticipated. Indeed there should be a trade-off between the economic outcome and allowable minimum risk of catastrophic failures. The selection of a residual strength factor γ_d is different from that of a reserve strength factor γ_i, and the procedure is complicated by the fact that the residual strength factor γ_d is also directly related to the location, extent and probability of damage in a structure. Since this type of damage is usually initiated by blunders, the occurrence of such damage cannot yet be expressed statistically for a particular marine structure. However, an approach similar to the damage survivability assessment of ships[9] may be used to estimate the residual strength factor γ_d. On the basis of available damage statistics (as shown in Figs. 6 and 7), some probability density curves can be obtained to determine the acceptable residual strength factor γ_d. Fig. 6 shows the distribution of the vertical extent of damage a_1, and Fig. 7 shows the distribution of the vertical location of damage p_1. The question of how to relate a_1 and p_1 to the residual strength factor γ_d needs to be investigated further. The basic trend of γ_d can perhaps be expressed as

$$\gamma_d \propto a_1^m p_1^n \tag{6}$$

where m and n are functions of location and extent of damage. From this expression, the larger the product $a_1^m p_1^n$, the greater the residual strength factor needs to be to sustain the initial damage. γ_d would also naturally depend on the consequence of a given damage. A target value for the residual strength factor can be given[2] as

$$0.6 \leq \gamma_d \leq 0.8 \tag{7}$$

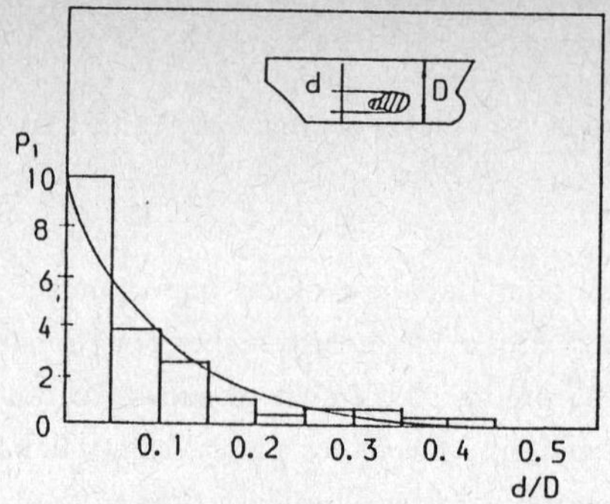

Fig. 6 Damage vertical extent distribution

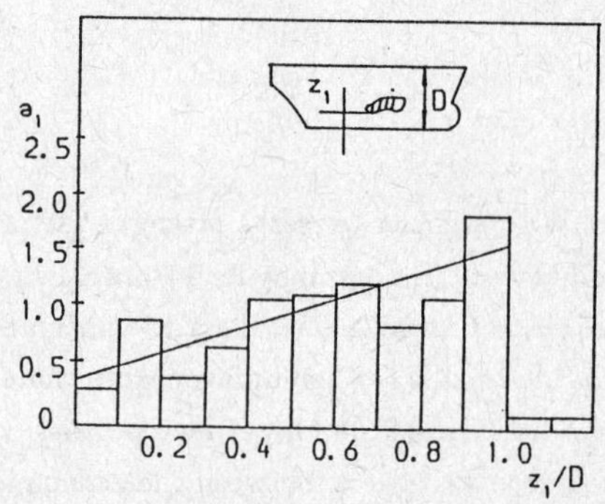

Fig. 7 Damage vertical location distribution

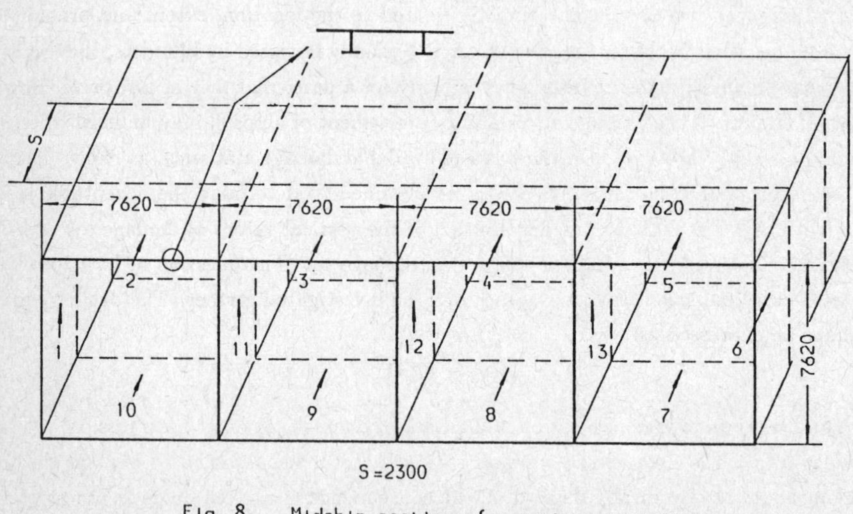

Fig. 8 Midship section of a pontoon structure

4. Structural Optimization Considering Damage Tolerance

Damage tolerance concepts can be incorporated in optimum structural design by introducing residual strength constraints for the overall structure for a range of damage conditions. A satisfactory design for a damage tolerant structure should have the following characteristics:-

- Sufficient reserve strength to avoid possible failure under some extreme loading states. Normal optimum structural design can take account of this by introducing reserve strength constraints.

- Sufficient residual strength to avoid catastrophic consequences due to initial damage. This provides alternative load paths to take the redistributed loads and augment the damage tolerance of the structure. This area has been ignored in many previous design applications. A conventional optimum structural design may result in a structure for which there is little or no likelihood of the existence of a feasible alternative load path.

- Favourable technical and economical efficiency. This may be judged in terms of multiple criteria of performance. Furthermore, damage tolerant design of structural systems lends itself to a multi-level treatment, with the structure being dealt with as a hierarchy of subsystems where the ultimate behaviour of a damaged structure is treated at the upper level and the detailed scantlings are determined at the lower levels. Hence, a multi-level, multiple criteria approach seems appropriate for such problems, in general[26].

In comparison with conventional optimum structural design, damage tolerant design involves the following additional considerations:-

- Identification of the basic damage states which are most likely to occur, in terms of the location and extent in the structure concerned.

- Selection of the appropriate reserve strength factor γ_i and residual strength factor γ_d, the latter being related to damage statistics.

- Estimation of the residual strength for each damage state, and formulation of the residual strength constraints.

A design system or subsystem can typically have several criteria of performance, and target values of such criteria of performance may not all be achievable. These can, therefore, be treated as goal constraints which for a general subsystem will be of the form

$$c_j(X_i, P_i, B_i) + d_j^- - d_j^+ = 0 \tag{8}$$

where X_i is a vector of design variables for the subsystem i (e.g. plate thickness, number of stiffening elements and other scantling details); P_i is a vector of parameters associated with the subsystem i (e.g. applied loads on the structure, material properties); B_i is a vector of behaviour variables input from outside the subsystem i (e.g. the ultimate overall load bearing capacity required); d_j^- and d_j^+ are the deviational variables for the jth goal constraint, restricted to values ≥ 0 and representing underachievement and overachievement respectively. If, for example, a particular c_j is required to be ≥ 0, then d_j^- is minimized only because an underachievement is to be avoided but overachievement is acceptable. If $c_j = 0$ is to be achieved, then d_j^- and d_j^+ are both minimized. By assigning priorities or weights to the deviational variables to reflect the relative importance of the associated goals, the target values of performance goals are achieved or approached as closely as possible.

The interaction conditions ensuring compatibility between the different levels and those between subsystems at the same level can also be handled as goal constraints, where achievement of the goals represents compatibility. The mathematical formulation of a typical two level damage tolerant design procedure is described below. This formulation is then used for the design of a simple barge structure. The goal constraints in this case arise out of a single criterion of performance, namely minimum weight, and the compatibility requirements of the structural subsystems.

Upper Level Subsystem Synthesis

The task at this level of design is to determine the upper level design variables Y defined as ultimate load bearing capabilities of the components. The behaviour variables B are determined for the subsequent design of lower level subsystems. The problem at this level is, therefore, to

find $\quad Y, \{D^-\}$ and $\{D^+\}$ $\quad (9a)$

(upper level design and deviational variables)

so as to minimize $\quad \{\alpha^-\}^T\{D^-\} + \{\alpha^+\}^T\{D^+\}$ $\quad (9b)$

(weighted sum of deviational variables)

subject to $\quad G_i(Y, P, B) \geq 0$ $\quad (9c)$

(reserve strength constraints)

$G_d(Y, P, B) \geq 0$ $\quad (9d)$

(residual strength constraints)

$G_s(Y, P, B) \geq 0$ $\quad (9e)$

(overall serviceability constraints)

$Q(Y, P, B) + \{D^-\} - \{D^+\} = 0$ $\quad (9f)$

(system interaction and performance goal constraints)

where, $\{\alpha^-\}$ and $\{\alpha^+\}$ are vectors of the weighting coefficients which reflect the designer's preference system; and $\{D^-\}$ and $\{D^+\}$ are the deviational variables for the upper level goal constraints. The goal constraints are normalised and the choice of weights is such that ensuring compatibility has a higher priority compared to minimizing weight.

Lower Level Subsystem Synthesis

For substructure l, all information concerning the influence of upper level design is passed to the lower level design by the behaviour variables B_l ($B_l \in B$). All of the detail scantlings X_l (e.g., plate thickness, number of stiffeners etc.) can be determined at this level. The problem at this level is, therefore, to find the values of lower level design variables X_l and lower level deviational variables so as to minimize the weighted sum of deviational variables for subsystem interaction and performance goal constraints as in the upper level problem. However, at this level, the local constraints are those due to ultimate and serviceability failure considerations.

Again, the behaviour variables B_l are calculated for information feedback to the upper level design. Iterations take place between the lower and upper design procedures until the convergence criteria are met. The optimization system used is a sequential linear programming algorithm incorporated in a software package GURU[26].

Damage Tolerant Design of a Barge

An offshore pontoon barge (see Fig. 8) has been designed by including residual strength constraints as described above. The principal dimensions of the barge are listed below:

Length	91.440 m
Breadth	30.480 m
Depth	7.6200 m
Draught (sea-going)	6.1670 m
Block coefficient	0.9460

The structure is assumed to support the normal service loads as given below:

Still water bending moment	M_S=189.470 MNm
Wave bending moment	M_W=296.058 MNm
Deck loading	q_d=0.1470 N/mm^2
Bottom loading	q_b=0.0747 N/mm^2
Side loading	q_s=0.0373 N/mm^2
Bulkhead loading	q_h=0.0373 N/mm^2

The transverse frame spacing is equal to 2.30 m. If the hull girder is holed or grossly deformed, the damaged area can be assumed to be ineffective under applied loads, and the cross section of the hull girder accordingly loses some effective area. However if the damage is less severe, this assumption may lead to a conservative design. Five damage states are considered for the design:-

Damage state 1 :	Exclusion of substructure 1
Damage state 2 :	Exclusion of substructure 2
Damage state 3 :	Exclusion of substructure 3
Damage state 4 :	Exclusion of substructure 9
Damage state 5 :	Exclusion of substructure 10

These five damage states are assumed to occur independently. Hogging and sagging conditions are studied and five residual strength constraints are formed for each condition.

The design model is decomposed into side, deck, bottom and bulkhead substructures. The ultimate load bearing capability of each substructure, which is defined as the ultimate stress multipled by the total cross section area, is taken as the upper level design variable so that the upper level design virtually re-distributes the ultimate load bearing capability throughout the hull girder. This assigned ultimate load bearing capability, through the behaviour variables, affects the lower level design procedure which optimizes the actual scantlings. The design cycles iterate between the lower and upper levels. The number of stiffeners in each substructure is modelled as an integer variable.

In this example, the effects of reserve and residual strength factors on the design of the tee stiffened hull girder were investigated. At the lower level, the safety margins for ultimate and serviceability failure limit states of substructures are 1.3 and 1.2 respectively. Firstly, normal structural design studies that excluded the residual strength constraints were carried out by changing the reserve strength factor γ_i from 1.5 to 2.0 for the whole structure. The total values of the weight per frame space thus obtained are indicated by the continuous line in Fig. 9. In damage tolerant structural design, the product of reserve strength factor γ_i and residual strength factor γ_d for the whole structure indicates the safety margin for the damaged structure. Damage tolerant designs for $\gamma_i \times \gamma_d$ values of 1.2 and 1.3 are shown by the broken line in Fig. 9. It can be observed from the numerical results that if the reserve strength factor γ_i is very high, the resultant structure will be able to support the applied loads with initial damage in the structure. This increase of the reserve strength factor γ_i takes no consideration of the residual strength. However, design with a high residual strength factor γ_d can lead to a very heavy weight penalty. The actual value of γ_d used would depend on the likelihood of various damage scenarios and the consequences thereof, and meaningful damage tolerant design would have to involve the identification of such damage situations.

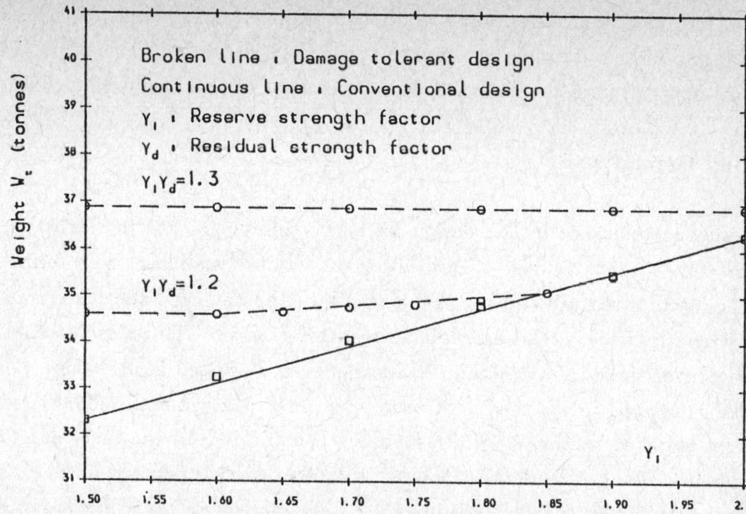

Fig. 9 Effects of reserve and residual strength factors on the design of a tee stiffened hull girder

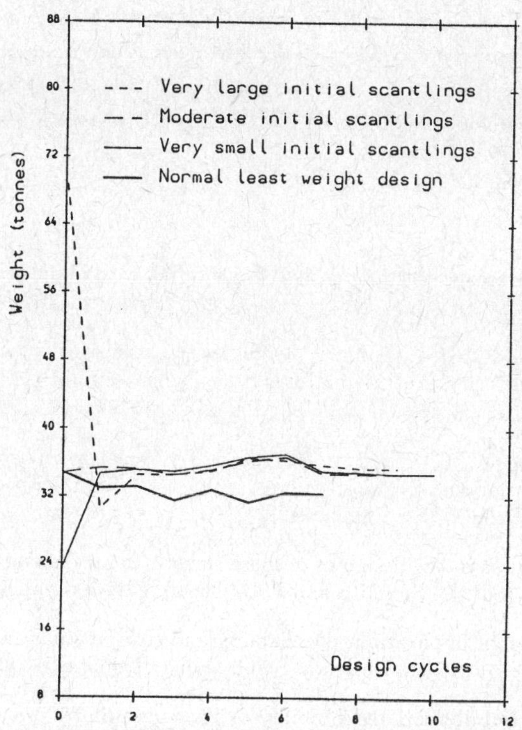

Fig. 10 Iteration history of normal and damage tolerant designs of a tee stiffened hull girders

Fig. 10 shows the convergence performance of the optimization algorithm. Such convergence checks for a range of design examples confirm the view that the design formulation, as described above, is robust and can form the basis for the development of practical design tools.

5. Concluding Remarks

The role of damage tolerance in the design for survivability of a marine structure has been recognised for some time. Damage tolerance can be explicitly addressed in terms of reserve strength of the intact structure and residual strength of the damaged structure, and measured by a reserve strength factor γ_i and a residual strength factor γ_d. The reserve strength factor is related to the conventional safety margin, which has been discussed in literature. The residual strength factor is related to the degree of seriousness of a damage state. The selection of an adequate target value of residual strength factor is complicated by the fact that the residual strength is damage-dependent and directional. The number of basic damage states for a structure can be very large, and there is thus a need to choose the damage states which are most likely to occur.

Damage tolerant design procedures can be formulated using a multi-level, multiple criteria optimization approach. In this approach, the upper level design problem is formulated to satisfy the damage tolerant requirements, while the lower level designs determine the actual scantlings of the substructure. The results indicate the relative influence of γ_i and γ_d, and also demonstrate that the proposed methodology can form the basis of a practical tool for optimum damage tolerant design.

6. References

[1] Haftka R. T., Damage tolerant design using collapse techniques, AIAA Journal, Vol. 21, No. 10, (1983)

[2] de Oliveira J. G. and R. A. Zimmer, Redundancy Considerations in the structural design of floating offshore platforms, An International Symposium on the Role of Design, Inspection and Redundancy in Marine Structural Reliability, Virginia, (1983)

[3] Lloyd J. R. and W. C. Clawson, Reserve and residual strength of pile founded offshore platforms, An International Symposium on the Role of Design, Inspection and Redundancy in Marine Structural Reliability, Virginia, (1983)

[4] Moan T., Advances in the design of offshore structures for damage tolerance in Advances in Marine Structures, (eds. C. S. Smith and J. D. Clark), Elsevier Applied Science, (1986)

[5] Frieze P. A., Design Implications of reliability and collapse analysis of mobile units, International Symposium on Practical Design in Shipbuilding, Trondheim, (1987)

[6] Regulations on Subdivision and Stability of Passenger Ships as an Equivalent to Part B of Chapter II of the International Convention for the Safety of Life at Sea, 1960, IMO Resolution A265 (VIII), London, (1974)

[7] Caldwell J. B., E. H. Hegazy and M. Pawlowsky, Safety of ships after collision, Second International Congress of Maritime Association of the East Mediterranean, (1981)

[8] Mistree F., Design of damage tolerant structural system, Engineering Optimization, Vol. 6, (1983)

[9] Sen P. and C. Konstantinidis, A time simulation approach to the assessment of damage survivability of RO/RO cargo ships, Trans. SNAME, (1987)

[10] Burke R. J., The consequences of extreme loadings on ship structures, Extreme Load Response Symposium, Arlington, (1981)

[11] ISSC, Proceeding of 9th International Ship Structures Congress, Genova, (1985)

[12] Faulkner D., Semi-probabilistic approach to the design of marine structures, Extreme Loads Response Symposium, Arlington, (1981)

[13] Ang H. S. and M. Amin, Safety factors and probability in structural design, ASCE, Vol. 95, No. ST7, (1969)

[14] Brown C. B., A fuzzy safety measure, ASCE, Vol. 105, No. EM5, (1979)

[15] Madsen H. O., S. Krenk and N. C. Lind, Methods of Structural Safety, Prentice-Hall, Inc., Englewood Cliffs, New Jersey, (1986)

[16] Thompson J. M. T. and G. H. Hunt, Dangers of structural optimization, Engineering Optimization, Vol. 1, (1974)

[17] Yao J. T. P., Safety and Reliability of Existing Structures, Pitman Publishing Inc., Boston, (1985)

[18] Frangopol D. M. and J. P. Curley, Effects of damage and redundancy on structural reliability, Journal of Structural Engineering, ASCE, Vol. 113, No. 7, (1986)

[19] ISSC, Proceeding of 7th International Ship Structures Congress, Paris, (1979)

[20] Ochi M. K., Principles of extreme value statistics and their application, Extreme Loads Response Symposium, Arlington, (1981)

[21] Majid K. I., Nonlinear Structures, Butterworths, London, (1972)

[22] Adamchak J. C., ULTSTR: A program for estimating the collapse moment of a ship's hull under longitudinal bending, DTNSRDC Report, 1982/076, (1982)

[23] Smith C. S., Structural redundancy and damage tolerance in relation to ultimate ship hull strength, International Symposium on the Role of Design, Inspection and Redundancy in Marine Structural Reliability, Virginia, (1983)

[24] Carlsen C. A. and A. J. Brown, Lifetime structural performance of mobile offshore units, Integrity of Offshore Structures-3, Glasgow, (1987)

[25] Faulkner D. and J. A. Sadden, Toward a unified approach to ship structural safety, RINA Spring Meeting, Paper No. 3, (1978)

[26] Shi W. B., P. Sen and J. B. Caldwell, Optimization of ship structural forms, International Conference on Computer Application in the Automation of Shipyard Operation and Ship Design, Shanghai, (1988)

DEVELOPMENT OF TECHNIQUES TO INVESTIGATE THE ULTIMATE LIMIT STATE OF TUBULAR FRAMED STRUCTURES

E. P. Shuttleworth, J. K. Ward & M. Lalani
The Steel Construction Institute, Silwood Park, Ascot, Berkshire

SUMMARY

Reserve strength, residual strength, redundancy and collapse mechanisms are important considerations when performing reliability assessments of structures. This paper describes work being undertaken at the Steel Construction Institute (SCI) to provide a means of quantifying these issues. Detailed strength modelling of structures, up to and beyond ultimate load, will be achieved by integrating component response characteristics for both member and joint behaviour, with provision for the interaction effects which occur in complex structures.

Included in this paper is a survey of recent background work and test programmes which have been carried out to determine the ultimate strength of tubular frames. Development work at SCI, under funding from oil and gas operators and the UK Department of Energy, covers testing of large scale tubular frames and components, and the production and calibration against the test data of a nonlinear analysis program for performing full collapse analysis, member redundancy studies and optimisation studies, as necessary in a reliability study. It is also intended that the system will be used in the decision making process for cost effective design, inspection, maintenance and repair of tubular steel structures.

INTRODUCTION

Offshore steel structures and other tubular framed structures are currently designed on the basis of traditional structural engineering practice. A combination of loads is applied to the structure and the internal forces in each brace member are established. Each member and joint is checked against permitted strengths given in design codes. The structure is considered to meet the selected code standards if all the individual components (i.e. members and joints) satisfy the code requirements. The allowable strengths are based on the large database of tests on isolated joints [2] and tubular beam column members [3-5].

However, it is recognised that, for example, fixed offshore structures are redundant and have a multiplicity of load paths to the extent that failure of one member can usually be tolerated without catastrophic collapse of the structure. This introduces the concept that redundant structures have a residual strength even if a primary member has failed. For undamaged structures, the conservatism in conventional design practice ensures that a substantial reserve strength exists beyond the design event.

To date, the effects of reserve strength, residual strength, redundancy and alternative load paths for structures as a whole, have essentially been treated in a qualitative manner or in terms of specific design examples. Recently, there has been an increased awareness of the need to quanitify these issues. This has been taken up by the regulatory bodies in the offshore industry who have started to consider the introduction of limit state codes for offshore structures [6,7].

In response to these moves, the Steel Construction Institute (SCI) recently started work on a research programme to address the issues of reserve and residual strength, redundancy and progressive collapse mechanisms for three dimensional steel tubular frames. The primary objective of the work is to develop a nonlinear numerical procedure for three dimensional frames which will be calibrated and substantiated by a series of tests on frames subjected to collapse loads. The interaction between component behaviour and global behaviour is described and an approach to the numerical treatment of the problem is addressed. A description of previous relevant work is presented.

Issues associated with the selection of environmental criteria and load generation are not considered in this paper which concentrates only on the resistance side of the design process.

Recent test data on tubular structures are surveyed, and examples are presented illustrating the nonlinear program's current capabilities.

COMPONENT AND GLOBAL FRAME BEHAVIOUR

The reliability of tubular structures is highly dependent on the approach adopted in calculating factors of safety. For tubular beam columns, a nominal factor of safety of 1.25 against first yield is inherently provided by design codes such as API RP2A/AISC [8]. Studies on the ultimate strength of tubular beam-columns [9,10] indicate a considerable reserve strength beyond first yield, as shown in Figure 1. For typical members designed to code requirements, the real safety factors for the design load case are closer to 2.0 based on the ultimate strength concept. The contribution of members will also depend on the failure modes of the members. Figure 2 presents typical failure modes of tubular members [11]. It can be seen that the maximum capacity of the tension and compression members can be reached at similar strain levels, and is dependent on the level of ductility inherent in the compression member (which is a function of the residual stresses, initial lack-of-straightness, slenderness ratio and diameter to thickness ratio).

The failure modes for tubular joints [2,12] show considerable reserve capacity beyond the point of first yield. This reserve is dependent on the joint geometry, configuration and load type. Typical load-deformation curves for axially loaded joints are shown in Figure 3. Joints subject to moment loads show similar characteristics except that the joint shows a lengthy plateau region at the ultimate load P_u, thereby demonstrating considerable plastic hinge rotation capacity without degradation of load.

As inferred above, the full plastic moment capacity of a tubular steel section is considerably larger than the moment at which yielding first occurs. Further, ductile yielding at a section or joint in a statically indeterminate structure will permit redistribution of internal forces so that larger loads can be applied. This redistribution is dependent on the interaction between joint nonlinearity and member nonlinearity, as well as the availability of alternative load paths. Indeed, loads can be increased until hinges occur at sufficient locations to create a collapse mechanism. The program discussed later in this paper will, in its fully enhanced state, provide interaction between member and joint nonlinearity so that the frame behaviour of connection-dominated collapse mechanisms may be assessed.

RECENT BACKGROUND WORK

Ultimate capacity of frames

Whilst a significant amount of research has been directed towards component behaviour [1-5], little work on the ultimate strength of frames has been carried out. Marshall [12] developed mathematically based engineering models to characterize the reliability of offshore platforms subjected to severe environmental loads. The models were based on simplified member strength assumptions and joint flexibility effects were ignored. Joint collapse was not considered.

Lloyd & Cawson [13] undertook a series of collapse analyses of an eight-legged structure with grouted and welded piles inside the leg sections. The analyses concentrated on member response to incremental loading and reserve factors in the range 2.43-3.48 for the welded pile model and 2.98-4.36 for the grouted model were obtained, depending on the direction of wave attack. The residual factors ranged from 0.81 to 1.00 for the structure considered and were dependent on the member removed. No attempt was made to model the post-ultimate stiffness of damaged members. In addition, for the analyses the joint collapse behaviour and stiffness characteristics were not considered.

A series of one-sixth scale two-dimensional frame tests was undertaken at the University of California [(14)] to investigate their behaviour when subjected to earthquake loads. The frames were designed to ensure that joint collapse did not take place. Energy absorption characteristics in the form of hysterisis loops were derived; failure was dominated by inelastic buckling of brace members.

Reference 15 presents some results obtained from four small scale frames (leg diameters less than 140 mm and brace diameters less 60 mm) subjected to earthquake loading. The joints were rigidly connected using external gusset plates and the tests were undertaken to provide an insight into energy absorption characteristics of frames dominated by member response.

Reference 16 presents results obtained from tests on six K-braced frames of realistic offshore sizes. In these tests, where there was no alternative load path following compression member buckling, it was still demonstrated that using the API RP2A/AISC [(8)] code there was a factor of safety of approximately 2.2 against ultimate collapse of the frame.

Reference 17 presents test results obtained from two small scale tests on three-dimensional X-braced frames which demonstrated the large residual strength and ductility evident in an indeterminate structure.

In Reference 18 a plane frame from a deepwater offshore jacket structure has been analysed using both the MARC and INTRA programs. Sensitivity studies were carried out to determine the effects of initial out-of-straightness, ovality and residual stresses, on the global frame response. Joints were assumed to be rigid.

Effects of joint flexibility

It has been concluded above that no studies have been conducted to date on the ultimate strength of frames with joint-dominated failures; all the limited numerical and experimental investigations have sought to identify member response rather than joint response.

The elastic analysis of fixed platforms assumes that the brace members are rigidly connected to the chord. The effects of chord wall flexibility are not normally taken into account. Examination of the load-deformation curves in Figure 3 reveals two primary characteristics:

(1) Tubular joints may exhibit local flexibility within the linear elastic range. This flexibility is a function of joint type, geometry and loading.

(2) In the elasto-plastic region tubular joints deform in a ductile manner, leading to significant flexibility in the nonlinear range.

The classic definition of joint flexibility is synonymous with (1) , and almost all the research effort to date has been concentrated in this area. Most design codes allow for the inclusion of joint flexibility but only DnV [19] give some formulae for the elastic flexibility of T joints subjected to brace moment loads. Notable research efforts on the effects of elastic flexibility are reported in References 20 and 21. Bouwkamp [20] used shell finite element models of joints to determine elastic flexibilities. These were accommodated in a two-dimensional frame analysis using substructuring techniques. The analysis results indicated that increases in deflections and leg forces and moments are likely. In addition, the results showed a reduction in brace axial forces and a modified distribution of pile loads. The UEG study [21] on the effects of elastic flexibilities encompassed the analyses of three two-dimensional frames. The flexibility coefficients used were based on experimental investigations of simple T, K/YT and KT joints. The results indicated changes similar to the Bouwkamp studies. In addition, changes in natural frequency of up to 82% were reported.

The only research work to date which addresses the effects of flexibility in the elasto-plastic region is reported in Reference 22. A series of analyses was undertaken on inverted K-braced two-dimensional frames subjected to lateral loads. Joint characteristics (utilizing nonlinear flexibilities) were represented using simplified load-deformation relationships for the various joints. The joints were modelled using a series of beam elements representing either beam-column behaviour, nonlinear joint behaviour or interaction behaviour (in the case of K joints). No experimental work was undertaken. The results indicated that ultimate capacity reduces as the ratio of chord radius to chord wall thickness increases for the frames considered, and nonlinear joint behaviour may have little effect on member collapse mechanisms but has a major effect on frame collapse behaviour.

DEVELOPMENTS IN NUMERICAL TECHNIQUES

Introduction

The ultimate limit state is when the structure becomes incapable of carrying the applied loads. For tubular framed structures such states can only be estimated using computerised numerical techniques which allow for geometric and material nonlinearity. Such methods are approximate, for simplifications and assumptions have to be made to satisfy computer resources. To justify the simplifications, the numerical techniques must be calibrated against experimental tests.

The ultimate capacity of tubular framed structures is governed mainly by beam-column behaviour and joint behaviour. Beam-column behaviour has been well researched, both experimentally and analytically. However, relatively little is known about the behaviour of multi-planar joints, for there are many variables to consider, including configuration, geometry, and loading. Experimental tests are expensive, leaving the dual approach of limited experimental tests and parametric analytical studies as the only way forward. The ultimate limit state of complex tubular joints can be studied analytically using nonlinear finite element analysis. It is only recently, however, that the method has started to be used, mainly because there are not many computer programs with a satisfactory and economical shell element formulation, or the automatic solution techniques to establish any reserve strength beyond the first peak-load. Further, such methods require calibration to give confidence in their use, for shell models only approximate the behaviour at welded tubular joint intersections.

To predict the reserve and residual strengths of tubular framed structures, including load redistribution and unloading beyond the ultimate load, both beam-column member buckling allowing for the spread of plasticity and initial imperfections, and nonlinear joint flexibility and strength, must be considered. Most nonlinear finite element programs contain the simple cubic beam element based on decoupled beam and column action. To model beam-column behaviour, each member must be subdivided and extra nodes introduced. Approximate methods have therefore been adopted. One program, INTRA, incorporates a phenomenological model which has only a single axial degree of freedom for which nonlinear properties, including the effective length factor, are defined by the user. Bending stiffness may be included by incorporating a bending element with no axial stiffness in parallel with the phenomenological element. The method has proved successful but it does require good knowledge of frame behaviour.

Research currently underway at The Steel Construction Institute includes the development of a nonlinear frame analysis program, including calibration of the program against large scale tubular frames and components. The analytical work consists of nonlinear finite element analysis of joints, to determine load-deflection curves and load interaction failure surfaces, and the implementation of these in the frame analysis program which includes a new element for beam-column behaviour.

Nonlinear analysis of joints

Calibration of the finite element program, FINAS, for ultimate load analysis of joints has been carried out. Current work involves the development of load-deflection curves and load interaction failure surfaces for the X, K and KT joints of the large scale frames to be tested to failure.

The FINAS program [23,24] was developed at Imperial College and contains doubly-curved 6 and 8 node isoparametric elements with six degrees of freedom per node. The element formulation includes transverse shear, but a penalty function procedure avoids 'shear locking' when the element is used as a thin shell. The automatic incremental solution techniques are not restricted to positive definite matrices, thus enabling unloading paths to be followed. The solution techniques include both load control and displacement control. When in displacement control, the freedom with the most rapid incremental change is monitored.

As part of the calibration exercise, the DT-joint shown in Figure 4 was analysed. Figures 5 and 6 illustrate the load-deflection behaviour of the joint. Figure 7 compares various analyses with an experimental test [25]. The experimental and analytical ultimate loads for both the coarse and a fine meshes, Figure 8, are in very good agreement. Agreement is to within 13% for the coarse mesh and 4% for the fine mesh. The analyses were performed using both the single-layer Ivanov criterion and the multi-layer approach monitoring the spread of plasticity through the plate thickness. The axial displacements at the peak load are not in good agreement, with the experimental displacements being larger. This is not unexpected, for the experimental result includes movement of the test frame, and the test specimen was made from seam-welded pipe, which has a lower stiffness due to residual stresses created in the fabrication process.

Analytical studies have also been carried out for the DT-joint under in-plane bending, with similar levels of agreement between experimental and analytical results. They give the necessary confidence in the use of nonlinear finite element analysis to predict complex joint behaviour, but with the proviso that the applied boundary conditions simulate the in-service behaviour of the joint.

Nonlinear analysis of frames

At present, most offshore structures are analysed linearly using simple cubic beam elements and assuming rigid connections. Linear analysis is only appropriate where the axial forces in the members are small so that the bending stiffness is not affected, or where members are dominated by axial load and bending is insignificant. In such cases, each member may be modelled by a single element with decoupled beam and column actions. In design, the approximation is allowed for in the various formulae in codes of practice for checking component strength.

In most offshore structures, beam-column action occurs. To predict the ultimate load behaviour both nonlinear and material properties and large deflection effects must be considered. Apart from ordinary beam elements, several of which must be used to represent each member, beam-column formulations with varying efficiencies have been developed [11,26,27]. At The Steel Construction Institute, an elastic formulation for beam-column behaviour has been developed which allows a member to be modelled by a single element. The formulation involves load incrementation and iterative procedures to accommodate beam-column nonlinearities. Elasto-plastic material behaviour, with strain hardening if required, is included by subdivision of an element and the introduction of a plastic hinge. However, an alternative formulation allows stresses and strains to be monitored at different points across the section and along a segment of the member.

The solution procedures include both load control and displacement control. The initial computer model need only consist of nodes at joints and single elements to model members, reducing data input and checking time to a minimum. As the load or a displacement is increased and plasticity occurs, the original elements are subdivided and extra nodes and elements introduced automatically. For solution, the frontal technique has been implemented since its efficiency depends on the element solution order. If the original element solution order is optimised, solution efficiency can be maintained provided each new element is spawned in series from its parent, when the increase in the frontal bandwidth will be only one node.

Several analyses have been undertaken to check the behaviour of the new element and the automatic solution strategies. Its behaviour has been found to be superior to other beam-column formulations. Four examples are presented:

- the solution for large displacement of a cantilever
- solutions for a perfect and imperfect simple truss
- analysis of Lee's frame
- analysis of a plane frame from an offshore structure.

Cantilever

Figure 9 shows the large deflection behaviour of a cantilever subjected to an increasing point load at its end. The results for a single element are compared with those using 8 elements of a cubic formulation available in the nonlinear analysis program, LUSAS. The results show very good agreement, and confirm the potential of the new element in fulfilling the requirement of using as few elements as possible to model nonlinear elastic beam-column behaviour in an offshore jacket structure.

Single Truss

The example shown in Figure 10 is of a very simple truss, for cases when the compression member is perfectly straight and when it has an imperfection of L/500. The results for the perfect compression member shows the deficiency of the beam formulation in LUSAS, since the buckling load is overestimated considerably when only 2 elements are used. However, the accuracy of the new element is extremely good, only 2 elements giving the same results as 6 LUSAS elements. The results for the imperfect strut reinforce the potential and efficiency of the new element for nonlinear analysis of frames.

Lee's Frame

Various analyses have been carried out for Lee's frame. Figure 11 shows the results using the open-section higher order element of FINAS, and 3 and 5 elements of the new beam-column element (SCI). The 5 element model (2 elements for the column, 3 elements for the horizontal beam) is in excellent agreement with Frey's solution for this 'theoretical' large displacement problem solved using displacement control.

Plane Frame

Figure 12 shows an end frame of a recent offshore structure. The jacket has four transverse frames, of the type shown, and two longitudinal frames. The jacket is piled at its corners and stands in a water depth of 155m.

The jacket was analysed under constant vertical loads and increasing horizontal loads. The applied vertical loads represented the buoyant self-weight and topside operating weight, and the basic horizontal loads (load factor = 1) represented the loads applied by a storm wave of height 30.5 m and period 17.5 secs, for the maximum base shear condition. The flooded legs were assumed fixed at their base, where they were also stiffened to model the pile-sleeve connections and prevent the formation of unrealistic plastic hinges. The applied loads are given in Figure 13.

Figure 14 shows the load-displacement behaviour of the frame. Under increasing transverse loading, the first plastic hinge forms at a load factor of 1.74. To reach the maximum load factor of 2.19, a complex pattern of opening and closing of plastic hinges takes place. When it is reached, the load rapidly falls off to below the original load level, with transverse displacements in excess of 1.7m. The plastic hinges at various load levels are shown in Figure 15.

This example was undertaken to demonstrate the capabilities of the program. In the real three-dimensional structure, the frame is likely to sustain a higher maximum load factor due to the greater level of redundancy providing alternative load paths. As the load levels in the members may not reflect those in the true structure, it must not be inferred that the reserve factor of 2.19 is typical of an offshore frame.

CLOSURE

In the bid for more economic and efficient offshore structures, a better understanding of structural behaviour must be achieved if acceptable safety margins are to be maintained. The level of structural redundancy has decreased and our design philosophy is changing to limit state concepts, but do we have the design tools to accommodate these changes, do we understand total structural response in sufficient detail, and what consistent factors of safety must we apply to materials and loads to maintain serviceability and prevent collapse? In deducing these factors, all of the various facets of the use of the structure, the environment, and the design, construction and installation processes, must be considered.

The ultimate limit state of a structure can only be estimated using computerised numerical techniques which allow for geometric and material nonlinearity. Such methods are approximate, for simplifications and assumptions have to be made to satisfy computer resources. To justify the simplifications, the numerical techniques must be calibrated against experimental tests.

The example analysis of the DT-joint presented shows that nonlinear finite element analysis can be relied on to predict the ultimate load behaviour of tubular joints. The examples for the new beam-column element are very encouraging but its performance must be calibrated against large scale frame tests. The element, when combined with load-deflection curves and interaction curves defining joint behaviour obtained using nonlinear finite element analysis, will prove a useful tool for reliability assessment of framed structures.

A project [28] at The Steel Construction Institute, funded by oil and gas operating companies and the UK Department of Energy and which is open to all interested parties, is currently undertaking large scale tests on two-dimensional frames for calibrating the new nonlinear frame analysis program.

ACKNOWLEDGEMENTS

The software development described in this paper is being carried out by Bassam Izzudin under the supervision of Dr A S Elnashai and the second author. The authors acknowledge their contribution to the research programme. They also thank Nancy Baddou and Maha Dabbous of The Steel Construction Institute for undertaking the majority of the example analyses.

References

1 YURA, J.A. et al. 'Ultimate Capacity Equations for Tubular Joints'. Paper OTC 3690, Offshore Technology Conference, Texas, 1980.

2 Underwater Engineering Group. 'Design Guidance on Tubular Joints in Steel Offshore Structures'. UEG publication UR33, 1985.

3 CHEN, W.F., & Ross, D.A.. 'Tests on Fabricated Tubular Columns'. Journal of the Structural Division ASCE 103(3), 1977.

4 SMITH, C.S. et al. 'Buckling Strength and Post-collapse Behaviour of Tubular Bracing Members Including Damage Effects'. BOSS '79, London, Vol. 2.

5 ELLINAS, C.P. et al. 'Buckling of Offshore Structures'. Granada Publishing Limited, 1984.

6 API Draft Limit State Rules for Offshore Structures.

7 FRIEZE, P.A. & PLANE, C.A. 'Partial Safety Factor Evaluation for Some Aspects of Buckling of Offshore Structures. A Pilot Study Based on 5400: Part 3'. University of Glasgow for UK Department of Energy. HMSO, London, 1987.

8 American Institute of Steel Construction. 'Manual of Steel Construction', 8th edn, 1984.

9 FOWLER, D.W. et al. 'Ultimate Strength of Tubular Columns under Combined Bending and Axial Load'. ASCE NSE meeting, San Francisco, April 1973.

10 SHERMAN, D.R. & GLASS, A.M. 'Ultimate Bending Capacity of Circular Tubes'. Paper OTC 2119, Offshore Technology Conference, Texas, 1974.

11 MARSHALL, P.W. & BEA, R.G. 'Failure Modes of Offshore Platforms'. BOSS '76, Vol. 2, Trondheim, Norway, 1976.

12 BILLINGTON, C.J., et al. 'Background to New Formulae for the Ultimate Limit State of Tubular Joints'. Paper OTC 4189, Offshore Technology Conference, Texas, 1982.

13 LLOYD, J.R. & CLAWSON, W.C. 'Reserve and Residual Strength of Pile Founded, Offshore Platforms'. The Role of Design, Inspection and Redundancy in Marine Structural Reliability. National Research Council, National Academy Press, Washington DC., 1984.

14 ZAYAS, V.A. et al. 'Ultimate Strength of Steel Offshore Structures'. BOSS '82, Vol. 2, Massachusetts.

15 INOUE, K. et al. 'Buckling Strength and Post-Buckling behaviour of Tubular Truss Tower'. Welding of Tubular Structures, IIW, Boston, 1984.

16 GRENDA, et al. 'Large-scale Ultimate Strength Testing of Tubular K-braced Frames'. Paper OTC 5832, Offshore Technology Conference, Texas, 1988.

17 SOREIDE et al. 'The Idealised Structural Unit Method on Space Tubular Frames'. SINTEF, Report No. 611, December 1986.

18 STEWART, G. et al. 'Ultimate Strength and Integrity Assessment of Fixed Offshore Platforms'. BOSS '88, Trondheim, June 1988.

19 DET NORSKE VERITAS. Rules for the Design, Construction and Inspection of Fixed Offshore Structures. DNV (plus amendments), 1977.

20 BOUWKAMP, J.G. et al. 'Effects of Joint Flexibility on the Response of Offshore Towers'. Paper OTC 3901, Offshore Technology Conference, Texas, 1980.

21 UNDERWATER ENGINEERING GROUP. 'Node Flexibility and its Effect on Jacket Structures'. UEG Report UR22, 1984.

22 UEDA, Y. et al. 'Flexibility and Yield Strength of Joints in Analysis of Tubular Offshore Structures'. Offshore Mechanics and Arctic Engineering Conference, Vol. 2, Japan, 1986.

23 TRUEB, U. 'Stability Problems of Elasto-plastic Plates and Shell by Finite Elements'. PhD Thesis, Imperial College. University of London, 1983.

24 BATES, D.N. 'The Mechanics of Thin Walled Structures with Special Reference to Finite Rotations'. PhD Thesis, Imperial College, University of London, 1987.

25 BOONE, T.J. et al. 'Chord Stress Effects on the Ultimate Strength of Tubular Joints'. Report PMFSEL 82-1, University of Texas, 1982.

26 RASHED, S.M.H. 'Behaviour to Ultimate Strength of Tubular Offshore Structures by the Idealised Structural Unit Method'. Report No. SK/R51, The Norwegian Institute of Technology, Trondheim, 1980.

27 ZHANG, S.K. 'The Failure Load Calculation of Offshore Jacket by Incremental Limit Analysis'. OMAE, Japan, 1986.

28 The Ultimate Limit State of Tubular Frames. Offshore Research Focus, No. 63, February 1988.

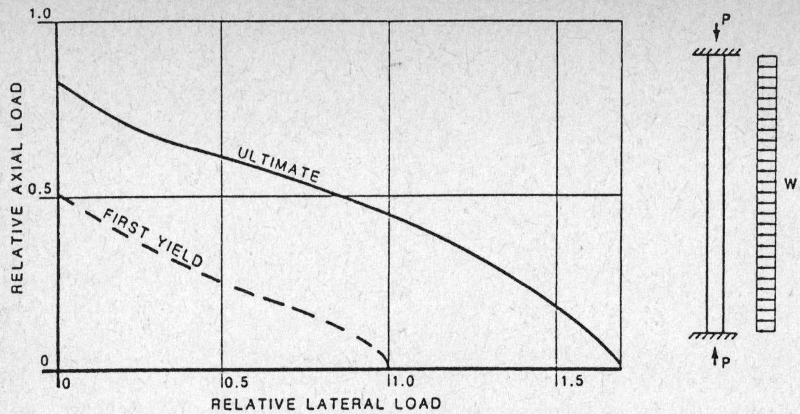

Figure 1. Tubular Beam-Columns with Fixed Ends-Failure Envelopes(9,10) Effective Length Factor = 0.7; L/D = 33.00.

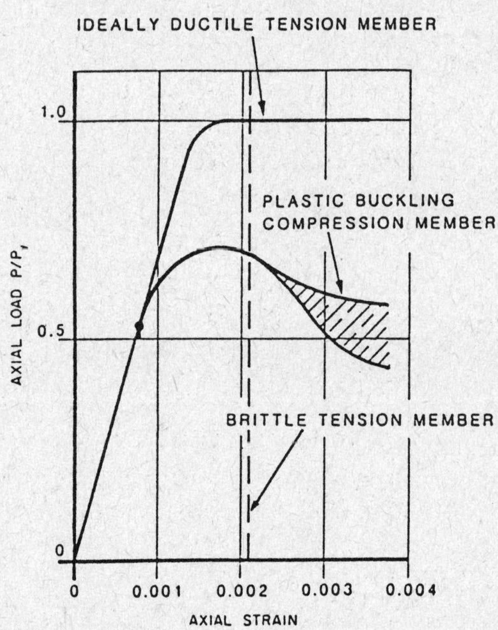

Figure 2. Typical Failure Modes for Tubular Members.

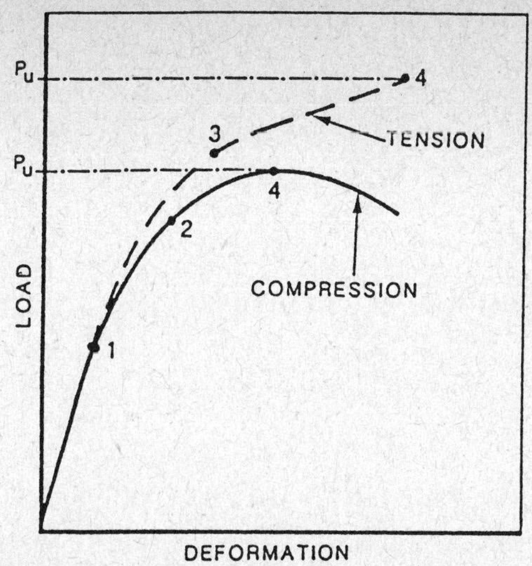

Figure 3. Typical Load-Deformation Curves for Axially Loaded Joints. Elastic Limit (1), Deformation Limit (2), Crack Initiation (3), and Ultimate Load (4).

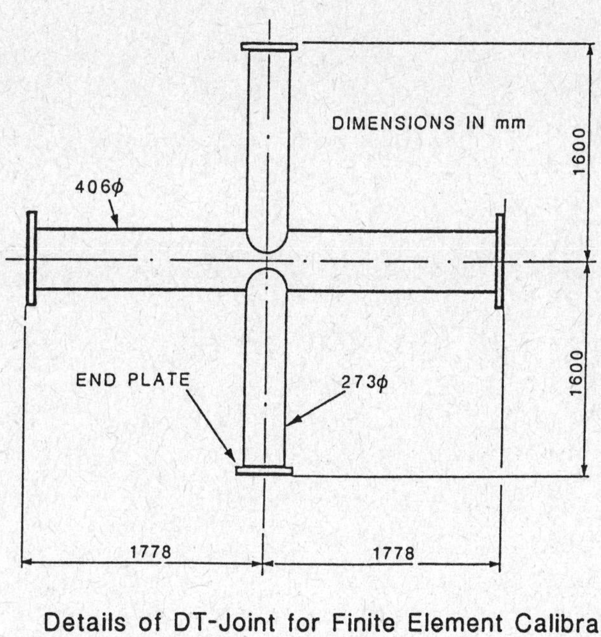

Figure 4. Details of DT-Joint for Finite Element Calibration.

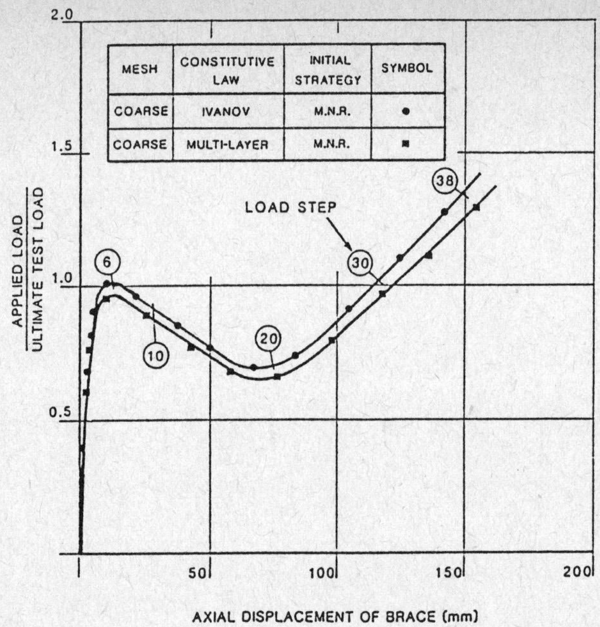

Figure 5. Load-Deflection Plot of Axially Loaded DT-Joint

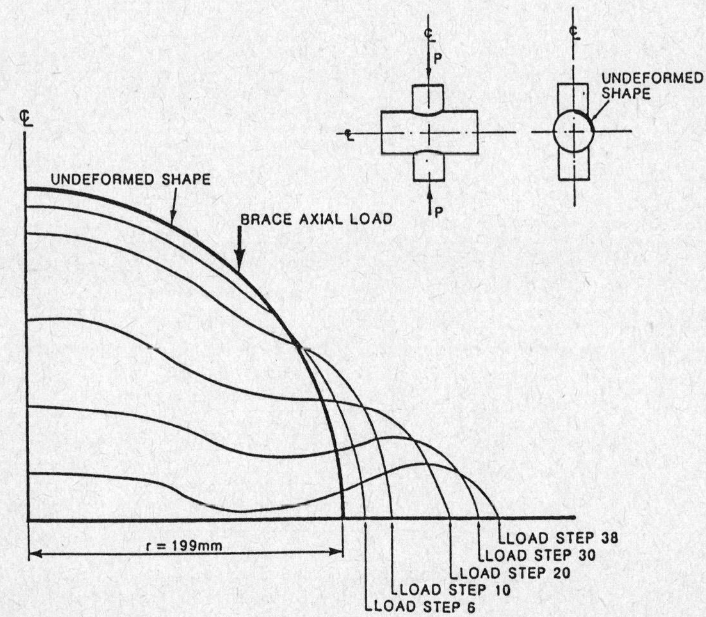

Figure 6. Deformed Shape of DT-Joint Chord Under Axial Load.

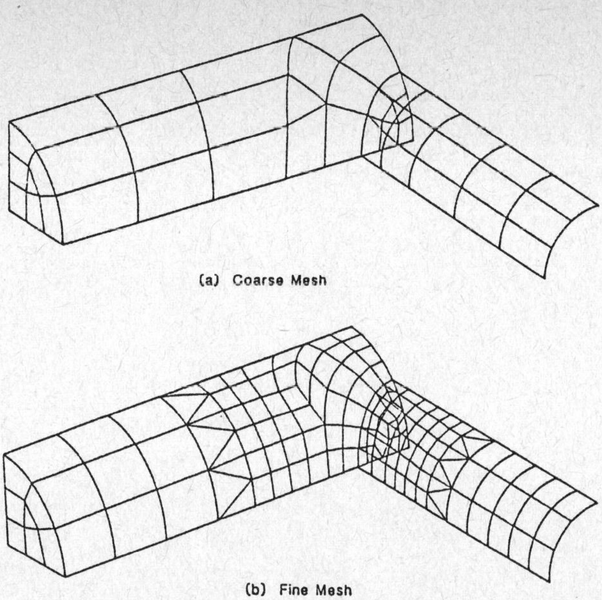

Figure 7. Finite Element Meshes for DT-Joint

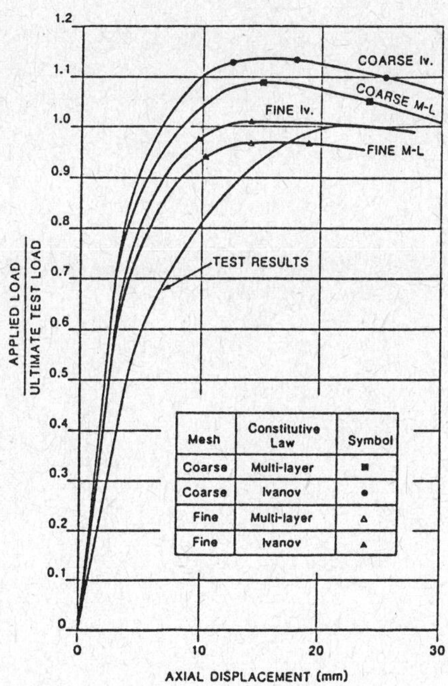

Figure 8. Load Displacement Plots for Axially Loaded DT-Joint

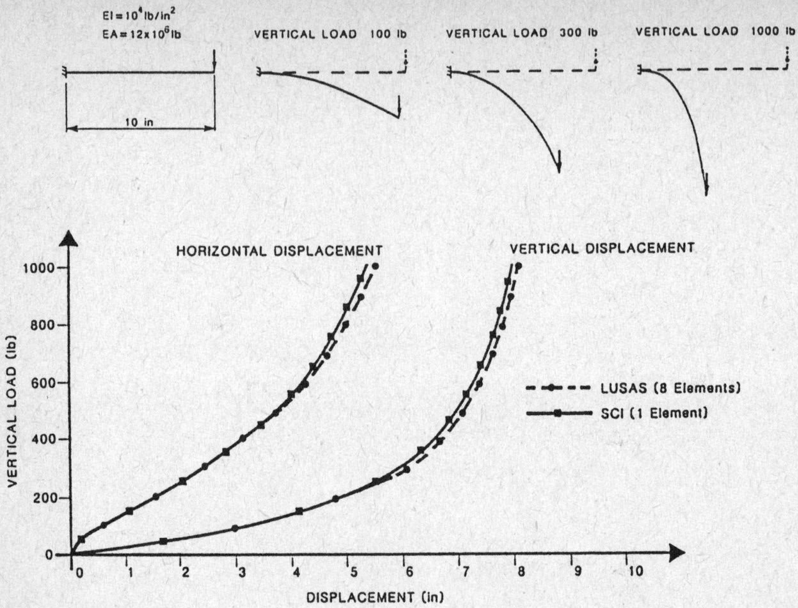

Figure 9. Large Displacement of a Cantilever

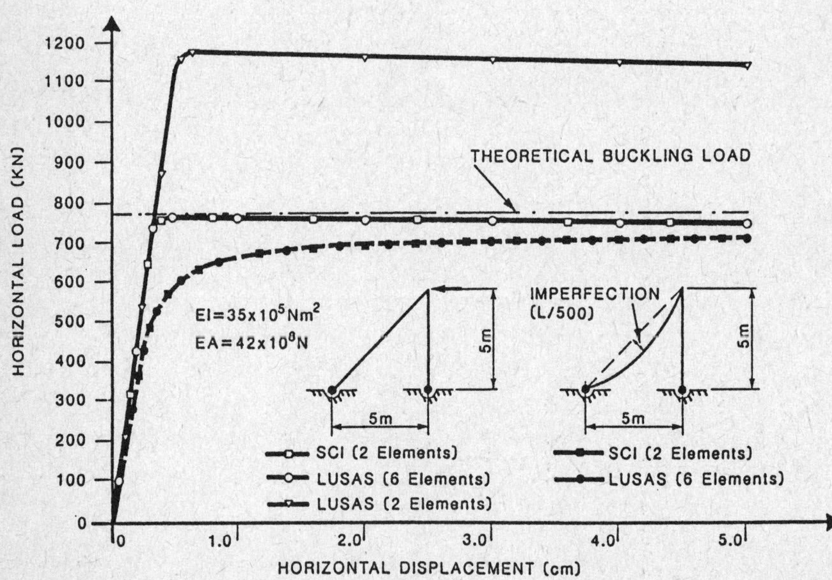

Figure 10. Analysis of Perfect and Imperfect Simple Truss.

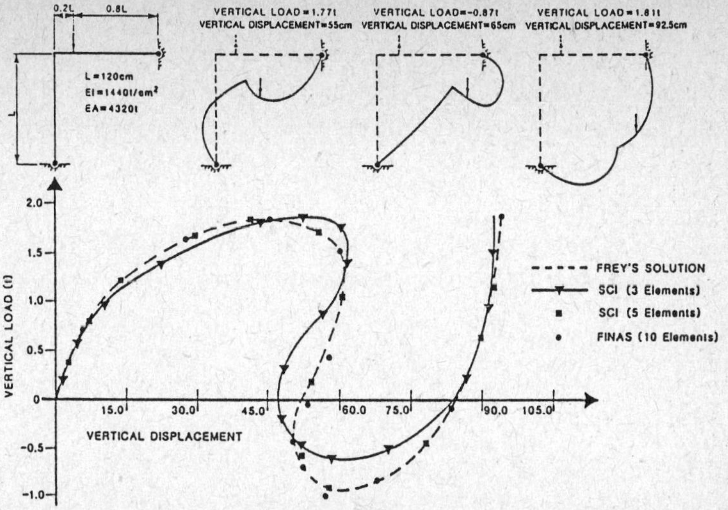

Figure 11. Solutions for Lee's Frame

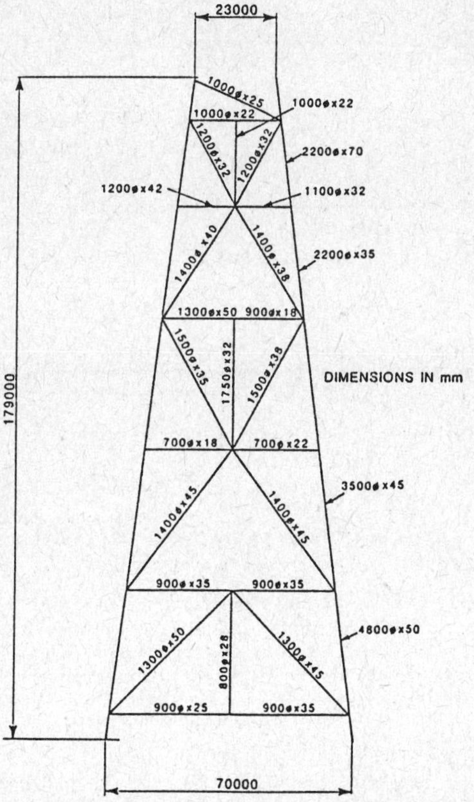

Figure 12. Plane Frame from Offshore Platform

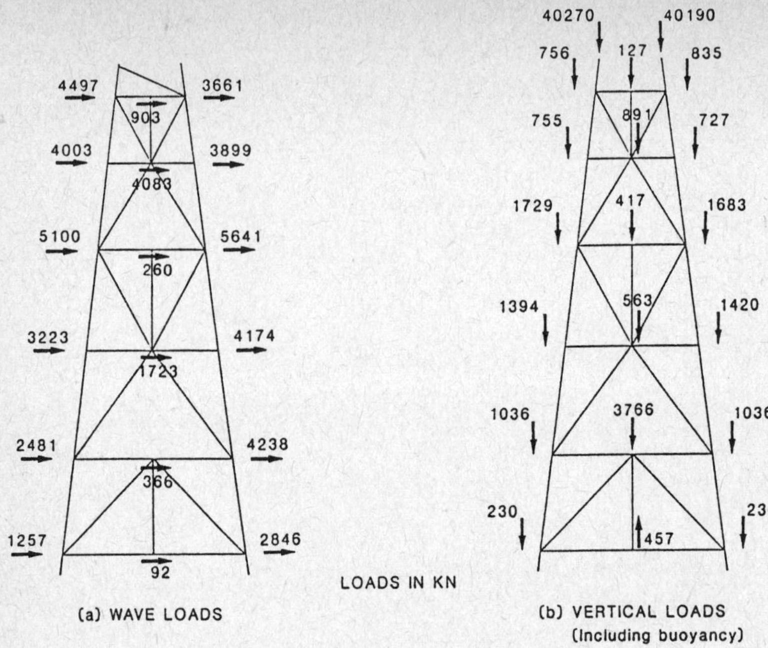

Figure 13. Loads Applied to Frame

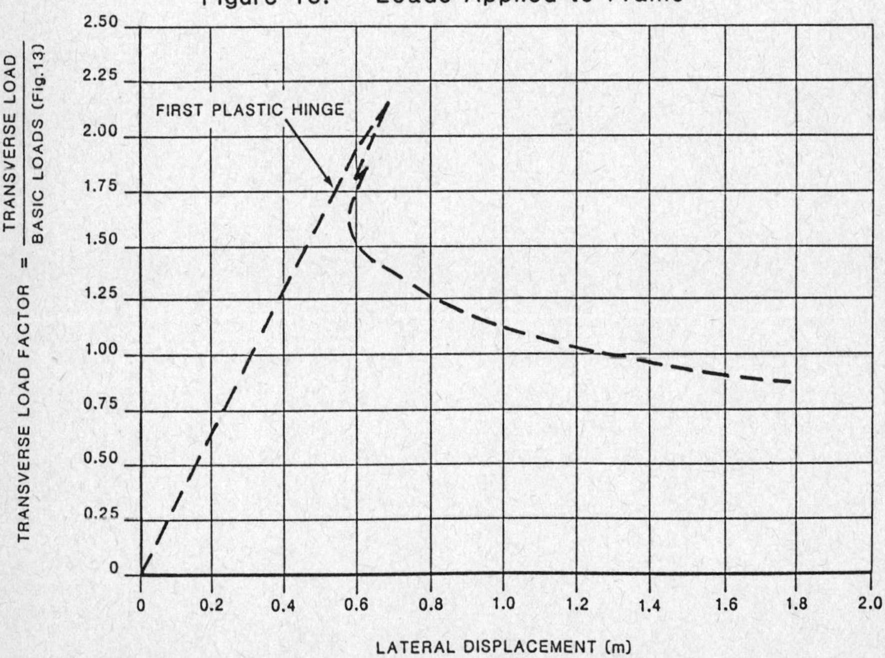

Figure 14. Lateral Load-Deflection Plot for Frame

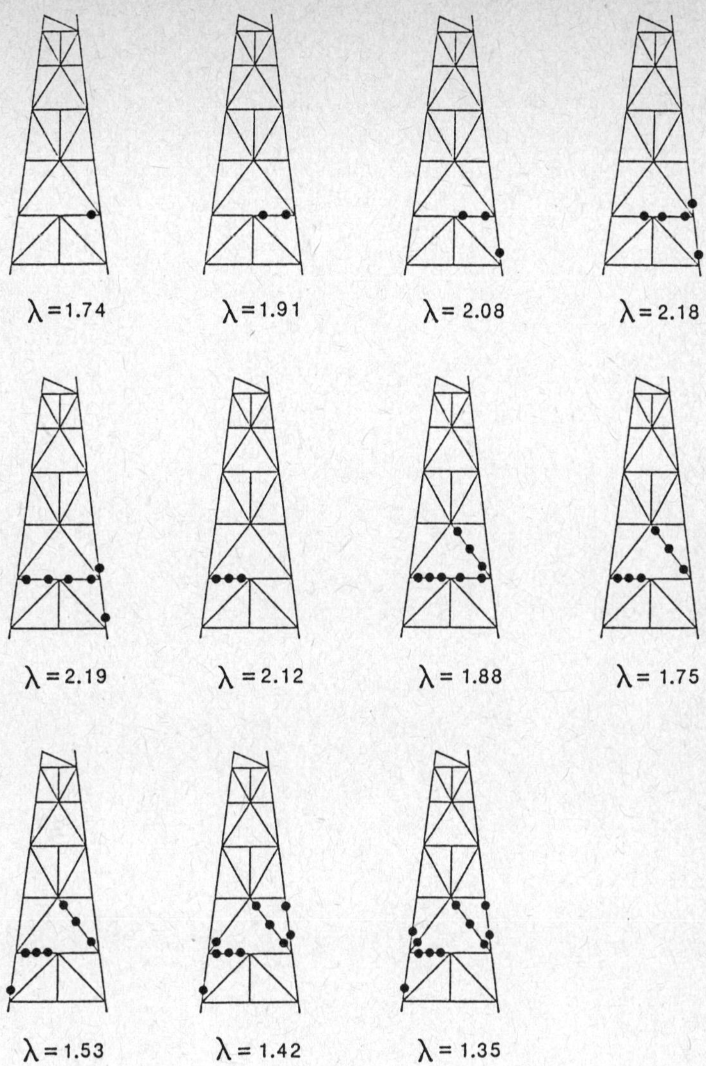

Figure 15. Development of Plastic Hinges at Various Transverse Load Factors (λ)

PROBABILISTIC FATIGUE OF OFFSHORE STRUCTURES

G. Sigurdsson
University of Aalborg
Sohngaardsholmsvej 57, DK-9000 Aalborg, Denmark

1. INTRODUCTION

Offshore structures of all types are generally subjected to cyclic loading from wind, current earthquakes and waves acting simultaneously, which cause time-varying stresses in the structure. The environmental quantities are of a random nature and are more or less correlated to each other through the generating and driving mechanism. Waves and earthquakes are generally considered to be the most important sources of the structural excitations. However, earthquake loads are only taken into account in the analysis of offshore structures close to or in tectonic offshore fields. For fixed offshore structures in deep water environments wind loads represent a contribution of about 5 % to the environmental loading [1]. Current loads are mostly considered to be unimportant in the dynamic analysis of offshore structures, because their frequencies are not sufficient to excite the structures. The reliability calculation of offshore structures is a difficult task due to the random nature of the loading, and also due to insufficient information of structural failure under these conditions. A stochastic assessment of the reliability analysis of structures is therefore inevitable. Dynamic loads, such as wave loads, produce stress fluctuations in the structural members and joints and are the primary cause of fatigue damages. A fatigue analysis of offshore structures can be described in general terms as a calculation procedure, starting from the waves and ending with a fatigue damage occurring in the material or in the joints. The links between the waves and the damage are formed by mathematical models for the wave forces, the structural behaviour and the material behaviour. I view of the stochastic and dynamic character of the waves it is an obvious choice to apply spectral fatigue analysis methods to the fatigue problem. In this paper, a stochastic reliability assessment for jacket type offshore structures subjected to wave loads in deep water environments is outlined. In the reliability assessment, structural and loading uncertainties are taken into account by means of some stochastic variables. To estimate statistical measures of structural stress variations the modal spectral analysis method is applied.

The analysis is divided into four steps:

I Description of the sea state

a) short-term model

b) long-term model

II Description of the wave loading

III Structural analysis

IV Fatigue analysis

2. DESCRIPTION OF THE SEA STATE

Short-term model:
The observed sea elevation at the fixed location $\overline{r}$ at the time t, $\eta(\overline{r},t)$, can be considered as a realization of non-stationary stochastic process, whose characteristic parameters vary slowly with time. Further, we assumed that for short-term periods (a few hours) the sea surface $\eta(\overline{r},t)$ can be considered as a realization of a stationary stochastic process. This process is assumed to be a zero-mean ergodic Gaussian process. The cross-spectral density of the sea surface at the points n and m (with spatial coordinates (x_n, y_n) and (x_m, x_m), respectively) can be written as:

$$S_{\eta_n\eta_m}(\omega) = S_{\eta\eta}(\omega) \int_\theta \psi(\theta)\; exp(-\epsilon\kappa(\omega)(\Delta x cos\theta + \Delta y sin\theta))d\theta \tag{1}$$

where $\epsilon = \sqrt{-1}$, $\Delta x = (x_n - x_m)$, $\Delta y = (y_n - y_m)$ and κ is the wave number, defined as:

$$\omega^2 = \kappa\; g\; tanh(\kappa d) \quad ; \quad \omega \geq 0.0 \quad , \quad \kappa \geq 0.0 \tag{2}$$

Serveral analytical expressions have been suggested for the spreading function $\psi(\theta)$. Generally, a cosine function is used:

$$\psi(\theta) = \begin{cases} K\; cos^{2n}(\theta - \overline{\theta}) & -\frac{\pi}{2} \leq (\theta - \overline{\theta}) \leq \frac{\pi}{2} \\ 0 & \text{elsewhere} \end{cases} \tag{3}$$

Where $\overline{\theta}$ denotes the average direction of wave propagation and K is a normalization factor defined so that the spreading function between $-\frac{\pi}{2}$ and $\frac{\pi}{2}$ is equal to one:

$$K = \frac{1}{\sqrt{\pi}} \frac{\Gamma(n+1)}{\Gamma(n+\frac{1}{2})} \tag{4}$$

where Γ is the Gamma function and n is a parameter defining the width of the distribution. For the limiting case $n = 0$ eq.(3) approaches the Dirac delta function corresponding to long-crested waves.

Here, the JONSWAP spectrum is adopted as a reasonable model of the sea surface, $S_{\eta\eta}$. This spectrum reads [6]:

$$S_{\eta\eta}(\omega) = \alpha\, g^2 \omega^{-5} exp(-\frac{5}{4}(\frac{\omega}{\omega_p})^{-4})\gamma^{exp\left(-\frac{1}{2}((\frac{\omega}{\omega_p}-1)/\sigma)^2\right)} \tag{5}$$

where

- ω is the frequency (rad/sec)
- α is the equlibrium range parameter
- g is the acceleration of gravity
- ω_p is the spectral peak ($= 2\pi/T_p$)
- γ is the spectral peak parameter
- σ is the spectral peak width parameter (here taken as 0.08)

Long-term model:
As mentioned earlier, it is assumed that the sea surface elevation at a fixed location for stort-term periods can be accurately modelled by a zero-mean ergodic Gaussian process. This process is completely characterized by the frequency spectrum $S_{\eta\eta}(\omega)$ which, for a given average direction of wave propagation $\overline{\theta}$, can be described by two parameters, namely by the significant wave height

H_s and the spectral peak periods T_p. The long-term probability distribution of the sea state is then given as a joint distribution of $\overline{\theta}, H_s$ and T_p , $p_{H_s T_p \overline{\theta}}(h,t,\theta)$. It is not possible to establish this distribution theoretically. The distribution has to be estimated from wave observations in the ocean area concerned or derived applying hidcasting models, i.e. the chosen analytical model has to be fitted in the best possible way to the data.

In most wave observations up till now no information of the mean direction of propagagion $\overline{\theta}$ has been included. If it is assumed that the joint distribution of H_s and T_p is independent of $\overline{\theta}$, the joint distribution $p_{H_s T_p \overline{\theta}}(h,t,\theta)$ can be written as:

$$p_{H_s T_p \overline{\theta}}(h,t,\theta) = p_{H_s T_p}(h,t)\, p_{\overline{\theta}}(\theta) \tag{6}$$

For our purpose the probability density function $p_{H_s T_p}(h,t)$ is conveniently written as:

$$p_{H_s T_p}(h,t) = p_{T_p|H_s}(t|h)\, p_{H_s}(h) \tag{7}$$

where $p_{H_s}(h)$ is the marginal probability density function for H_s and $p_{T_p|H_s}(t|h)$ is the conditional probability density function for T_p given H_s. $p_{H_s}(h)$ and $p_{T_p|H_s}(t|h)$ are fitted to the observations separately. The numerical values for $p_{H_s T_p}(h,t)$ are obtained by means of eq.(7). Here $p_{H_s}(h)$ is modelled by log-normal distribution for $H_s \leq v$ and by Weibull distribution for $H_s < v$, i.e.

$$p_{H_s} = \begin{cases} \frac{1}{\sqrt{2\pi}\sigma_{H_s} h}\, exp(\frac{-(ln\, h - \mu_{H_s})^2}{2\sigma^2_{H_s}}) & h \leq v \\ \frac{\xi}{\rho}(\frac{h}{\rho})^{\xi-1}\, exp(-\frac{h}{\rho})^{\xi} & \rho > 0\,, \xi > 0\,, h > v \end{cases} \tag{8}$$

where μ_{H_s} and $\sigma^2_{H_s}$ are the mean and variance of the variable $ln\,(H_s)$, respectively, and where continuity is required for $p_{H_s}(h)$ and $P_{H_s}(h)$ at $h{=}v$. The conditional distribution of T_p given H_s is approximated by the log-normal distribution, i.e.

$$p_{T_p|H_s}(t|h) = \frac{1}{\sqrt{2\pi}\sigma_{T_p} t}\, exp(\frac{-(ln\, t - \mu_{T_p})^2}{2\sigma^2_{T_p}}) \tag{9}$$

where μ_{T_p} and $\sigma^2_{T_p}$ are the mean and variance of the variable $ln\,(T_p)$, respectively.

The marginal probability density function $p_{\overline{\theta}}(\theta)$ will divide the circle into a certain number of sectors and associate each sector with a point probability $p_{\overline{\theta}}(\theta_k), k = 1,...,n$ where θ_k is the midpoint of sector no. k and n is the number of sectors.

3. DESCRIPTION OF THE WAVE LOADING

In chapter 2 the statistical nature of the waves was dealt with. Now the consequential loading on a structural element is considered. It is well known that the force on a vertically placed circular cylinder subjected to wave action consists of a drag as well as an inertia component. It is assumed that Morison's equation can be applied to a cyndrical member oriented in a random manner. The total wave force per unit length of cylinder of the diameter D at the position $\overline{r}_o = (x_o, y_o, z_o)$ at the time t is:

$$\overline{f}_n(\overline{r}_o,t) = \begin{bmatrix} f_{nx}(\overline{r}_o,t) \\ f_{ny}(\overline{r}_o,t) \\ f_{nz}(\overline{r}_o,t) \end{bmatrix} = \overline{f}_{nD} + \overline{f}_{nI} \tag{10}$$

where

$\overline{f}_{nD} \quad = K_D \, |\overline{\dot{u}}_n(\overline{r}_0,t)| \, \overline{\dot{u}}_n(\overline{r}_0,t)$

$\overline{f}_{nI} \quad = K_I \, \overline{\ddot{u}}_n(\overline{r}_0,t)$

$K_D \quad = 1/2\, C_D \, \rho D$

$K_I \quad = 1/4\, C_M \, \pi \, \rho \, D^2$

$\overline{\dot{u}}(\overline{r}_0,t)$ is the horizontal water particle velocity at the position $\overline{r}_0$ at the time t.

$\overline{\ddot{u}}(\overline{r}_0,t)$ is the horizontal water particle acceleration at the position $\overline{r}_0$ at the time t.

x,y,z is the global coordinate system.

C_D is the drag coefficient.

C_M is the inertia coefficient.

ρ is the density of water.

and where the subscript n refers to the normal direction of the cylinder.
The non-linear drag term $\overline{f}_{nD}$ in eq.(10) makes the computations for correlations and spectral densities extremely difficult and intractable, and therefore, a recourse to linearization of the drag term in eq.(10) is made. The "minimum square error linearization method" [9] is used here. The linearized version of the drag term $\overline{f}_{nD}$ becomes :

$$\overline{f}_{DL} = K_D \, \overline{\overline{L}} \, \overline{\dot{u}}_n(\overline{r}_0,t) = K_D \begin{bmatrix} l_{11} & l_{12} & l_{13} \\ l_{21} & l_{22} & l_{23} \\ l_{31} & l_{32} & l_{33} \end{bmatrix} \begin{bmatrix} \dot{u}_{nx} \\ \dot{u}_{ny} \\ \dot{u}_{nz} \end{bmatrix} \tag{11}$$

where $\overline{\overline{L}}$ is the linearization coefficient matrix which is given in appendix A.
Eq.(11) can now be written as:

$$\overline{f}_n(\overline{r}_0,t) = \begin{bmatrix} f_{nx}(\overline{r}_0,t) \\ f_{ny}(\overline{r}_0,t) \\ f_{nz}(\overline{r}_0,t) \end{bmatrix} = K_D \, \overline{\overline{L}} \, \overline{\dot{u}}_n(\overline{r}_0,t) + K_I \overline{\ddot{u}}_n(\overline{r}_0,t) \tag{12}$$

The normal vectors $\overline{\dot{u}}_n$ and $\overline{\ddot{u}}_n$ in eq.(12) can be expressed in terms of a unit vector $\overline{c} = (c_x, c_y, c_z)$ along the cylinder axis as follows (see Appendix A):

$$\overline{\dot{u}}_n = \overline{c} \times (\overline{\dot{u}^T} \times \overline{c}) = \overline{\overline{C}} \, \overline{\dot{u}}$$

$$\overline{\ddot{u}}_n = \overline{c} \times (\overline{\ddot{u}^T} \times \overline{c}) = \overline{\overline{C}} \, \overline{\ddot{u}} \tag{13}$$

where

$$\overline{\overline{C}} = \begin{bmatrix} (1-c_x^2) & -c_x c_y & -c_x c_z \\ & (1-c_y^2) & -c_y c_z \\ \text{sym.} & & (1-c_z^2) \end{bmatrix} = \left[\{\overline{C}\}_x \; \{\overline{C}\}_y \; \{\overline{C}\}_z \right]$$

$$\overline{\dot{u}} = \begin{bmatrix} \dot{u}_x \\ \dot{u}_y \\ \dot{u}_z \end{bmatrix} \quad ; \quad \overline{\ddot{u}} = \begin{bmatrix} \ddot{u}_x \\ \ddot{u}_y \\ \ddot{u}_z \end{bmatrix}$$

Now eq.(12) can be rewritten as:

$$\overline{f}_n(\overline{r}_0,t) = \begin{bmatrix} f_{nx}(\overline{r}_0,t) \\ f_{ny}(\overline{r}_0,t) \\ f_{nz}(\overline{r}_0,t) \end{bmatrix} = K_D \, \overline{\overline{L}} \, \overline{\overline{C}} \, \overline{\dot{u}}(\overline{r}_0,t) + K_I \, \overline{\overline{C}} \, \overline{\ddot{u}}(\overline{r}_0,t) \tag{14}$$

Two points in the wave field are considered, i.e. point l with the coordinates $\overline{r}_l = (x_l, y_l, z_l)$ and point m with the coordinates $\overline{r}_m = (x_m, y_m, z_m)$. l denotes a circular cylindrical element L with the diameter D_L and the unit vector $\overline{c}_L = (c_{x_L}, c_{y_L}, c_{z_L})$ along the cylinder axis and point m denotes a circular cylindrical element M with the diameter D_M and the unit vector $\overline{c}_M = (c_{x_M}, c_{y_M}, c_{z_M})$ along the cylinder axis. The cross-covariance function for the various combinations of wave force components at l and m can now be expressed as a function of the covariance functions of $\dot{\overline{u}}$ and $\ddot{\overline{u}}$. The cross-spectral densities between various components can be found by deriving the Fourier transforms of the corresponding cross-covarince functions. The cross-spectral density between the forces f_{nil} and f_{nim} becomes (subscripts nil and njm $(i, j = x, y, z)$ denote the force perpendicular to the elements in the directions i and j at the points l and m):

$$
\begin{aligned}
S_{f_{nil} f_{njm}}(\omega) &= K_{D_L}\, K_{D_M} [(\{\overline{B}_l\}_i\ \{\overline{B}_m\}_j)]\, [S_{\dot{u}_l \dot{u}_m}] \\
&+ K_{D_L}\, K_{I_M} [(\{\overline{B}_l\}_i\ \{\overline{C}_M\}_j)]\, [S_{\dot{u}_l \ddot{u}_m}] \\
&+ K_{I_L}\, K_{D_M} [(\{\overline{C}_L\}_i\ \{\overline{B}_m\}_j)]\, [S_{\ddot{u}_l \dot{u}_m}] \\
&+ K_{I_L}\, K_{I_M} [(\{\overline{C}_L\}_i\ \{\overline{C}_M\}_j)]\, [S_{\ddot{u}_l \ddot{u}_m}]
\end{aligned}
\tag{15}
$$

where

$$
\begin{aligned}
[\{\overline{B}_l\}_x\ \{\overline{B}_l\}_y\ \{\overline{B}_l\}_z] &= \overline{\overline{L}}_l\, \overline{\overline{C}}_L \\
[\{\overline{B}_m\}_x\ \{\overline{B}_m\}_y\ \{\overline{B}_m\}_z] &= \overline{\overline{L}}_m\, \overline{\overline{C}}_M
\end{aligned}
$$

$$
\overline{\overline{C}}_L = [\{\overline{C}_L\}_x\ \{\overline{C}_L\}_y\ \{\overline{C}_L\}_z] = \begin{bmatrix} (1 - c_{x_L}^2) & -c_{x_L} c_{y_L} & -c_{x_L} c_{z_L} \\ & (1 - c_{y_L}^2) & -c_{y_L} c_{z_L} \\ \text{sym.} & & (1 - c_{z_L}^2) \end{bmatrix}
$$

$$
\overline{\overline{C}}_M = [\{\overline{C}_M\}_x\ \{\overline{C}_M\}_y\ \{\overline{C}_M\}_z] = \begin{bmatrix} (1 - c_{x_M}^2) & -c_{x_M} c_{y_M} & -c_{x_M} c_{z_M} \\ & (1 - c_{y_M}^2) & -c_{y_M} c_{z_M} \\ \text{sym.} & & (1 - c_{z_M}^2) \end{bmatrix}
$$

$\overline{\overline{L}}_l$ and $\overline{\overline{L}}_m$ are the linearization coefficient matrices for point l and point m, respectively.

$$
[S_{\dot{u}_l \dot{u}_m}] = \begin{bmatrix} S_{\dot{u}_{x_l} \dot{u}_{xm}}(\omega) & S_{\dot{u}_{x_l} \dot{u}_{ym}}(\omega) & S_{\dot{u}_{x_l} \dot{u}_{zm}}(\omega) \\ S_{\dot{u}_{y_l} \dot{u}_{xm}}(\omega) & S_{\dot{u}_{y_l} \dot{u}_{ym}}(\omega) & S_{\dot{u}_{y_l} \dot{u}_{zm}}(\omega) \\ S_{\dot{u}_{z_l} \dot{u}_{xm}}(\omega) & S_{\dot{u}_{z_l} \dot{u}_{ym}}(\omega) & S_{\dot{u}_{z_l} \dot{u}_{zm}}(\omega) \end{bmatrix}
$$

$$
[S_{\dot{u}_l \ddot{u}_m}] = \begin{bmatrix} S_{\dot{u}_{x_l} \ddot{u}_{xm}}(\omega) & S_{\dot{u}_{x_l} \ddot{u}_{ym}}(\omega) & S_{\dot{u}_{x_l} \ddot{u}_{zm}}(\omega) \\ S_{\dot{u}_{y_l} \ddot{u}_{xm}}(\omega) & S_{\dot{u}_{y_l} \ddot{u}_{ym}}(\omega) & S_{\dot{u}_{y_l} \ddot{u}_{zm}}(\omega) \\ S_{\dot{u}_{z_l} \ddot{u}_{xm}}(\omega) & S_{\dot{u}_{z_l} \ddot{u}_{ym}}(\omega) & S_{\dot{u}_{z_l} \ddot{u}_{zm}}(\omega) \end{bmatrix}
$$

$$
[S_{\ddot{u}_l \dot{u}_m}] = \begin{bmatrix} S_{\ddot{u}_{x_l} \dot{u}_{xm}}(\omega) & S_{\ddot{u}_{x_l} \dot{u}_{ym}}(\omega) & S_{\ddot{u}_{x_l} \dot{u}_{zm}}(\omega) \\ S_{\ddot{u}_{y_l} \dot{u}_{xm}}(\omega) & S_{\ddot{u}_{y_l} \dot{u}_{ym}}(\omega) & S_{\ddot{u}_{y_l} \dot{u}_{zm}}(\omega) \\ S_{\ddot{u}_{z_l} \dot{u}_{xm}}(\omega) & S_{\ddot{u}_{z_l} \dot{u}_{ym}}(\omega) & S_{\ddot{u}_{z_l} \dot{u}_{zm}}(\omega) \end{bmatrix}
$$

$$
[S_{\ddot{u}_l \ddot{u}_m}] = \begin{bmatrix} S_{\ddot{u}_{x_l} \ddot{u}_{xm}}(\omega) & S_{\ddot{u}_{x_l} \ddot{u}_{ym}}(\omega) & S_{\ddot{u}_{x_l} \ddot{u}_{zm}}(\omega) \\ S_{\ddot{u}_{y_l} \ddot{u}_{xm}}(\omega) & S_{\ddot{u}_{y_l} \ddot{u}_{ym}}(\omega) & S_{\ddot{u}_{y_l} \ddot{u}_{zm}}(\omega) \\ S_{\ddot{u}_{z_l} \ddot{u}_{xm}}(\omega) & S_{\ddot{u}_{z_l} \ddot{u}_{ym}}(\omega) & S_{\ddot{u}_{z_l} \ddot{u}_{zm}}(\omega) \end{bmatrix}
$$

The brackets [...][...] in eq.(15), and later in eq.(18), are not matrix multiplications in the conventional sense, they are used here only to denote a row-to-column multiplication. After one row-to-column multiplication a sum is made and added to the sum of the second row-to-column multiplication and so on, so that the final result is only a single term. The cross-spectral densities of the water particle velocity $[S_{\dot{u}_l \dot{u}_m}]$ may be expressed in terms of the one-dimensional wave spectral density $S_{\eta\eta}(\omega)$ by using eq.(1) as, [2]:

$$[S_{\dot{u}_l \dot{u}_m}] = S_{\eta\eta}(\omega) \int_\theta \omega^2 \overline{\overline{\Lambda}}(\omega, z_l, z_m)\psi(\theta) exp(-\epsilon\kappa(\omega)(\Delta x\, cos\theta + \Delta y\, sin\theta))d\theta \qquad (16)$$

where

$$\overline{\overline{\Lambda}}(\omega, z_l, z_m) = \overline{A}(\overline{\kappa}(\omega), z_l)\overline{A}^{*T}(\overline{\kappa}(\omega), z_m)$$

$$\overline{A}(\overline{\kappa}(\omega), z) = \frac{1}{sinh(\kappa(\omega)d)} \begin{bmatrix} cos\theta\, cosh(\kappa(\omega)z) \\ sin\theta\, cosh(\kappa(\omega)z) \\ \epsilon\, sinh(\kappa(\omega)z) \end{bmatrix}$$

*	denotes complex conjugated
ϵ	$= \sqrt{-1}$
d	is the water depth
Δx	$= x_l - x_m$
Δy	$= y_l - y_m$
$\kappa(\omega)$	is the wave number, defined in eq.(7)
$\psi(\theta)$	is the spreading function, defined in eq.(19)

The z-coordinates are measured from the bottom positive upwards.

The cross-spectral densities of water particle accelerations $[S_{\ddot{u}_l \ddot{u}_m}]$, acceleration and velocity $[S_{\ddot{u}_l \dot{u}_m}]$ and velocity and acceleration $[S_{\dot{u}_l \ddot{u}_m}]$ can be obtained using the properties of the derived processes. This gives:

$$[S_{\ddot{u}_l \ddot{u}_m}] = \omega^2 [S_{\dot{u}_l \dot{u}_m}]$$

$$[S_{\dot{u}_l \ddot{u}_m}] = -[S_{\ddot{u}_l \dot{u}_m}] = \epsilon\omega\, [S_{\dot{u}_l \dot{u}_m}] \qquad (17)$$

Applying eq.(17), eq.(15) can be rewritten as:

$$S_{f_{nil} f_{njm}}(\omega) = [K_{D_L}\ K_{D_M}(\{\overline{B}_l\}_i \{\overline{B}_m\}_j) + \omega^2 K_{I_L}\ K_{I_M}(\{\overline{C}_L\}_i \{\overline{C}_M\}_j)$$

$$+ \epsilon\omega(K_{D_L} K_{I_M}(\{\overline{B}_l\}_i\ \{\overline{C}_M\}_j) - K_{I_L} K_{D_M}(\{\overline{C}_L\}_i\ \{\overline{B}_m\}_j))]\ [S_{\dot{u}_l \dot{u}_m}] \qquad (18)$$

4. STRUCTURAL ANALYSIS

It is assumed that the structure can be modelled as a space frame of three-dimensional beam elements connected by nodal points, where each structural member in the structure has one or more elements. If the structural system is modelled by a linear system and by a finite number of degrees of freedom then the dynamic equations may written as:

$$\overline{\overline{M}}\, \ddot{\overline{x}} + \overline{\overline{C}}\, \dot{\overline{x}} + \overline{\overline{K}}\, \overline{x} = \overline{F} \qquad (19)$$

where

$\overline{x}$ is the displacement vector

$\overline{\overline{M}}$ is the mass matrix

$\overline{\overline{C}}$ is the damping matrix

$\overline{\overline{K}}$ is the stiffness matrix

$\overline{F}$ is the load vector which varies with time

The matrix equation (19) represents a finite number of coupled differential equations. In this paper it is chosen to use a "modal analysis" to transform the coupled system into an uncoupled system. The uncoupled system becomes:

$$\ddot{q}_j + 2\zeta_j\omega_j\dot{q}_j + \omega_j^2 q_j = f_j \tag{20}$$

where q_j is the modal coordinates , ζ_j is the damping ratio and ω_j is the jth natural frequency. The stresses $\overline{s}$ at internal points in the structure may be found as:

$$\overline{s} = \overline{\overline{T}}\,\overline{q} \tag{21}$$

where the components T_{ij} in the matrix $\overline{\overline{T}}$ indicate the stress at point i due to displacement in mode j, and $\overline{q}$ is the solution of eq.(20). The cross-spectral density of the stresses at points k and l may be written as (see eq.(21)):

$$S_{s_k s_l}(\omega) = \sum_{i=1}^{n}\sum_{j=1}^{n} T_{ki}T_{lj}S_{q_i q_j}(\omega) \tag{22a}$$

where

$$S_{q_i q_j}(\omega) = H^*_{q_i f_i}(\omega)H_{q_j f_j}(\omega)S_{f_i f_j}(\omega) \tag{22b}$$

$$H_{q_i f_i}(\omega) = \frac{1}{\omega_i^2 - \omega^2 + 2\epsilon\zeta_i\omega\omega_i} \tag{22c}$$

$$S_{f_i f_j}(\omega) = \sum_{r=1}^{m}\sum_{s=1}^{m} \phi_{ir}\phi_{js}S_{F_r F_s}(\omega) \tag{22d}$$

and where n is the number of mode shapes, $*$ denotes the complex conjugate, $\epsilon = \sqrt{-1}$, ϕ_{ir} is the (i,r) element in the mode shape matrix and m is the number of degrees of freedom.
The cross-spectral density of the load at the points r and s, $S_{F_r F_s}$, may be found as shown in chapter 3 (see eq.(18)).
For the fatigue analysis in chapter 5 the cross-spectral density of the stresses as a whole is not interesting. However, three characteristics, namely the area m_0, the second moment m_2 and the fourth moment m_4 of the auto-spectral density (i.e. $k = l$ in eq.(22a)) are of interest. The area of the auto-spectral density can be derived as:

$$m_0(s_k) = \sigma^2(s_k) = \int_0^\infty S_{s_k s_k}(\omega)d\omega \tag{23}$$

and the second and the fourth moment of the auto-spectral density as:

$$m_2(s_k) = \int_0^\infty \omega^2 S_{s_k s_k}(\omega) d\omega \tag{24}$$

$$m_4(s_k) = \int_0^\infty \omega^4 S_{s_k s_k}(\omega) d\omega \tag{25}$$

5. FATIGUE ANALYSIS

In chapter 4 it was shown how the spectral densities of the stresses in a given hot spot in the structure can be estimated. Here fatigue damage is defined as a result of cumulative damage because of stress fluctuations (the stress amplitude).

For a short-term period the sea state is assumed to be a zero-mean ergodic Gaussian process, see chapter 2. By using the linearized version of the Morison equation (see chapter 3), and by modelling the structural system linearly (see chapter 4) the stress response S becomes a zero-mean ergodic Gaussian process too.
A number of cycles counting algoritms have been proposed, see i.e. [20]. Two of the counting methods, namely the range count method (RC) and the rainflow count method (RFC) are generally recognized as the method which produces the best results, and they will be included here. Both methods give the same result for an ideal narrow-band stress history, but for wide-banded stress history the result can be very different.

In section 5.1 it will be shown how the probability of fatigue failure can be estimated by using Miner's rule and the so-called S-N approach.

5.1 Reliability Analysis

The relationship between the stress fluctuation and the damage can be found by using Miner's rule, witch states in essence that every stress cycle i results in the degree of damage D_i equal to

$$D_i = \frac{1}{N_i} \tag{26}$$

where N_i is the number of cycles to failure, if the same stress cycle is repeated over and over again. The most commonly used model to determine N_i is the so-called $S-N$ approach [14]:

$$N_i = \left(\frac{2S_i}{K}\right)^{-m} \tag{27}$$

where S_i is the stress amplitude and K and m are constants which can be determined by constant-amplitude test. To allow for the scatter in the number of cycles to fatigue and to allow for a model uncertainty by using Miner's rule, K and m are modelled as random variables. To allow for the uncertainty in the estimation of the stress amplitude a new random variable B will be introduced into eq.(27) as:

$$N_i = \left(\frac{2S_iB}{K}\right)^{-m} \tag{28}$$

Under constant amplitude loading failure occurs by definition when the total degree of damage $D_{tot} = \sum D_i$ attains the value D_{fail} equal to 1. However, with variable-amplitude random loading the influences due to the load history may cause failure at the value D_{fail} different from 1. To take into account the uncertainty of the failure defintion, D_{fail} will be modelled as a random variable.

In fatigue analysis of jackets, the analysis will primarily focus on the welded joint between the members. When considering a fatigue failure in tubular joints the geometry of the whole nodal point becomes very important, since stress concentrations will occur due to the non-uniform stiffness of the chord wall and the brace. The locations, or points at which the highest stress occurs, are called hot spots. In welded joints two different hot spots for each brace in the joint are found, one at the weld toe on the brace side, the other on the chord side, i.e. for K-joints there are four hot spots. The stress concentration factor (SCF) is defined as the ratio of the hot spot stress σ_{max} to the nominal stress σ_N in the brace, i.e.

$$SCF = \frac{\sigma_{max}}{\sigma_N} \tag{29}$$

The SCF's for a given joint geometry and loads can be estimated either by full-scale tests or by a FEM-analysis. Here the SCF's are estimated by using some empirical formulas suggested by Kuang, [12], [13], which are based on thin shell FEM-analysis of different joint-geometry and loads. Two modes of fatigue failure, called failure elements, are defined to occur for each brace in a tubular joint, cracking at the hot spot toe of the weld jointing the brace to the chord (brace fatigue) and cracking at the hot spot in the wall of the chord itself (punching shear fatigue). The locations in the chord/brace intersection, where the hot spot stresses occur, depend on the external loads. In [12] it is recommended to check the 8 points along the brace/chord intersection to locate the hot spots, see figure 1.

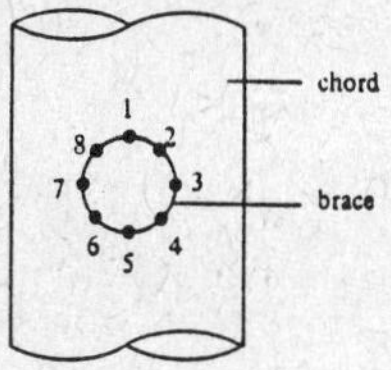

Figure 1. Points in the brace/chord intersection where the stress concentrations are checked.

For a given sea state, the mean fatigue damage of a failure element under consideration per stress amplitude, $\overline{D}_i$ can be determined by substituting eq.(26) and eq.(28) as:

$$\overline{D}_i = \int_0^{\infty} D_i(s)\ p_S(s) ds \tag{30}$$

where $p_S(s)$ is the distribution function of the stress amplitudes, see section 5.1.

The total damage in the failure element, D_{tot} is obtained by summing up the mean damage $\overline{D}_i$ over the service life of the structure, taking account of the long-term distribution of the sea states (see chapter 2):

$$D_{tot} = \int_{\overline{\theta}} \int_{H_s} \int_{T_p} \frac{T_L}{T_{mp}(t,h,\theta)}\ \overline{D}_i(t,h,\theta)\ p_{T_p|H_s}(t|h)\ p_{H_s}(h)\ p_{\overline{\theta}}(\theta)\ dt\, dh\, d\theta \tag{31}$$

where $p_{H_s}(h)$ is the marginal probability density function of the significant wave height H_s, $p_{T_p|H_s}(t|h)$ is the conditional probability density function of the wave spectral peak periods T_p, given H_s, $p_{\overline{\theta}}(\theta)$ is the probability density function of the mean direction of the wave propagation, $\overline{\theta}$, T_L is the total service life and T_{mp} is mean period of a stress cycle within the sea state.
The fatigue failure mode for the failure element is described by a safety margin M, defined as:

$$M = D_{fail}\ -\ D_{tot} \tag{32}$$

and the probability of failure P_f is

$$P_f = P(M \leq 0) \tag{33}$$

For a narrow-banded stress process $p_S(s)$ becomes Rayleigh distributed. $\overline{D}_i$ can be written as:

$$\overline{D}_i = \int_0^\infty B^m \left(\frac{2s}{K}\right)^m \frac{s}{\sigma_S^2}\, exp\left(-\frac{s^2}{2\sigma_S^2}\right) ds$$

$$= B^m \frac{\sigma_S^m}{K^m} \left(2\sqrt{2}\right)^m \Gamma(1+\frac{m}{2}) \tag{34}$$

where $\Gamma(\cdots)$ is the gamma function.

And the total degree of damage D_{tot} can be written as:

$$D_{tot} = \int_{\overline{\theta}} \int_{H_s} \int_{T_p} \frac{T_L B^m}{T_{mp}(t,h,\theta)} \frac{\sigma_S^m(t,h,\theta)}{K^m} \left(2\sqrt{2}\right)^m \Gamma(1+\frac{m}{2})$$

$$p_{T_p|H_s}(t|h)\; p_{H_s}(h)\; p_{\overline{\theta}}\; dt\, dh\, d\theta \tag{35}$$

where T_{mp} can be estimated as:

$$T_{mp} = 2\,\pi\,\sqrt{\frac{m_0}{m_2}} \tag{36}$$

where m_0 and m_2 are the area and the second moment of the auto-spectral density of the stress spectra in the failure element.

In general, the safety margin M, as in eq.(32), is a function of the number of correlated non-normally distributed random variables $\overline{X} = (X_1,\ldots,X_n)$ called basic variables (where $\overline{X} = (D_{fail}, B, m, K)$), i.e. $M = f(\overline{X})$. $f(\overline{x})$ is called the failure function defined in such a way that it divides the n-dimensional basic variables space ω into two regions, namely a safe region ω_s, where $f(\overline{x}) > 0$, and an unsafe region ω_f, where $f(\overline{x}) \leq 0$. By a suitable transformation the correlated and non-normally distributed variables $\overline{X}$ are transformed into uncorrelated and standardized normally distributed variables $\overline{Z}$. By this transformation the failure surface is given by $f(\overline{z}) = 0$ in the corresponding z-space. In the n-dimensional z-space the reliability index β is defined as the shortest distance from the origin to the failure surface, i.e.

$$\beta = \min_{\overline{z}\in\partial\omega} \left(\sum_{i=1}^{n} z_i^2\right)^{\frac{1}{2}} \tag{37}$$

It can be shown that the probability of failure P_f in eq.(33) can be determined with good approximation from

$$P_f \approx \Phi(-\beta) \tag{38}$$

where $\Phi(\cdot)$ is the standard normal distribution function.

Until now a probability of failure for one failure element has been considered. When the probability of fatigue failure of the whole structure is considered the probabilities of failure for every failure element must be evaluated. If the largest of these probabilities of failure is used as a measure of the probability of fatigue failure of the structure this is called reliability modelling at level 0. A more satisfactory estimate of the probability of fatigue failure of the structure P_f^s is based on a systems approach (see [15]). This is called reliability modelling at level 1. The systems probability of fatigue failure can be estimated as:

$$P_f^s \approx 1 - \Phi_n(\overline{\beta}, \overline{\overline{\rho}}) \tag{39}$$

where n is the number of failure elements in the structure, Φ_n is the n-dimensional standardized normal distribution function, $\overline{\beta}$ $(= \beta_1, \ldots, \beta_n)$ the reliability indices for the failure elements and $\overline{\overline{\rho}}$ the correlation matrix for the safety margins. In real structures the number of failure elements n is very often very large, but usually a large number of failure elements is not significant for the systems reliability which means that evaluation of the systems probability of fatigue failure in eq.(39) becomes much less complicated.

6. APPLICATION

In the above chapters a method for estimating the probability of fatigue failure is briefly described. To make this method applicable a new computer package "SAOFF" (Stochastic Analysis Of Fatigue Failure) has been made. The program package which is written in FORTRAN, consists of five calculation blocks, namely:

1) STIFFMAS
This program reads the structural data and creates the global stiffness and mass matrices for the structure.

2) EIGEN
This program evaluates the n smallest eigenfrequencies and corresponding eigenvectors (mode shapes), where n is defined by the user.

3) MODAL
This program is the most complex and time-consuming part of the whole program package. Here the 0-,2- and 4-moments of the cross-spectral density of the modal displacements are evaluated, namely (see chapters 3 and 4 for more details):

0-moment:

$$\int_0^\infty S_{q_i q_j}(\omega) d\omega$$

2-moment:

$$\int_0^\infty \omega^2 S_{q_i q_j}(\omega) d\omega$$

4-moment:

$$\int_0^\infty \omega^4 S_{q_i q_j}(\omega) d\omega$$

4) SIGMA
In this program the auto-spectral densities and their moments for the hot spot stresses in joints defined by the user are evaluated. (Here the SCFs are taken into account in the calculation of T_{ki} and T_{kj}).

5) RELIA
In this program the probability of fatigue failure of failure elements in the joints (which was defined in SIGMA) is estimated. Here the user can choose between 3 different estimates of the distribution of the stress amplitudes $p_S(s)$, namely :
1) Rayleigh distribution (narrow-banded approach)

2) Distribution defined by the RFC-method (simulation)
3) Distribution defined by the RC-method (analytical estimation or simulation)
And the systems probability of fatigue failure is estimated by using Hohenbichler approximation (see chapter 5 for more details)

6.1 Example

Consider the model of a steel jacket offshore platform in figure 2. All structural elements are tubular beam elements made of steel with modulus of elasticity $E = 0.205 \cdot 10^9$ kN/m^2 and density $\rho = 7800$kg/m^3.
The cross-sectional diameters and thickness are shown in table 1. The foundation is modelled as elastic springs with horizontal stiffness equal to $1.2 \cdot 10^5$kN/m, vertical stiffness equal to 10^6 kN/m and rotational stiffness equal to $1.2 \cdot 10^6$kNm/rad. The total mass of the deck is assumed to be $4.8 \cdot 10^6$kg. The service life of the structure is taken as 25 years.
The calculation is carried out by considering 1 direction of wave propagation $\overline{\theta}$, namely $\overline{\theta}_1 = 0^o$ (x-direction), with the probability $p_{\overline{\theta}_1} = 1.0$ where $\overline{\theta}$ is defined in figure 2. Long crested waves are assumed ($n = 0$ in eq.(3)).

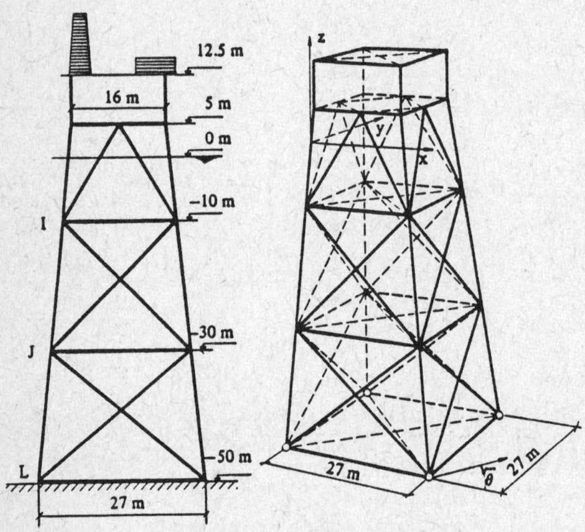

Figure 2. Steel jacket offshore platform.

The parameter in the long-term probability density function of the significant wave height H_s, $p_{H_s}(h)$, and conditional probability density function of the wave spectral peak periods T_p given H_s, $p_{T_p|H_s}(t|h)$ are estimated by fitting observations from the northern part of the North Sea in the period 1980-1983 (8222 observations), see [19] for more details. The parameters are (see eq.(8) and eq.(9)):

$$\sigma^2_{H_s} = 0.376, \qquad \mu_{H_s} = 0.836, \qquad v = 3.27 \text{ m}, \qquad \rho = 2.822, \qquad \xi = 1.547$$

$$\mu_{T_p} = 1.59 + 0.42\ ln(h + 2)$$

$$\sigma^2_{T_p} = 0.005 + 0.85\ exp\left(-0.13\ h^{1.35}\right)$$

Members	Diameter (m)	Thickness (m)
deck legs	2.00	0.050
jacket legs	1.20	0.016
braces (vertical plane)	1.20	0.016
braces (horizontal plane):		
level +5	0.80	0.008
level -10	1.20	0.014
level -30	1.20	0.014
level -30 (diagonal)	1.20	0.016
level -50	1.20	0.014

Table 1. Cross-sectional data for structural elements.

The total damage calculation for each failure element is carried out by considering 15 sea states, see table 2.

H_s (m)	p_{H_s}	T_p (sec)	$p_{T_p\|H_s}$
0.8	0.30924	5.8	0.366318
		7.9	0.442132
		11.5	0.191550
2.5	0.42741	7.1	0.306097
		9.2	0.447116
		12.6	0.246787
4.3	0.22634	8.9	0.331933
		10.6	0.439235
		14.0	0.228832
7.9	0.03621	11.2	0.293697
		12.8	0.432785
		14.9	0.273518
12.0	0.00080	13.6	0.312393
		14.9	0.452909
		16.7	0.234698

Table 2. The sea states under consideration and their probabilities.

In figure 2 two joints in the structure are considered, namely joints I and J (two TK-joints which give 12 failure elements). Detailed data and numbering of failure elements for the joints under consideration are shown in figure 3. The location of failure elements in the chord/brace intersection is determined by checking 8 points along the chord/brace intersection, see figure 1. The stochastic variables D_{fail} and B (see chapter 5.1) are assumed to be uncorrelated, but they are the assumed to be fully correlated between failure elements and with the same statistical characteristics, respectively. m and K are assumed to have correlation coefficient equal to -0.44 for each failure element, but uncorrelated between failure elements. The statistical characteristics for the stochastic variables are shown in table 3.

Basic variable	Variable	Distri-buted	Expected value	Standard deviation
X_1	D_{fail}	N	1.0	0.1
X_2	B	LN	1.0	0.2
$X_3, \ldots X_{18}$	$K_1, \ldots K_{16}$	LN	6400N/mm^2	1024N/mm^2
$X_{19}, \ldots X_{34}$	$m_1, \ldots m_{16}$	N	3.8	0.095

Table 3. Statistical characteristics for the stochastic variables (N: normal, LN: log-normal).

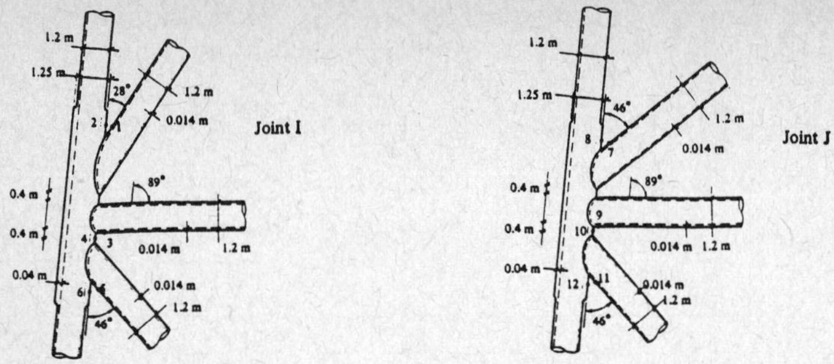

Figure 3. Detailed data and location of failure elements for joints under consideration (• means failure element).

The drag coefficient C_D in Morison's equation is taken as 1.3, but the coefficient of inertia C_M is assumed to vary as [17]:

$$C_M = \begin{cases} 2 & \text{for } 0 \le x \le 0.6 \\ 2(1.65\ exp(-0.8974\ x)) & \text{for } 0.6 < x < 2.0 \\ 2(0.798/\sqrt{x^3}\ ) & \text{for } x \ge 2.0 \end{cases}$$

where $x = \frac{D}{2g}\ \omega^2$ in which D denotes a member diameter, g denotes the acceleration of gravity and ω is the frequency.

The number of eigenfrequencies (and mode shapes) in the modal analysis is taken as 3 and the damping ratio ζ is taken as 1 % for all mode shapes. The significant failure elements are defined as the failure elements which have safety indices less than $\beta_{min} + 2.1$, where β_{min} is the lowest safety index for the failure elements. The three lowest eigenfrequencies are obtained as:

$$\omega_1 = 3.01\ rad/sec$$
$$\omega_2 = 3.01\ rad/sec$$
$$\omega_3 = 6.48\ rad/sec$$

A run of the program SIGMA showed that the irregularity factor, α ($=\frac{m_2}{\sqrt{m_0 m_4}}$, where m_i is the ith moment of the stress spectra), of the stress spectra in the failure elements is 0.37-0.6 for most of the sea states which mean broad banded stress spectrum.
In figure 4 a typical normmalized stress spectrum (normalized as $m_0 = 1.0$) for the failure elements ($H_s = 4.3$m, $T_p = 10.6$ sec, $\alpha = 0.51$).

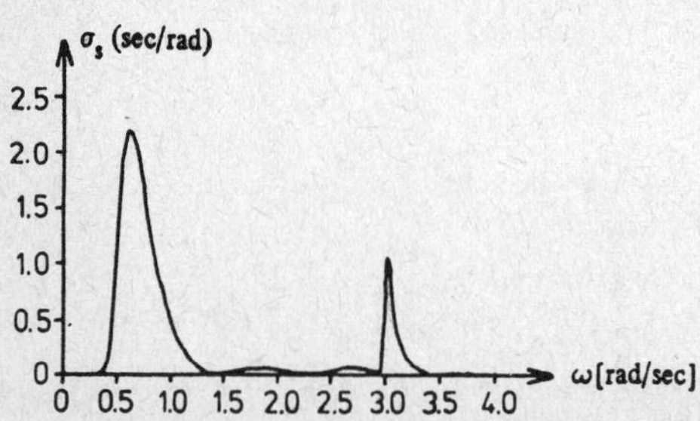

Figure 4. Typical normalized stress spectrum for the failure elements.

The distribution of stress amplitudes will be estimated by :

1 : Rayleigh distribution.

2 : Distribution defined by the RFC-method.

3 : Distribution defined by the RC-method.

Four not fully correlated significant failure elements are identified. They are given in table 4.

Failure element i	1	7	3	9	System reliability index β^s
Rayleigh β_i	1.70	1.71	3.46	3.80	1.45
RFC β_i	1.89	1.89	3.64	3.98	1.62
RC β_i	2.43	2.44	4.19	4.53	2.16

Table 4. Safety indices for the significant failure elements and Hohenbichler approximation of the system reliability index.

The correlation coefficient matrix of the linearized safety margins of the significant failure elements is:

$$\bar{\bar{\rho}} = \begin{bmatrix} 1.0 & 0.65 & 0.65 & 0.65 \\ & 1.0 & 0.65 & 0.65 \\ \text{sym.} & & 1.0 & 0.65 \\ & & & 1.0 \end{bmatrix}$$

As we can see from table 4, there are significant differences between the safety indices of a failure element dependent on how the distribution of the stress amplitudes is estimated and it is especially interesting to see the significant difference between the probability density functions estimated by the RFC and the RC methods, see figure 5.

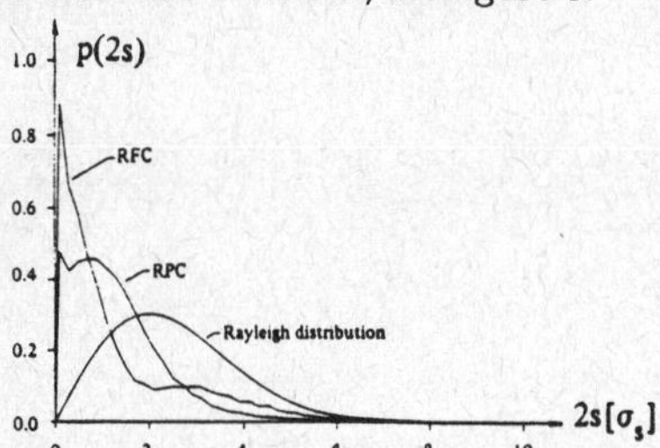

Figure 5. Estimation of the distribution density function of the stress amplitude based on simulation and using RFC- and the RC-methods.

7. CONCLUSIONS

- A method to estimate the reliability of offshore structures subjected to wave loads in deep water environments is presented.
- Failure modes corresponding to fatigue failure are used.
- The reliability is estimated using a first-order reliability method.
- The statistically measured of structural stress variations is estimated by using modal spectral analysis method.

- The damage in the failure elements is estimated by using Miner's rule and S-N approach.
- The distribution function of the stress amplitudes, for a given sea state, is estimated by Rayleigh distribution, Rice distribution, RFC-method and RC-method.
- For most of the sea states under consideration the stress process became broad-banded and therefore, the results by using Rayleigh distribution cannot be expected to give satisfactory results.
- The RFC and RC methods give a different estimation of the distribution function of the stress amplitudes, resulting in a major difference in the estimate of the safety indices.
- This means that the order and definition of the stress amplitudes must be taken into account in the damage accumulation model.
- For this purpose more experimental and theoretical work are needed.

REFERENCES

[1] Watt, B. J.: *Basic Structural System - A Review of Their Design and Analysis Requirements.* Numerical Methods in Offshore Engineering, 1978.

[2] Sigbjørnsson, R.: *Stochastic Theory of Wave Loading Processes.* Eng. Struct., Vol. 1, January 1979, pp.58-64.

[3] Sigbjørnsson, R. & E. K. Smith: *Wave Induced Vibrations of Gravity Platforms: A Stochastic Theory.* Applied Mathematical Modelling, Vol. 4, June 1980, pp.155-165.

[4] Sarpkaya, T. & M. Isaacson: *Mechanics of Wave Forces on Offshore Structures.* van Nostrand Reinhold Co., 1981.

[5] Haver, S.: *Long-Term Response Analysis - Advantages and Present Limitations.* Paper presented at the "Deep Water Jacket Seminar", Statoil, Trondheim, August 1985.

[6] Sigbjørnsson, R., K. Bell & I. Holand: *Dynamic Response of Framed and Gravity Structures to Waves.* Numerical Methods in Offshore Engineering, 1978.

[7] Olufsen, A., K. A. Farnes & D. Fergestad: *FAROW - A Computer Program for Dynamic Response Analysis and Fatigue Life Estimation of Offshore Structures Exposed to Ocean Waves - Theoretical Manual.* SINTEF, Report STF71 A86040, ISBN No:82-595-4318-4, 1986.

[8] Haver, S.: *Wave Climate of Northern Norway.* Applied Ocean Research, Vol. 7, No. 2, 1985, pp.85-92

[9] Atalik, T. S. & S. Utku: *Stochastic Linearization of Multi-Degree-of-Freedom Non-Linear System.* Earthquake Engiveering and Structural Dynamics, Vol. 4, 1976, pp. 411-420.

[10] Langen, I. & R. Sigbjørnsson: *Dynamisk Analyse av Konstruksjoner* (in Norwegian). Tapir Publishers, Trondheim, 1979.

[11] Cronin, D. J., P. S. Godfrey, P. M. Hook & T. A. Wyatt: *Spectral Fatigue Analysis for Offshore Structures.* Numerical Methods in Offshore Engineering, 1978.

[12] Almar-Næs, A.(ed).: *Fatigue Handbook.* Tapir Publishers, Trondheim, 1985.

[13] Dansk Standard DS 449. *Danish Code of Practice for Pile-Supported Offshore Steel Structures.* Teknisk Forlag, april 1983. ISBN 87-571-0838-2.

[14] Baker, M. J.: *Supplementary Notes on Fatigue and Fracture Reliability of Offshore Structures.* Structural Safety and Reliability: Theory and Practice, Short Course, September 1985.

[15] Thoft-Christensen, P. & Y. Murotsu: *Application of Structural Systems Reliability Theory.* Springer-Verlag, 1986.

[16] Lin, Y. K.: *Probabilistic Theory of Structural Dynamices.* McGraw-Hill, 1967.

[17] Karadeniz, H.: *Stochastic Analysis Program for Offshore Structures (SAPOS).* Report, Department of Civil Engineering, Delft University of Technology, Delft, Netherlands, May 1985.

[18] Hohenbichler, M.: *An Approximation to the Multivariate Normal Distribution.* DIALOG 6-82, Danish Academy of Engineers, Lyngby, 1982, pp.79-100.

[19] Haver, S. & K. A. Nyhus.: *Wave Climate Elevation for Design Purposes.* 5th OMAE, Tokyo, April 1986.

[20] Wirsching P. H. & A. M. Shehata.: *Fatigue Under Wide Band Stresses Using the Rain-Flow Method.* Journal of Engineering Materials and Technology, July 1977, pp.205-211.

APPENDIX A

This appendix deals with three-dimensional linearization of drag forces by the "minimum mean square error linearization method" [9]. A circular cylinder as shown in figure A1 is considered, and it is assumed that Morison's equation may be applied to a cylindrical member in a random manner. The non-linear term in Morison's equation may be written as:

$$\overline{f}_D = \begin{bmatrix} f_{Dx} \\ f_{Dy} \\ f_{Dz} \end{bmatrix} = K_D \, |\overline{\dot{u}}_n| \, \overline{\dot{u}}_n = K_D \, (\dot{u}_{nx}^2 + \dot{u}_{ny}^2 + \dot{u}_{nz}^2)^{\frac{1}{2}} \begin{bmatrix} \dot{u}_{nx} \\ \dot{u}_{ny} \\ \dot{u}_{nz} \end{bmatrix} \tag{A1}$$

where

$$\overline{\dot{u}}_n = \begin{bmatrix} \dot{u}_{nx} \\ \dot{u}_{ny} \\ \dot{u}_{nz} \end{bmatrix} = \overline{c}(\overline{\dot{u}}^T \times \overline{c}) = \begin{bmatrix} 1-c_x^2 & -c_x c_y & -c_x c_z \\ & 1-c_y^2 & -c_y c_z \\ \text{sym.} & & 1-c_z^2 \end{bmatrix} \begin{bmatrix} \dot{u}_x \\ \dot{u}_y \\ \dot{u}_z \end{bmatrix} \tag{A2}$$

$\overline{c} \; = (c_x, c_y, c_z)$

$\overline{\dot{u}}^T = (\dot{u}_x, \dot{u}_y, \dot{u}_z)$

$\dot{u}_x$, $\dot{u}_y$ and $\dot{u}_z$ are components of the water particle velocity in the x, y and z-direction, respectively, and $\overline{c}$ is a unit vector along the cylinder axis.

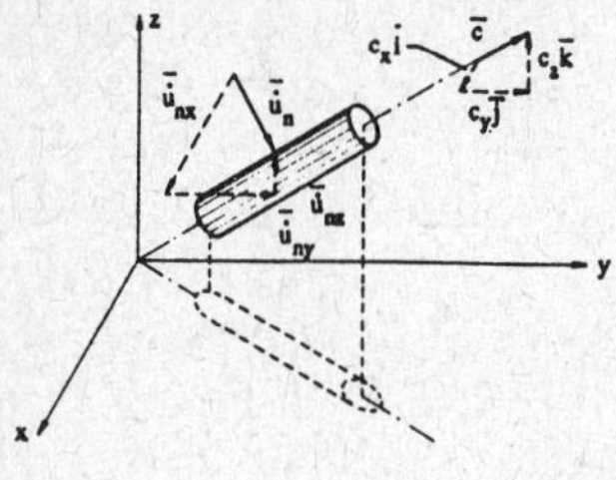

Figure A1. $\overline{i}$, $\overline{j}$ and $\overline{k}$ represent the base vectors in the x, y, z-coordinate system.

The linearized version of equation (A1) is

$$\overline{f}_{DL} = K_D \, \overline{\overline{L}} \, \overline{\dot{u}}_n \tag{A3}$$

where $\overline{\overline{L}}$ is the linearization coefficient matrix, expressed as:

$$\overline{\overline{L}} = \begin{bmatrix} l_{xx} & l_{xy} & l_{xz} \\ l_{yx} & l_{yy} & l_{yz} \\ l_{zx} & l_{zy} & l_{zz} \end{bmatrix}$$

The error introduced by using equation (A3) instead of equation (A1) is defined as:

$$\overline{e} = (\overline{\overline{L}}\, \overline{\dot{u}}_n \; - \; \overline{g}(\overline{\dot{u}}_n)) \tag{A4}$$

where

$$\overline{g}(\overline{\dot{u}}_n) = \begin{bmatrix} g_x(\overline{\dot{u}}_n) \\ g_y(\overline{\dot{u}}_n) \\ g_z(\overline{\dot{u}}_n) \end{bmatrix} = (\dot{u}_{nx}^2 + \dot{u}_{ny}^2 + \dot{u}_{nz}^2)^{\frac{1}{2}} \begin{bmatrix} \dot{u}_{nx} \\ \dot{u}_{ny} \\ \dot{u}_{nz} \end{bmatrix}$$

The criterion that the mean square value of the error $\overline{e}$ is at a minimum is expressed as:

$$E[\overline{e}\,\overline{e}^T] \;\rightarrow \text{minimum}$$

where $E[\cdots]$ denotes the expected value.

The coefficients l_{ij} in the linearization matrix $\overline{\overline{L}}$ may be written as [9],[10]:

$$l_{ij} = E[\; \frac{\partial g_i(\overline{\dot{u}}_n)}{\partial \dot{u}_{nj}}] \tag{A5}$$

The matrix $\overline{\overline{L}}$ may now be expressed as:

$$\overline{\overline{L}} = E \begin{bmatrix} \frac{2\dot{u}_{nx}^2 + \dot{u}_{ny}^2 + \dot{u}_{nz}^2}{|\dot{u}_n|} & \frac{\dot{u}_{nx}\dot{u}_{ny}}{|\dot{u}_n|} & \frac{\dot{u}_{nx}\dot{u}_{nz}}{|\dot{u}_n|} \\ & \frac{\dot{u}_{nx}^2 + 2\dot{u}_{ny}^2 + \dot{u}_{nz}^2}{|\dot{u}_n|} & \frac{\dot{u}_{nz}\dot{u}_{ny}}{|\dot{u}_n|} \\ \text{sym.} & & \frac{\dot{u}_{nx}^2 + 2\dot{u}_{ny}^2 + \dot{u}_{nz}^2}{|\dot{u}_n|} \end{bmatrix} = E[\overline{\overline{M}}]$$

where $|\overline{\dot{u}}_n| = (\dot{u}_{nx}^2 + \dot{u}_{ny}^2 + \dot{u}_{nz}^2)^{\frac{1}{2}}$.

When the water particle velocity is assumed to be a zero-mean Gaussian process then the following expression is obtained

$$E[\overline{\overline{M}}] = \int_{-\infty}^{\infty}\int_{-\infty}^{\infty}\int_{-\infty}^{\infty} \overline{\overline{M}} N_3(0, \overline{\overline{\Sigma}}_{\dot{u}_x\dot{u}_y\dot{u}_z}) d\dot{u}_x d\dot{u}_y d\dot{u}_z \tag{A6}$$

where $N_3(0, \overline{\overline{\Sigma}}_{\dot{u}_x\dot{u}_y\dot{u}_z})$ is a three-dimensional normal density function defined by:

$$N_3(0, \overline{\overline{\Sigma}}_{\dot{u}_x\dot{u}_y\dot{u}_z}) = \frac{1}{(2\pi)^{3/2}(det(\overline{\overline{\Sigma}}_{\dot{u}_x\dot{u}_y\dot{u}_z}))^{1/2}} exp(-\frac{1}{2}\overline{\dot{u}}^T \overline{\overline{\Sigma}}_{\dot{u}_x\dot{u}_y\dot{u}_z}^{-1}\; \overline{\dot{u}})$$

where the covariance matrix $\overline{\overline{\Sigma}}_{\dot{u}_x\dot{u}_y\dot{u}_z}$ is defined by:

$$\overline{\overline{\Sigma}}_{\dot{u}_x\dot{u}_y\dot{u}_z} = \begin{bmatrix} \int_0^\infty S_{\dot{u}_x\dot{u}_x}(\omega)d\omega & \int_0^\infty S_{\dot{u}_x\dot{u}_y}(\omega)d\omega & \int_0^\infty S_{\dot{u}_x\dot{u}_z}(\omega)d\omega \\ \int_0^\infty S_{\dot{u}_y\dot{u}_x}(\omega)d\omega & \int_0^\infty S_{\dot{u}_y\dot{u}_y}(\omega)d\omega & \int_0^\infty S_{\dot{u}_y\dot{u}_z}(\omega)d\omega \\ \int_0^\infty S_{\dot{u}_z\dot{u}_x}(\omega)d\omega & \int_0^\infty S_{\dot{u}_z\dot{u}_y}(\omega)d\omega & \int_0^\infty S_{\dot{u}_z\dot{u}_z}(\omega)d\omega \end{bmatrix}$$

$S_{\dot{u}_i\dot{u}_j}(\omega)$ is the cross-spectral density of the water particle velocity $\dot{u}_i$ and $\dot{u}_j$ $(i, j = x, y, z)$. Thus,

$$l_{xx} = E[m_{11}] = \int_{-\infty}^{\infty}\int_{-\infty}^{\infty}\int_{-\infty}^{\infty} \frac{2\dot{u}_{nx}^2 + \dot{u}_{ny}^2 + \dot{u}_{nz}^2}{|\bar{\dot{u}}_n|} \frac{1}{(2\pi)^{3/2}(det(\bar{\bar{\Sigma}}_{\dot{u}_x\dot{u}_y\dot{u}_z}))^{1/2}}$$

$$exp(-\frac{1}{2}\bar{\dot{u}}^T \bar{\bar{\Sigma}}^{-1}_{\dot{u}_x\dot{u}_y\dot{u}_z} \bar{\dot{u}}) d\dot{u}_x d\dot{u}_y d\dot{u}_z$$

where $\dot{u}_{nx}$, $\dot{u}_{ny}$ and $\dot{u}_{nz}$ can be expressed as functions of $\dot{u}_x$, $\dot{u}_y$ and $\dot{u}_z$ (see eq. (A2)).

BAYESIAN PREDICTION OF DESIGN WAVE HEIGHTS

C. Guedes Soares
Naval Architecture and Marine Engineering Section
Technical University of Lisbon
Av. Rovisco Pais, I.S.T., 1096 Lisboa, Portugal

ABSTRACT: A brief review is made of the methods in current use to predict design wave heights. Predictions of various methods are compared using wave data from FAMITA which allows an assessment of their model uncertainty. A method is proposed to combine information from the different predictions into a single design value.

1. INTRODUCTION

Prediction of wave loading on offshore structures is obviously one of the initial phases of a design procedure. However, before this load distribution can be calculated it is necessary to define the wave conditions that will generate it [1].

In general, local wind generated wave fields are random in nature and are well described by a wave spectrum which indicates how the energy is distributed along the frequency domain. Theory is well established to calculate the response of any marine structure to such wave field.

The high waves are associated with large wave lengths and with high wave velocities which implies that with time they will be separated from the initial wave system, propagating as swell which looks like a two dimensional regular wave system.

The extreme loading condition in many offshore platforms occurs when they are subjected to such a wave system. This is why the so called design wave approach has become popular for shallow and moderate depth structures at least.

System reliability analysis of jacket structures has adopted a design wave as the load model [2] since at present the method hardly handles other more complicated representations.

Whenever the structure is sensitive to dynamic response the spectral approach is often adopted. However, even in an irregular sea state, the larger wave loads are associated with the higher waves and thus the quantification of their effect on the structure becomes of interest.

In a recent study [3] a simulation was made in which an irregular sea state was used to excite the response of a jacket structure and the corresponding wave loads were calculated. Afterwards the individual high waves in the wave systems were identified and fitted by regular wave models. The structure was loaded by such single waves and the wave induced forces in this case compared well with the larger forces that resulted from the simulation study.

This result confirms the usefullness of using the design wave approach giving relevance to the problem of deriving that wave, which is the main subject of this paper. A method is proposed hereafter to derive a design wave height using all the available information from different prediction methods.

2. LONG-TERM EXTRAPOLATION OF WAVE HEIGHTS

The derivation of the design waves is normally based on a probabilistic model that represents the distribution of wave heights over periods of time comparable to the life of the structure. The design wave is then chosen as the most probable wave in that period of time.

Often the bivariate distribution of heights and periods of the waves is considered as a product of the conditional distribution of periods and the marginal distribution of heights:

$$f(h,t) = f(t|h)\, f(h) \tag{1}$$

Thus the major concern is the fitting and extrapolation of the wave height distribution function. Having obtained the design wave height, the associated period is then chosen.

Various methods are available to derive a characteristic value of the design wave. The most common one weights the short-term distribution of wave heights by the long-term distribution of wave parameters.

In a short-term period of stationarity the wave height is well represented by a Rayleigh distribution, which is conditional on the variance of surface elevation or on the significant wave height (H_s) [4]:

$$Q_s(h|H_s) = \exp - \frac{2\,h^2}{H_s^2} \tag{2}$$

in which $Q_s(h)$ is the probability of exceeding the height h in the short-term. The Rayleigh distribution was originally derived as the distribution of wave amplitudes, which can be approximated by h/2 specially in narrow band situations. It has been shown recently [5] that this distribution can still be applied to sea states with finite band width if the value of H_s is corrected accordingly.

The long-term distribution of wave heights is obtained by integrating the short term conditional distribution multipied by the marginal distribution of H_s and a factor $n(H_s)$ which is proportional to the percentage of time the considered sea state lasts [6,7]:

$$Q_L(h) = \int n(H_s)\, Q_s(h|H_s)\, f(H_s)\, dH_s \tag{3}$$

The factor $n(H_s)$ depends directly on the average period of the sea state, T_m, but in turn T_m is a function of H_s.

The design wave is the most probable one in the return period considered. Often return periods of 50 years or 100 years are sought, leading to probability levels around 10^{-8}.

Usually one uses wave statistics of $f(H_s)$ to construct $Q_L(h)$ which is only assessed with confidence to levels up to 10^{-5} or 10^{-6} corresponding to the existing data bases. An extrapolation becomes necessary for the desired lower probability levels.

The common procedure is to calculate various values of $Q_L(h)$ to which a theoretical probability distribution is fitted and is used as a basis for extrapolation. Experience has shown that a log-normal distribution is adequate for low and moderate wave heights [6] but a Weibull distribution fits better the tails of the distribution [7]. A suggestion has even been made of using both distributions to model the whole range of wave heights [8].

Another method of extrapolation uses the most probable maximum wave height in each sea state and makes extrapolations based on these values [9].

A comparable approach, avoids the problem of correlation between successive sea states which may occur in the previous case, is the use of annual maxima of wave height, which are then fitted by a Gumbel distribution. In fact one year is the period of climatological and environmental processes which implies that yearly maxima should be statistically independent [10-12].

Still a different procedure is to extrapolate the sea state parameters, based on the assumption that the largest wave should occur during the most severe sea state. Once the sea state parameters H_s and T_m are determined, the largest wave to occur in that sea state is determined from a short-term model [11].

A last model to be considered here is the extrapolation based on storms. Storms in a given location are assumed to be independent events occurring with a constant arrival rate, which is described by a Poisson process. A storm is modelled as a succession of sea states with increasing H_s up to a maximum value after which they decrease again [13]. In this case the extrapolation is made associating the Poisson model of storm occurrence with the maximum wave height in each storm [14].

A detailed account of the various methods is given by Nolte [10]. Predictions of design waves at the weather station FAMITA (57° 30'N, 03° 00'E) made by different authors using the various methods are summarized in Table 1 which is taken from [16]. It can be observed that the predictions range from 22.5m to 34.9m, which involves a high degree of uncertainty in the methods of extrapolation.

Table 1. Results of different prediction methods applied to data of weather station FAMITA (57°30'N, 03°00'E).

1. Visual data 1959-1969 - 100 years return period [10]	
a. Individual wave model	27.0 m
b. Most prob. max. wave height model	25.2 m
c. Storm model	24.7 m
d. Annual extremes	25.0 m
2. Visual data 1959-1974 - 100 years return period [11]	
a. Max significant wave height model	22.5 m
b. Annual extreme	25.6 m
c. As a. using hindcast data	28.4 m
3. Visual data 1959-1977 - 100 years return period [12]	
a. Individual wave model	32.7 m
b. Most prob. max. wave height model	24.0 m
c. Max. significant wave height model	27.0 m
d. Storm model	27.2 m
e. Annual extremes	34.9 m
4. Instrumental data 1969-1974 - 50 years period [15]	
a. Most prob. max. wave height model	28.4 m
b. Storm model	32.0 m

3. BAYESIAN APPROACH TO UNCERTAINTY MODELLING

Probabilistic methods have been developed to deal with random variables and stochastic processes which model the natural variability of physical processes, like for example sea surface elevation. However, the state of incomplete knowledge about a phenomenon or the uncertainty about its outcome can also be modelled by a random variable and can be manipulated with the same probabilistic methods. This uncertainty can concern the parameters that describe a random variable or even the corresponding probabilistic model as discussed with some detail in [17].

Consider a probability distribution of a variable which is described by n parameters p_i (i=1,2,...,n). The distribution of X is in reality conditional on the values of p_i that have been determined from the sampled data. Thus, the Bayesian distribution of X is obtained by weighting the conditional distribution of X with the marginal distribution of the parameters:

$$f_X(x) = \int_{\underline{p}} f_X(x|\underline{p}) \cdot f_{\underline{p}}(\underline{p}) d\underline{p} \qquad (4)$$

where $\underline{p}$ is the vector of all parameters of the distribution.

The distribution of the parameters is the prior distribution or the updated one depending on the situation. In this work $f_P(P)$ will be postulated on the basis of expert judgement and the question of its updating will not be dealt with here because there is almost no data available on waves of the size of design waves.

In the previous expression, the Bayesian distribution is associated with a particular model M_i and in fact it is conditional on that model being the correct one. If various models are possible candidates, the Bayesian distribution that incorporates both parameter and model uncertainty is [17]:

$$f_X(x) = \int_{\underline{M}} f_X(x|M)\ f_{\underline{M}}(\underline{m})\ d\underline{M} \qquad (5)$$

where the conditional distribution on this expression is in fact the distribution given by eqn. 4, i.e. it is the result of integrating over all possible values of parameters for that model.

In practice models are discrete and a probabilistic description of the state of knowledge about their adequacy is a probability mass function. Thus combining the two previous expressions leads to:

$$f_X(x) = \sum_{i=1}^{n} p[m_i] \int_{\underline{p}} f_i(x|\underline{p})\ f(\underline{p})\ d\underline{p} \qquad (6)$$

where $p[m_i]$ is the probability assigned to model i and $f_i(x|\underline{p})$ is the conditional probability distribution associated with that model.

The rational behind Bayesian methods is the minimization of the expected losses under uncertainty. In principle only one of the models M_i in the above formulation is the correct one. However in the state of uncertainty about which one is correct, the use of all methods minimizes the expected losses that would result from choosing the prediction of a wrong method.

In the case of sea waves, the significant wave height of a sea state is a derived parameter which is thus subjected to uncertainty due to sampling variability. In modelling the long-term distribution of significant wave heights or of individual wave heights the log-normal distribution was sometimes better than the Weibull [18] but the opposite has occurred on occasions [6]. This means that there is also uncertainty about which probabilistic model is more correct in a new situation.

Still other types of extrapolation models are the different ones that were described in the previous section to lead to the design wave. They all have a theoretical justification behind but they lead to different results, as indicated in Table 1.

The Bayesian approach allows one to take all these uncertainties into account. It considers that the parameters estimated from data are themselves a random variable which can be included in the analysis. In addition, probabilities are interpreted as degrees of belief which can quantify subjective evaluations. These probabilities can be combined with the ones that describe the natural randomness to yield statements about the total state of knowledge about a certain situation [19,20].

The Bayesian approach is extremely valuable in cases that deal with rare events, like for example design waves, since in this case the amount of statistical data available to test the hypothesis about probabilistic models is very scarce and do not allow definite conclusions to be obtained from classical statistics alone.

Bayesian methods can be used to combine existing knowledge about a variable with new evidence in the form of data i.e., to update distributions or parameters. However, because existing information about waves of the size of the design waves is extremely scarce, very little would be added to the existing knowledge by this updating procedure. Thus the main aim of this work is to show how one can use a Bayesian point of view to incorporate model uncertainties in the derivation of design waves.

4. NUMERICAL EXAMPLE

To illustrate the concepts presented in the previous sections example calculations will be performed to predict a design wave height at FAMITA weather station.

In [21] predictions were made of the 100 years significant wave height at Famita based on 17 data sets that are one year long and then on 10 data sets that are 10 years long. In the first case the predictions varied between 9.02m and 14.11m, while in the second they ranged from

10.49m to 12.79m. Using all that information leads to mean values and coefficients of variation of 11.66m and 0.14 from the yearly data sets and 11.76m and 0.06 from the 10 years data set. These results show that the average values determined in both cases are consistent but the statistical uncertainty associated with the predictions is larger for the smaller data sets, as expected.

Haver [8] used Weibull distributions with different parameters to model the tail of a long-term distribution of wave heights, which were then used to make extrapolations. The results indicated in his figure 4.16 show that the 100 years wave predicted by the four Weibull distributions he adopted are 1.15, 1.26 and 1.38 times larger than the smaller value. All of these fits were considered adequate, i.e. they passed a Chi-square test of fit at 5% significance level, and the spreading in the predictions indicates the effect of parameter uncertainty. If all models are considered to have an equal probability of being correct, an average value of 1.20 and a coefficient of variation of 0.13 are obtained.

This uncertainty could also be interpreted as probabilistic model uncertainty since one is really using different Weibull distributions. However, probabilistic model uncertainty would be more correctly associated with the results of extrapolations with a Weibull and with a log-normal distribution for example, which would lead to even larger differences.

Physical model uncertainty in the present context could be associated with the use of the various extrapolation models described in section 2 to make the design wave predictions. Table 1 indicates the results of these methods applied to the same data set by the same author. Weighting all the models equally would lead to the results indicated in Table 2.

A possible model of the design wave height is

$$H_d = B_s \cdot B_p \cdot B_m \cdot \tilde{H}_d$$

where B is a correction factor that accounts for the statistical uncertainty (s), the probabilistic model uncertainty (p) and the phisical model uncertainty (m). $\tilde{H}_d$ is the Bayesian best estimate of the design wave height based on the available information. It should be noted that this is the usual representation of model uncertainty [22] which is in accordance with Ditlevsen's generalization [23].

Table 2. Bayesian Predictions of Design Wave Height from Data Sets of Table 1 with Equal Weights.

	1959-69	1959-74	1959-77	1969-74	All
Mean(m)	25.5	24.1	29.2	30.2	27.3
COV	.04	.09	.15	.08	.09

Since $\tilde{H}_d$ is already the best estimator, the correction factors B_i have a mean value of 1. Their purpose is to quantify the uncertainty of the predictions. Based on the results previously discussed, the uncertainty of each parameter as expressed by its COV is assumed to be $V_s = .06$, $V_p = .13$ and $V_m = .09$. Assuming that those sources are independent, the uncertainty of H_d is:

$$V_d = V_s^2 + V_p^2 + V_m^2 = .17$$

The best estimate of the design wave height could be obtained using different criteria. Considering the prediction of each author in Table 1 equally important would make $p[m_i]=1/n$ and the average of all values in Table 2 would be 27.3m as indicated.

One could argue that the weight of the values should be proportional to the size of the data base in number of years. In this case

$$p[\underline{m}] = [.21, .31, .38, .10]$$

and the result of applying eqn.(6) would be $H_d = 26.9$ m.

Instead one could argue that more importance should be given to the predictions that are the result of more individual methods. In this case one would have

$$p[\underline{m}] = [.31, .15, .39, .15]$$

and the result would be $H_d = 27.4$ m.

If one would now say that both the size of the data set and the number of methods used are important and that they are equally important, the last two sets of information can be combined leading to:

$$p[\underline{m}] = [.24, .17, .54, .05]$$

resulting now in $H_d = 27.5$ m.

It is apparent that whenever using all the available information the Bayesian predictions of the design wave show a much smaller variability than one would have using predictions of individual methods. The Bayesian predictions range between 26.9 m and 27.5 m while individual extrapolations they go from 22.5 m to 34.9 m.

The spread of the individual extrapolations is larger than one would expect from a consistent application of the methods available. In this respect this work does not intend to provide the best estimate of the design wave height at Famita nor does it aims at evaluating the correctness of each of the five extrapolation methods.

Furthermore the various methods have been develloped for different situations and ocean areas and thus their relative weights must vary in accordance with them. The predicted values and the weights assigned to each method are only intended to illustrate the concepts proposed here.

5. CONCLUSIONS

There is no theoretical justification to choose or to discard totally any method of representing the long-term distribution of wave heights among the existing proposals. Furthermore there is no theoretical justification to prefer any probability distribution to fit the long-term distribution of individual wave heights and there is not enough data to make meaningful tests of fit in the low probability range.

However, when adopting the various methods available a large spread in predictions is obtained. A Bayesian approach allows one to use the information from all prediction methods, combining it in one value and thus minimising the expected error.

Depending how the various predictions are weighted different final values will be obtained but their range of variation is much smaller than in the case of individual predictions. The adoption of this method would allow more consistent design wave predictions to be made.

6. ACKNOWLEDGEMENTS

This work was performed in the scope of the project on "Development of Codes for the Design of Marine Structures", which is partially funded by INIC, the National Institute for Scientific Research through CEMUL, the Centre for Mechanics and Materials of the Technical University of Lisbon.

7. REFERENCES

1. Guedes Soares, C., "Reliability of Marine Structures", Reliability Engineering, A. Amendola and A. Saiz de Bustamante (Eds), Kluwer Acad. Pub., Dordrecht, 1988, pp. 513-559.
2. Thoft-Christensen, P., and Murotsu, Y., Application of Structural Systems Reliability Theory, Springer-Verlag, 1986.
3. Dogliani, M., and Cazzulo, R., "Stochastic Wave Loads for Reliability Analysis of Jacket Structures", 2nd Working Conference on Reliability and Optimization of Structural Systems, Imperial College, London, September 1988.
4. Longuet-Higgins, M.S., "On the Statistical Distribution of the Height of Sea Waves", J. Marine Research, Vol. 11, 1952, pp. 245-266.
5. Longuet-Higgins, M.S.,"On the Distribution of the Heights of Sea Waves: Some Effects of Non-Linearity and Finite Band Width", J. Geophysical Research, Vol. 85, 1980, pp. 1519-1523.
6. Jaspers, N.H., "Statistical Distribution Patterns of Ocean Waves and of Wave Induced Ship Stresses and Motions with Engineering Applications", Transactions Soc. Naval Architects and Marine Engineers (SNAME), Vol. 64, 1956, pp. 375-432.
7. Battjes, J.A., "Long-Term Wave Height Distributions at Seven Stations Around the British Isles", Deutsche Hydrograph. Zeitsch., Vol. 25, 1972.
8. Haver, S., "Analysis of Uncertainties Related to the Stochastic Modelling of Ocean Waves", Report UR-80-09, Division of Marine Structures, The Norwegian Institute of Technology, 1980

9. Draper, L., "Derivation of a Design Wave from Instrumental Records of Sea Waves", Proc. Inst. Civil Engrs, Vol. 26, 1963, pp. 291-304.

10. Nolte, K.G., "Statistical Methods for Determining Extreme Sea States", Proc. 2nd Int. Conf. on Port and Ocean Engineering, Under Artic Conditions, Reykjavik, 1973, pp. 705-742.

11. Houmb, O.G., Mo, K., and Overvik, T., "Reliability Tests of Visual Wave Data and Estimations of Extreme Sea States", Report No. 5, Div. Port and Ocean Engineering, Norwegian Institute of Technology, 1978.

12. Eide, L.I., "Variations of Estimates of Extreme Wave Heights", Report No. 79-0049, Det Norske Veritas, 1979.

13. Borgman, L.E., "Probabilities for Highest Waves in a Hurricane", J. Waterways, Port and Ocean Div., ASCE, Vol. 99, 1973, pp. 185-207.

14. Petraukas, C., and Aagard, P.M., "Extrapolation of Historical Storm Data for Estimating Design Wave Heights", Proc. 2nd Offshore Technology Conf., 1970, Paper OTC 1190.

15. Saerthe, H.J., "On High Wave Conditions in the Northern North Sea", Oceanology International, 1975.

16. Guedes Soares, C., and Moan, T., "On the Uncertainties Related to the Extreme Environmental Loading on a Cylindrical Pile", Reliability Theory and its Application in Structural and Soil Mechanics, P. Thoft-Christensen (Ed), Martinus Nijhoff Pub., The Hague, 1983, pp. 351-364.

17. Guedes Soares, C., "Uncertainty Modelling in Systems Reliability Analysis", ISPRA Course on Advanced Systems Reliability Modelling, Madrid, 19-23 September 1988.

18. Guedes Soares, C., Costa, M.D.S., and Lopes, L.C., "Wave Climate Modelling for Engineering Purposes", Computer Modelling in Ocean Engineering, Schrefler, B.A., and Zienkiewicz O.C., (Eds), A.A. Balkema, Rotterdam, 1988, pp.169-175.

19. Cornell, C.A., "Bayesian Statistical Decision Theory and Reliability-Based Design", Proc. Int. Conf. on Structural Safety and Reliability, A.M. Freudenthal (Ed.), Pergamon Press, New York, 1972, pp. 47-68.

20. Martz, H.F., and Waller, R.A., Bayesian Reliability Analysis, J. Wiley & Sons, New York, 1982.

21. Guedes Soares, C., "Assessment of the Uncertainty in Visual Observations of Wave Height", Ocean Engineering, Vol. 13, 1986, pp. 37-56.

22. Ang, A. H-S., and Cornell, C.A., "Reliability Bases of Structural Safety and Design", J. Struct. Div. ASCE, Vol. 100, 1975, pp. 1755-1769.

23. Ditlevsen, O., "Model Uncertainty in Structural Reliability", Structural Safety, Vol. 1, pp. 73-86.

INSPECTION STRATEGIES FOR CONCRETE BRIDGES

J. D. Sørensen & P. Thoft-Christensen
University of Aalborg
Sohngaardsholmsvej 57, DK-9000 Aalborg, Denmark

ABSTRACT
In this paper an optimal inspection strategy for concrete bridges based on periodic routine and detailed inspections is presented.

The failure mode considered is corrosion of the reinforcement due to chlorides. A simple modelling of the corrosion and of the inspection strategy is presented. The optimal inspection strategy is determined from an optimization problem, where the design variables are time intervals between detailed inspections and the concrete cover. The strategy is illustrated on a simple structure, namely a reinforced concrete beam.

1. INTRODUCTION
Periodic inspection intervals for structures like bridges are in general determined on the basis of experience and practical convenience rather than rational methods. However, in the last decade there has been a growing interest in deriving inspection strategies on a more rational basis. In this paper optimal inspection strategies for bridges are investigated on the assumption that two different types of inspections are used, namely routine inspections and detailed inspections. Routine inspections typically take place at one year intervals while the detailed inspections are only performed every five years or so.

Only failure due to corrosion of the reinforcement is considered. A model for corrosion is derived and the proposed strategy is applied to a simple structural element of a concrete bridge. The corresponding optimal inspection strategy is obtained from a complicated integer optimization problem. The object function is the estimated total cost in the expected lifetime of the bridge and the constraints are reliability based.

2. INSPECTION REGULATIONS
Traditionally inspection regulations for concrete bridges are based on three types of inspection:

1. Routine inspections.
2. Detailed inspections.
3. Special inspections.

The special inspections are only used when routine or detailed inspections show that there is a need for a more complete (and expensive) inspection. The special inspections are therefore not included in the regulations as inspections that are repeated within fixed intervals. It is therefore not relevant to include them in the optimal inspection strategies derived in the present paper.

The time intervals between routine and detailed inspections vary from country to country according to existing inspection regulations. To illustrate this a number of regulations will be briefly described in this section. Most of the information shown here is obtained from the proceedings edited by Nowak & Absi [1].

Lemarié [2] reports that technical guidelines for the inspection and maintenance of bridges were published in 1979 in *France*. According to these guidelines there are two types of inspection, namely permanent inspections performed by local agents and periodical inspections. The periodical inspections include a systematic yearly visit (relatively superficial) to all bridges more than 10 m long and a detailed inspection every five years for at least all bridges exceeding 120 m.

In *Belgium* a computer assisted bridge management scheme was developed in 1977 (see De Buck [3]). In Belgium the maintenance of bridges involves two activities: the control of the bridges and the repair and maintenance works. The control is based on three levels of inspection:

1. A routine inspection and annual survey.
2. "Type A" general inspection every 3 years.
3. "Type B" general specialized inspection when the "type A" inspection reveals the need for it.

In *Denmark* a bridge management system developed for the Danish State Railways consists of three modules (see Rostam [4]). Module B is the inspection module which comprises the following activities:

1. A superficial inspection.
2. A principal inspection (visual registration and evaluation of damage).
3. A special inspection (specially qualified inspection to be made when needed).

The observation and inspection of bridges in *Germany* is governed by the German code DIN 1076 (see Zichner [5]). It specifies:

1. Visual inspection of the general conditions without special equipment 4 times a year.
2. General inspection every three years.
3. Main inspection every six years.
4. Special survey in the event of accident or natural disaster that may affect the short-term safety of the structure.

In *Switzerland* it has been proposed to split bridge surveilance up into (see Favre [6]):

1. Routine inspections (every 15 months).
2. Periodic inspections (every 5 years).
3. Special inspections (according to needs).

In *Italy* the bridge structure supervision activities are regulated by the Ministry for Public Works Circular n.6736/61 (see Malisardi & Nebbia [7]). It is prescribed that an inspection should be carried out at least once every 3 months. At least once a year a more specialized inspection should take place.

It is seen from the data presented above that a routine type inspection is prescribed for every 3 to 15 months in the countries mentioned. Therefore, it seems reasonable to fix the intervals between routine inspections to 1 year. The 1 year interval is also convenient from a climate point of view.

The data for the detailed inspections deviate more, but intervals from 3 to 5 years seem to be the most commonly used. The time between detailed inspections is the unknown to be determined by the optimization problem formulated in this paper.

3. MODELLING OF CORROSION

As mentioned in the introduction only a single structural element in a concrete bridge is considered, viz. the T-beam shown in figure 1.

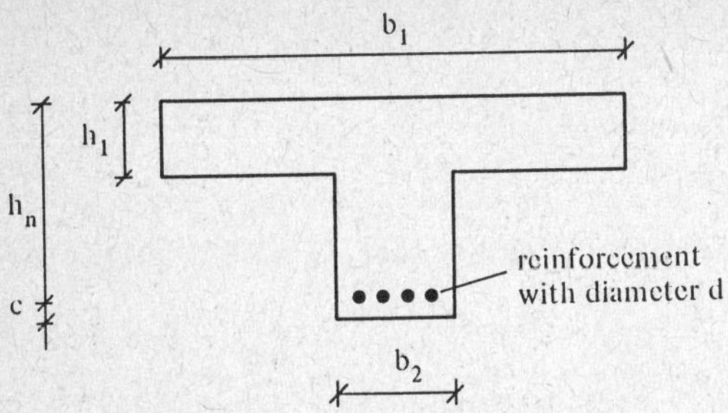

Figure 1. Model of concrete beam.

The beam is loaded by a bending moment kP, where P is an external load and k is a coefficient of influence. The concrete cover is modelled by c. The reinforcement is assumed to consist of n bars each with the diameter d. Let the total reinforcement area $A(t)$ be modelled as a function of the time t. Then a safety margin $M_F(t)$ corresponding to failure of the beam is (in the normal case where the neutral line distance is less than h_1)

$$M_F(t) = Z_1(h_n A(t)\,\sigma_y - \frac{A(t)^2 \sigma_y^2}{2\,b_1 \sigma_c}) - kP \tag{1}$$

where

σ_y is the yield stress of the reinforcement,

σ_c is the yield stress of the concrete in compression,

Z_1 is the uncertainty variable modelling the uncertainty connected with estimating the moment capacity.

A reinforced concrete beam in a bridge structure is exposed to a large number of hazards, e.g. alternating load, extreme load, frost, acids, chlorides, de-icing salts, pollution and alkali aggregates. Several of these hazards have the effect that the reinforcement is exposed to corrosion. In this paper one of the most important sources of corrosion is considered, namely chlorides which have a de-passivating effect on the steel if the concentration is high enough. Chlorides that are in the concrete initially are found to be of much less importance than chlorides coming from rock salt used for de-icing roads, see Vassie [8].

A number of physical factors is of importance for corrosion of reinforcement generated by chlorides, namely low depth of concrete cover, leaking joints, faulty drainage, absence of waterproof membranes and the number and intensity of freeze-thaw cycles. However, in this paper only such cases are considered where the chloride concentration outside the concrete, the initial chloride concentration and the critical chloride concentration (starting corrosion) are given. Clearly the concentration of chlorides outside the concrete beam is highly dependent on the location of the beam in the structure.

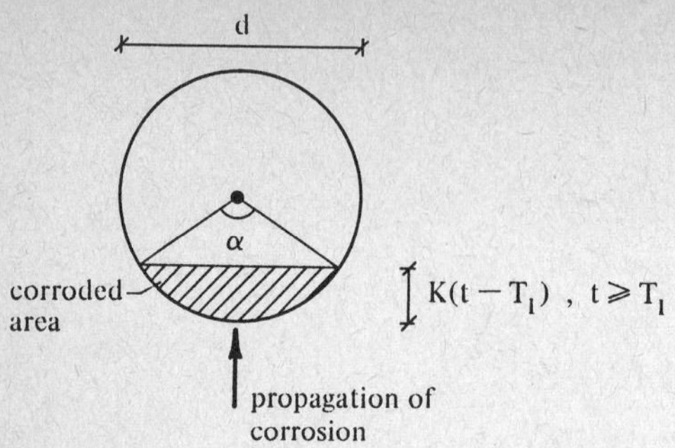

Figure 2. Model of local corrosion propagation.

The *time of corrosion initiation* T_I depends on the chloride concentration outside the beam, how fast the chlorides penetrate the concrete cover and on the critical chloride concentration. If the penetration process is modelled as a diffusion process T_I can be estimated from (see [9])

$$T_I = \frac{c^2}{4a}[erf^{-1}(\frac{c_c - c_\infty}{c_0 - c_\infty})]^{-2} \tag{2}$$

where

a is the coefficient of diffusion (= 400 mm^2/year for water/cement ratio = 0.55)

c_c is the critical chloride concentration (= 0.15)

c_0 is the initial chloride concentration (= 0.04)

c_∞ is the outer chloride concentration (= 0.2)

erf is the error function

The values shown in parenthesis are values used by Andersen & Lyck [9].

After initiation of corrosion in the reinforcement it is assumed that it propagates as local corrosion. All bars are assumed to corrode with the same velocity starting at the same time. The corrosion model shown in figure 2 is used in this paper.

The propagation velocity K is assumed to be constant

$$K = G\,k_t \tag{3}$$

where

G = 200 μm/year for chloride initiated corrosion

k_t = 1 at normal temperatures.

The total reinforcement area as a function of the time t then becomes

$$A(t) = \begin{cases} n\,\pi\frac{d^2}{4} & t \le T_I \\ n(\pi\frac{d^2}{4} - \frac{d^2}{8}(\alpha - sin\alpha)) & T_I \le t \le T_I + \frac{d}{K} \\ 0 & T_I + \frac{d}{K} \le t \end{cases} \tag{4}$$

where

$$\alpha = 2\,arc\,cos\,(1 - \frac{2K(t - T_I)}{d}) \tag{5}$$

4. MODELLING OF INSPECTION STRATEGY

As described in section 2 most national inspection programmes for concrete bridges consist of two types of inspection, namely routine inspections at fixed time intervals Δt and detailed inspections at longer intervals (being a multiple of the Δt). If the routine inspections are performed with inspection quality q_1 and the detailed inspections with quality q_2 the inspection programme can be modelled as shown in figure 3. $T_i, i = 1, \dots, N$, are the times of detailed inspections and T is the expected lifetime of the beam.

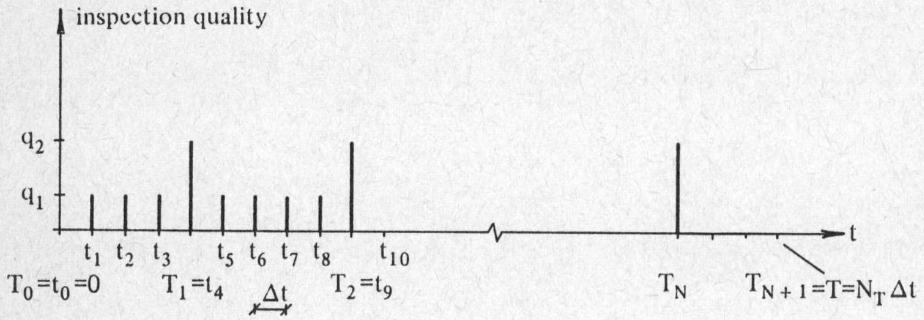

Figure 3. Inspection plan.

The detailed inspections are assumed to be performed at the time intervals

$$\Delta t_i = T_i - T_{i-1} = N_i \Delta t \tag{6}$$

where $N_i, i = 1 \dots, N$, is a positive integer. The inspection method used for the detailed inspections are assumed to provide an estimate of the chloride concentration variation with the depth from the concrete surface and to give a picture of where corrosion (general and local) occurs. The inspection methods include such methods as visual inspection, measurements of half-cell potentials, resistance, chloride content and concrete quality.

A number of repair strategies can be chosen. In this paper it is assumed that the beam is repaired when corrosion of the reinforcement is detected. The repair is assumed to be immediate and complete (both concrete and reinforcement is replaced, if necessary). The routine inspections are assumed to be visual inspections. If corrosion has started and has developed to a certain degree the concrete cover will flake off. The structure is assumed to be repaired if peelings are discovered during the inspections.

An optimal inspection and repair strategy can be determined from an optimization problem where the optimization variables are the total number of detailed inspections N, the time intervals between detailed inspections $N_1, N_2, \dots, N_N$ and the concrete cover c. In Thoft-Christensen &

Sørensen [10] a similar problem to determine an optimal inspection and repair strategy for a fatigue sensitive element (hot spot) in an offshore jacket structure is considered. The optimization problem considered here is:

$$\min_{\substack{N \\ N_1 N_2 \ldots N_N \\ c}} \quad C(N, N_1, \ldots, N_N, c) \tag{7}$$

$$s.t. \quad \beta(t) \geq \beta^{min} \quad , \quad 0 \leq t \leq T \tag{8}$$

$$\sum_{i=1}^{N} N_i \Delta t \leq T \tag{9}$$

$$c^{min} \leq c \leq c^{max} \tag{10}$$

The objective function is modelled as

$$C(N, N_1, \ldots, N_N, c) = C_I(c) + C_{IN}(N, N_1, \ldots, N_N, c) \\ + C_R(N, N_1, \ldots, N_N, c) + C_F(N, N_1, \ldots, N_N, c) \tag{11}$$

where

$$C_I(c) = C_{IO} + C_{I1} c \tag{12}$$

is the initial cost,

$$C_{IN}(N, N_1, \ldots, N_N, c) = \sum_{i=1}^{N} [(N_i - 1)\, C_{IN1} + C_{IN2}] + C_{IN1}\, (N_T - \sum_{i=1}^{N} N_i - 1) \tag{13}$$

is the inspection cost,

$$C_R(N, N_1, \ldots, N_N, c) = C_{RO} \sum_{i=1}^{N_T} E[R_i] \tag{14}$$

is the repair cost, and

$$C_F(N, N_1, \ldots, N_N, c) = C_{FO} \sum_{i=i}^{N_T} P_F(i\Delta t) \tag{15}$$

is the cost of failure. C_{IO}, C_{I1}, C_{IN1}, C_{IN2}, C_{RO} and C_{FO} are constants.

$E[R_i]$ is the expected number of repairs at the time $i\Delta t$ and $P_f(i\Delta t)$ is the probability of failure at the time $i\Delta t$. In the optimization problem the influence on the cost from the rate of interest is neglected.

The reliability index $\beta(t)$ is determined from

$$\beta(t) = -\Phi^{-1}(P_f(t)) \tag{16}$$

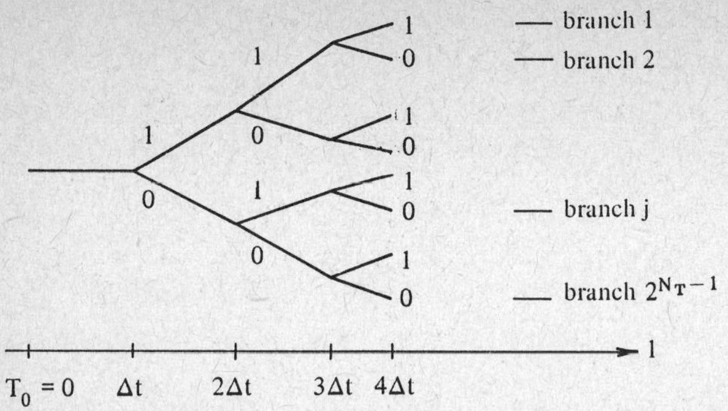

Figure 4. Repair realizations. 0 signifies non-repair and 1 signifies repair.

The total number of different repair courses (branches) are 2^{N_T-1}, see fig. 4. The safety margin corresponding to the event that the beam is repaired after a detailed inspection is written

$$M_D(t) = c_c - C_M \tag{17}$$

where C_M is the measured concentration of chlorides at the depth of the reinforcement. C_M is estimated from

$$C_M = [(c_0 - c_\infty)\, erf\,(\frac{c}{2\sqrt{at}}) + c_\infty]\, Z_D \tag{18}$$

where Z_D is a measuring uncertainty variable.

Correspondingly the safety margin for the event that repair is performed after a routine inspection is written

$$M_R(t) = Z_R(A(t) - c_1 A(0)) \tag{19}$$

where c_1 is a constant, $c_1 \in [0,1]$ and Z_R is a measuring uncertainty variable.

The probability of failure at the time t is determined as follows.

For $0 \le t \le t_1$:

$$P_F(t) = P(M_F(t) \le 0) \tag{20}$$

For $t_1 < t \le t_2$:

$$P_f(t) = P(B_1^1 \cap M_F^0(t) \le 0) + P(B_1^2 \cap M_F^1(t) \le 0) \tag{21}$$

where $M_F^1(t)$ is the failure safety margin at the time t given repair at the time t_1. B_1^1 and B_1^2 are the events that branches 1 and 2 occur, respectively. For example

$$B_1^2 = \{M_R(t_1) \le 0\} \tag{22}$$

For $t_2 < t \le t_3$:

$$\begin{aligned} P_f(t) = & P(B_2^1 \cap M_F^{00}(t) \le 0) + P(B_2^2 \cap M_F^{01}(t) \le 0) \\ & + P(B_2^3 \cap M_F^{10}(t) \le 0) + P(B_2^4 \cap M_F^{11}(t) \le 0) \end{aligned} \tag{23}$$

where for example

$$B_2^3 = \{M_R(t_1) \leq 0 \cap M_R^1(t_2) > 0\} \tag{24}$$

(23) is used if the second inspection is a routine inspection. If instead it is a detailed inspection $P_f(t)$ is determined from

$$\begin{aligned} P_f(t) = P(B_2^1 \cap M_F^{00}(t) \leq 0) + P(B_2^2 \cap M_F^{01}(t) \leq 0) \\ + P(B_2^3 \cap M_F^{10}(t) \leq 0) + P(B_2^4 \cap M_F^{11}(t) \leq 0) \end{aligned} \tag{25}$$

where for example

$$B_2^3 = \{M_R(t_1) \leq 0 \cap M_D^1(t_2) > 0\} \tag{26}$$

Correspondingly the expected number of repairs can be determined from

$$E[R_1] = P(M_R(t_1) \leq 0) \tag{27}$$

$$E[R_2] = P(B_1^1 \cap M_R^0(t_2) \leq 0) + P(B_1^2 \cap M_R^1(t_2) \leq 0) \tag{28}$$

$$\begin{aligned} E[R_3] = P(B_2^1 \cap M_R^{00}(t_3) \leq 0) + P(B_2^2 \cap M_R^{01}(t_3) \leq 0) \\ + P(B_2^3 \cap M_R^{10}(t_3) \leq 0) + P(B_2^4 \cap M_R^{11}(t_3) \leq 0) \end{aligned} \tag{29}$$

or if e.g. the first detailed inspection is performed at the third inspection

$$\begin{aligned} E[R_3] = P(B_2^1 \cap M_D^{00}(T_1) \leq 0) + P(B_2^2 \cap M_D^{01}(T_1) \leq 0) \\ + P(B_2^3 \cap M_D^{10}(T_1) \leq 0) + P(B_2^4 \cap M_D^{11}(T_1) \leq 0) \end{aligned} \tag{30}$$

In order to be able to evaluate $P_f(t)$ and $E[R_1]$ within a reasonable computer time the following simplifications are used in the example in section 5.

The inspections are grouped such that the routine inspections preceding a detailed inspection are treated as one group of inspections. Further, it is assumed that the probability of more than one repair is negligible. The probability of repair in $[0, T_1]$ is then approximated by

$$E[R_{T_1}] \cong (N_1 - 1)\, P(M_R(T_1) \leq 0) + P(M_D(T_1) \leq 0) \tag{31}$$

The probability of repair in $]T_1, T_2]$ is approximated by

$$\begin{aligned} E[R_{T_2}] \cong (N_2 - 1)\, P(M_D(T_1) > 0 \cap M_R^0(T_2) \leq 0) \\ + P(M_D(T_1) > 0 \cap M_D^0(T_2) \leq 0) \end{aligned} \tag{32}$$

The probability of repair in $]T_2, T_3]$ is approximated by

$$\begin{aligned} E[R_{T_3}] \cong (N_3 - 1)\, P(M_D^0(T_2) > 0 \cap M_R^{00}(T_3) \leq 0) \\ + P(M_D^0(T_2) > 0 \cap M_D^{00}(T_3) \leq 0) \end{aligned} \tag{33}$$

etc.

The probability of failure is approximated by

$$P_f(t) \cong \begin{cases} P(M_F(t) \leq 0) & \text{for} \quad 0 \leq t \leq T_1 \\ P(M_D^0(T_1) > 0 \cap M_F^0(t) \leq 0) & \text{for} \quad T_1 < t \leq T_2 \\ P(M_D^0(T_2) > 0 \cap M_F^{00}(t) \leq 0) & \text{for} \quad T_2 < t \leq T_3 \\ \text{etc.} \end{cases} \tag{34}$$

The Constraint (8) in the optimization problem is approximated by

$$\beta(T_i) \geq \beta^{min} \quad , \quad i = 1, 2, \ldots, N, N+1 \tag{35}$$

The probabilities and the sensitivity factors with respect to the optimization variables are determined using FORM (First Order Reliability Methods).

5. EXAMPLE

Consider the structural beam as shown in figure 1 with $b_1 = 1500$ mm, $b_2 = 400$ mm, $h_1 = 300$ mm, $h_n = 840$ mm, $d = 15$ mm and $n = 7$. Let the beam be subjected to a bending moment kP, where $k = 1500$ mm and where the load P is modelled as a log-normally distributed stochastic variable: P: LN(60,000 N, 12,000 N).

Further, let the yield stresses σ_y and σ_c be log-normally distributed:

σ_c: LN(36 N/mm^2, 7.5 N/mm^2)

σ_y: LN(600 N/mm^2, 60 N/mm^2)

The variables in eq. (2) are modelled as

a: LN(400 mm^2/year, 80 mm^2/year)

$c_c =$ 0.15

$c_0 =$ 0.04

$c_\infty =$ 0.2

The propagation velocity K is modelled as:

K: LN(0.2 mm/year, 0.04 mm/year)

The model uncertainty variables are modelled as:

Z_1: LN(1, 0.10)

Z_R: N(1, 0.10), N means normal distribution

Z_D: LN(1, 0.05)

The constant c_1 in eq. (19) is put equal to $c_1 = 0.5$.

In the optimization problem (7) - (10) the parameters are chosen as

β^{min} = 4.0

T = 70 years

Δt = 1 year

c^{min} = 30 mm, c^{max} = 60 mm

and (the costs are in 10^3 DKK).

$C_{I0} = 2000$, $C_{I1} = 200$, $C_{IN1} = 10$, $C_{IN2} = 100$, $C_{R0} = 200$ and $C_{F0} = 1000$.

The optimization problem (7) - (10) is a mixed integer/pseudo-integer/real variable problem. It is solved sequentially for fixed N treating $N_1, \ldots, N_N$ as real variables and using the NLPQL algorithm developed by Schittkowski [11]. The final values for $N_1, \cdots, N_N$ are chosen as the integers closest to the final values.

The result is: $N = 5$, $N_1 = 14$, $N_2 = 13$, $N_3 = 14$, $N_4 = 13$, $N_5 = 8$ and $c = 30$ mm.

This result is illustrated in figure 5.

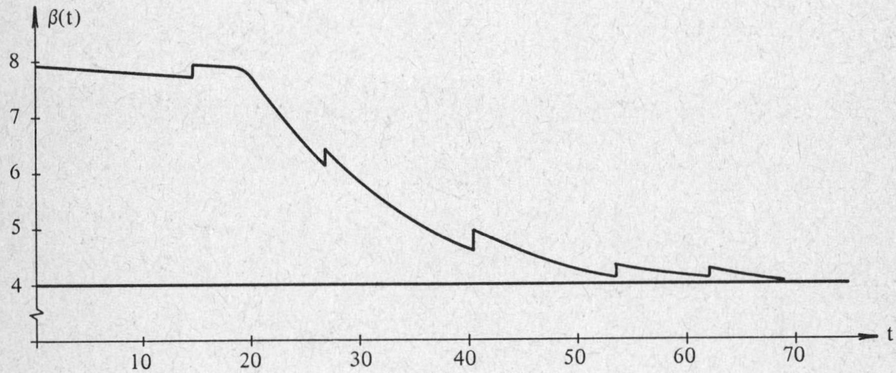

Figure 5. Reliability index β as a function of time.

The figure shows that the reliability index is only decreasing slightly for the first approximately 20 years (corresponding to the expected value of the initiation time T_I for the chloride to penetrate the concrete). Further, it is seen that each inspection (the detailed inspections and the routine inspections are grouped) gives only a small increase in the reliability index β, i.e. the effect of the inspection on the reliability is in this example not significant.

The long time intervals between detailed inspections ($\approx$ 13 - 14 years) in this example seem to be unrealistic. The reason for these large values could be that only corrosion due to chloride penetration is taken into account. The time intervals are highly sensitive to the input data.

6. CONCLUSION

A method to determine an optimal inspection strategy for concrete bridges is presented. The failure mode considered is corrosion of the reinforcement due to chloride penetration.

The inspection strategy is modelled on the basis of the current practice in many countries, where routine inspections are performed approximately once a year and detailed inspections are performed at longer intervals (three to five years).

The optimal inspection strategy is determined from an optimization problem, where the objective function is the expected total cost, and the constraints ensure that the reliability is satisfactory at any time. The design variables are the number of inspections, the time intervals between inspections and a design parameter (the concrete cover).

As an example a single reinforced concrete beam is considered. Using some simplifications with regard to determination of probabilities of failure and repair the optimal inspection strategy is determined. The example shows that inspections with the present modelling have only a small significance for the reliability index. However, the example indicates that the proposed procedure can be applied to more general examples where the total structural system is considered.

7. REFERENCES

[1] *Bridge Evaluation, Repair and Rehabilitation.* Proceedings of the 1st US-European Workshop. Editors A. S. Nowak & E. Absi. The University of Michigan, Ann Arbor, Michigan, USA, 1987.

[2] Lemarié, P. *Bridge Management and Maintenance in France.* In [1], pp. 51-57.

[3] De Buck, J. *Bridge Maintenance in Belgium.* In [1], pp. 63-98.

[4] Rostam, S. *Bridge Monitoring and Maintenance Strategy.* In [1], pp. 146-153.

[5] Zichner, T. *Disorder - Repair - Strengthening of Prestressed Concrete Bridges based on the Inspection of 43 Valley-Bridges.* In [1], pp. 168-179.

[6] Favre, R. *Methodology for the Monitoring of Bridges.* In [1], pp. 197-208.

[7] Malisardi, L. & Nebbia, G. *Supervision of Bridge and Tunnel Structures.* In [1], pp. 219-233.

[8] Vassie, P. *Reinforcement Corrosion and the Durability of Concrete Bridges.* Proc. Instn. Civ. Engrs., Part 1, Vol. 76, 1984, pp. 713-723.

[9] Andersen, K. & Lyck, S. *Optimal Repair Strategies for Concrete Bridges* (in Danish). M.Sc. Thesis, University of Aalborg, 1988.

[10] Thoft-Christensen, P. & Sørensen, J. D. *Integrated Reliability-Based Optimal Design of Structures.* Proc. IFIP WG 7.5 Conference on Reliability and Optimization of Structural Systems. Springer-Verlag, Lecture Notes in Engineering, Vol. 33, 1987, pp. 385-398.

[11] Schittkowski, K. *NLPQL : A FORTRAN Subroutine Solving Constrained Non-linear Programming Problems.* Annals of Operations Research, 1986.

THE VIRTUAL DISTORTION METHOD APPLIED TO THE RELIABILITY ANALYSIS OF OFFSHORE STRUCTURES

R. C. Turner* **, J. T. Gierlinski**, G. M. Zintilis**
M. J. Baker* & J. Holnicki-Szulc***

*Imperial College of Science, Technology and Medicine, London, UK
**W. S. Atkins Engineering Sciences, Epsom, UK
***Institute of Fundamental Technological Research, Warsaw, Poland

ABSTRACT

The recently developed virtual distortion method of nonlinear structural analysis is an efficient analysis tool ideally suited to repetitive analysis. The method is briefly described in the first part of the paper, and is used to analyse the progressive collapse of a jacket-type truss structure under cyclic loading.

The second part of the paper concerns the use of VDM within reliability procedures. Some problems were encountered in using advanced structural analysis techniques in conjunction with simple reliability theory - these problems concern FOSM/SORM system analysis methods and are of general interest. The problems are discussed, and a number of potential solutions are offered. Although one approach has been shown to be feasible, the topic is still under investigation to find a more satisfying solution. Suggestions are welcome.

INTRODUCTION

In many areas, including the offshore industry, there is a rapidly growing interest in probabilistic methods and reliability assessment. Structural reappraisal, and the optimization of inspection, maintenance and repair strategies are particular areas of interest. They are also characteristically complex problems which require careful modelling. Taking account of failure due to both extreme loading and fatigue, structural life is sensitive to structural behaviour and the loading processes. Thus, realistic and rational models should be used if engineers are to have confidence in the results, and if reliability-based methods are to be accepted.

In the past, structural behaviour in system reliability procedures has been idealised and simplified. Often elastic-perfectly plastic, or elastic-brittle member behaviour has been assumed. Such assumptions are inappropriate for modelling offshore structures which have complex framing and which are generally composed of semi-slender members. Procedures based on the member replacement/modification approach, which involve structural re-analysis, have been developed for first-order reliability methods to model a more generalised member behaviour. However, the behaviour is still idealised, and the methods can be criticised because of their treatment of the loading process or load path. Conventional nonlinear analysis methods are, in general, too expensive to use with reliability methods. However, a recently developed method of nonlinear structural analysis - the Virtual Distortion Method - offers significant improvements in efficiency, especially for repetitive analysis. The method uses the principle of superposition and is closely related to the principle of virtual work. Its application with reliability methods is the main topic of this paper.

The work presented in this paper is closely related to the BRITE P1270 project set up and partly sponsored by the Commission of European Communities. The project, whose title is "Reliability Methods for the Design and Operation of Offshore Structures", started in December 1986 and is scheduled to finish in December 1990. The research work, with a total value of 2,470K ECU, is shared by eleven contractors with five industrial sponsors, as listed in the acknowledgement at the end of this paper. The project is separated into four sub-projects:

I. Statistics of basic variables, linear component modelling and environmental loading,

II. Nonlinear behaviour of damaged or intact components and jacket progressive collapse,

III. System reliability,

IV. Inspection, maintenance and repair.

The project covers a number of topics, and this paper is only concerned with some of the work in sub-projects II and III. One of the important aspects of this BRITE project is the development of the Virtual Distortion Method (VDM) for nonlinear structural analysis.

PROGRESSIVE COLLAPSE ANALYSIS USING VDM

Concept of the Virtual Distortion Method

The virtual distortion method is based on the concept of superimposing linear and virtual states of a structure. Linear states are obtained from a linear solution, and virtual states are generated through the introduction of virtual distortions into selected finite elements within the structure. To explain this idea, consider a single loading step, and assume that S_L is the original linear state of a structure, and S_M is the final modified state. The fundamental postulate of the method states that

$$S_M = S_L + S_V \tag{1}$$

where S_V is the virtual state determined as a linear function of virtual distortions $\overset{0}{\epsilon}$. In the case of an element of a skeletal structure, distortions $\overset{0}{\epsilon}$ will have the form of member shortening/elongation, curvature, etc. An arbitrary state S of the structure is defined (known) if ϵ and σ are known, where ϵ are generalised strains and σ are generalised stresses. Equation (1) can thus be replaced by a system of equations for strains $\overset{M}{\epsilon}$ and stresses $\overset{M}{\sigma}$ in the modified structure

$$\overset{M}{\epsilon} = \overset{L}{\epsilon} + \overset{V}{\epsilon} \tag{2}$$

$$\overset{M}{\sigma} = \overset{L}{\sigma} + \overset{V}{\sigma} \tag{3}$$

where $\overset{L}{\epsilon}$ and $\overset{L}{\sigma}$ are strains and stresses in the original structure due to the external load,
$\overset{V}{\epsilon}$ and $\overset{V}{\sigma}$ are virtual strains and virtual stresses in the original structure caused by the virtual distortions $\overset{0}{\epsilon}$.

Strains $\overset{V}{\epsilon}$ and stresses $\overset{V}{\sigma}$ are defined as follows:

$$\overset{V}{\epsilon} = D\overset{0}{\epsilon} \tag{4}$$

$$\overset{V}{\sigma} = E\,(D-I)\overset{0}{\epsilon} \tag{5}$$

where D is the influence matrix of virtual distortions
I is the unit matrix
E is the vector of elastic properties.

Substitution of definition (4) for virtual strains and (5) for virtual stresses into the postulated equations (2) and (3) respectively, gives the final form of definitions for strains $\overset{M}{\epsilon}$ and stresses $\overset{M}{\sigma}$ which are generated in the modified structure.

$$\overset{M}{\epsilon} = \overset{L}{\epsilon} + D\overset{0}{\epsilon} \tag{6}$$

$$\overset{M}{\sigma} = \overset{L}{\sigma} + E\,(D-I)\overset{o}{\epsilon} \qquad (7)$$

Imposing constraints on the final strains $\overset{M}{\epsilon}$ or stresses $\overset{M}{\sigma}$ provide the equations necessary to calculate the virtual distortions $\overset{o}{\epsilon}$. These constraints can, in general, be presented as inequalities

$$\overset{M}{\epsilon} \leqslant f\left[\,\bar{\epsilon},\ \bar{\sigma},\ \overset{L}{\epsilon},\ \overset{o}{\epsilon}\,\right]$$

or

$$\overset{M}{\sigma} \leqslant f\left[\,\bar{\epsilon},\ \bar{\sigma},\ \overset{L}{\epsilon},\ \overset{o}{\epsilon}\,\right]$$

where: $\bar{\epsilon}$ and $\bar{\sigma}$ denote the limit value of strain and stress components, respectively.

In the case of the progressive collapse analysis of skeletal structures, a simple form of the second group of inequalities is employed. Namely, that for the overloaded structural members, the equality

$$\overset{M}{\sigma} = f\left[\,\bar{\epsilon},\ \bar{\sigma},\ \overset{L}{\epsilon},\ \overset{o}{\epsilon}\,\right] \qquad (8)$$

is satisfied, where $\bar{\sigma}$ is equal to a yield stress, or to zero if the member is fractured, and $\bar{\epsilon}$ is the strain fracture limit.

Restricting, for simplicity, further considerations to truss–type structures, equation (8) becomes

$$\overset{M}{\sigma} = \left[\,1 - \alpha\,\right]\bar{\sigma} + \alpha E\,\left[\,\overset{L}{\epsilon} + D\,\overset{o}{\epsilon}\,\right] \qquad (9)$$

for plastic members, and

$$\overset{M}{\sigma} = 0 \qquad (10)$$

for fractured members. In equation (9), α denotes the hardening/softening slope on the stress–strain diagram. From the above, the corresponding equations for virtual distortions have the following form

$$\left[\,D\left[\,1-\alpha\,\right] - I\,\right]\overset{o}{\epsilon} = -\,E^{-1}\left[\,\overset{L}{\sigma} - \bar{\sigma}\,\right]\left[\,1-\alpha\,\right] \qquad (11)$$

for plastic members, and

$$(D-I)\overset{o}{\epsilon} = -\,E^{-1}\,\overset{L}{\sigma}. \qquad (12)$$

for fractured members.

The equations (11) and (12) constitute the fundamental set of equations of the virtual distortion method. Solution of these equations provides the virtual distortions $\overset{o}{\epsilon}$, which, when substituted into (6) and (7) define the state S_M of the modifed structure. As mentioned above, the virtual distortions $\overset{o}{\epsilon}$ are introduced only into the overloaded members - plastic or fractured. Thus the number of equations included in (11) or (12) is usually very small in comparison with the size of the system of equilibrium equations. This fact is the source of the considerable computational efficiency of the virtual distortion method.

If the load is applied in an incremental manner, strains, stresses and virtual distortions have to be accumulated at the end of each increment. The corresponding algorithm of computational steps with for the assumption of perfect plasticity was discussed in detail by Holnicki and Gierlinski (1988); other areas for potential applications of VDM were also emphasised. Generalisation of the algorithm to hardening/softening plasticity will be presented elsewhere. Below, an example of an offshore jacket under non-proportional loading and with members exhibiting hardening/softening behaviour is presented.

Progressive Collapse Analysis of an Offshore Jacket

To illustrate the application of VDM, the progressive collapse analysis of an offshore jacket modelled as a pin-jointed structure (shown in Figure 1) is considered. It is assumed that the stress-strain diagram for each individual member is non-symmetric, with hardening on the tension side and softening on the compression side to imitate the buckling phenomenon (Figure 2). The derivation of a more accurate nonlinear member model to be used with VDM will be presented elsewhere. Two independent sets of loads are acting on the jacket; vertical dead load, and horizontal load to model wave loading. The load history, describing the changes to each load set, is shown in Figure 3. The structural response was monitored at the end of each load increment. Initially, during increments 1 and 2, the jacket remains elastic. Then plastic loading and unloading follow, with an increasing number of damaged members at subsequent peaks of horizontal load, until total collapse occurs during increment 9. The horizontal displacement at the top is shown in Figure 4, and the sequence of member failures is given in Table 1. The position of the damaged members is also shown in Figure 5.

VDM APPLIED TO RELIABILITY METHODS

The efficiency of VDM, especially for repetitive analysis, makes it an ideal tool for use with reliability methods. VDM can be used with simulation techniques to estimate the system reliability directly, and it can be used with FOSM (or SORM) methods to evaluate the probability of individual failure sequences.

Nonlinear FE programs have been used to evaluate component reliabilities for a number of years, and it is not the first time that they have been used with system reliability methods. Edwards *et al* (1985) used a general nonlinear program with a simulation technique to analyse a 2-D idealisation of an offshore structure; because of the expense of the analysis they only used 60 trials. Amdahl *et al* (1987) combined an advanced progressive collapse analysis program with first-order reliability methods to analyse space frames. By analysing a simple structure they found that it was necessary to control the formation of plastic hinges. This often precludes standard nonlinear programs from use with failure tree enumeration methods because it is not usually possible to control the sequence of member failures.

However, because the VDM program can be modified to accept a prescribed failure sequence, it was not anticipated that there would be any serious difficulties in applying VDM to system reliability methods. Unfortunately, some difficulties were encountered with failure tree enumeration methods. These problems are general in nature and stem from the basic assumptions of the system reliability methods.

Review of Basic Failure Tree Enumeration Theory

From basic reliability theory, the probability of failure of the system $(P_f)_{sys}$ may be evaluated by taking the union of the failure probabilities for all the sequences F_i leading to system failure.

$$(P_f)_{sys} = P\,\{F_1 \cup F_2 \cup \ldots\ldots\, F_m\} \tag{13}$$

Generally, it is impractical to evaluate every potential sequence, so a strategy is used to evaluate the stochastically significant sequences. For special cases, when the member behaviour is idealised and system failure is defined by structural collapse, the events F can be evaluated and manipulated directly. However, a more general approach is to use a failure tree enumeration strategy. The probability for a failure sequence F_i is then computed from the joint probabilities of the consecutive failure events.

$$P\,\{F_i\} = P\,\{E_a \cap E_{b|a} \cap E_{c|a,b} \cdots\cdots\} \tag{14}$$

The event E_a represents the failure of member **a**, and $P\{E_a\}$ is the probability that at least member **a** has failed. It is evaluated from the results of an elastic analysis for the intact structure using a safety margin Z_a:

$$Z_a = g\,(\bar{R}_a, \bar{S}) \tag{15}$$

The failure function is generally defined explicitly, and although the failure surface given by $Z_a = 0$ is only expressed in terms of the loading variables, $\bar{S}$, and member resistance variables, $\bar{R}_a$, it can be assumed to represent a hypersurface in the space of all the basic variables.

The event $E_{b|a}$ represents the failure for member **b** given that member **a** has failed, i.e. it is a conditional event.

The early methods of structural systems analysis were for one dimensional failure functions and elastic-perfectly plastic or elastic brittle member behaviour. Then the safety margin was derived from successive elastic analysis and the Artificial Load Concept. That is, the structure was re-analysed with the failed member(s) removed for brittle behaviour, or else replaced by unit artificial loads - the effects of which could then be factored to represent the plastic capacity of the damaged member(s).

Hence $$Z_{b|a} = g\ (\bar{R}_b, \bar{R}'_a, \bar{S}) \qquad (16)$$

where $\bar{R}'_a$ represent the post-failure resistance variables for member **a**

Again, the failure surface given by $Z_{b|a} = 0$ represents a hypersurface in basic variable space. The safety margin $Z_{b|a}$ is defined explicitly, and the probability of failure for the conditional event can be evaluated using a suitable calculation procedure.

With this approach for strain independent member post-failure behaviour it is important to realise two points. Firstly, the safety margin can be derived independently of any load path assumptions. Thus, although the deterministic behaviour is load path dependent, a state in probability space is load path independent. Secondly, it is assumed that failed members do not unload, i.e. strain reversal does not cause failed members to revert to elastic behaviour.

In practice, member behaviour generally differs considerably from idealised elastic plastic or elastic brittle behaviour. So the basic approach has been extended to treat bi-linear member behaviour by replacing the failed member with its post-failure stiffness and re-analysing the structure elastically (Moses and Rashedi (1983), Melchers and Tang (1984), and others). Thoft-Christensen and Murotsu (1986) have also applied the method to a linearised moment/axial force interaction, and recently Turner and Baker (1987) have shown that multi-dimensional failure criteria and the associated flow rule can be considered.

However, when the member post-failure behaviour is strain dependent the safety margin can only be defined by assuming a load path; often the load is assumed to be applied proportionally in basic variable space.

There is a serious consequence of this load path requirement, since now for an event $E_{b|a}$ the failure space defined by $Z_{b|a} \leq 0$ includes regions which cannot physically occur. The event $E_{b|a}$ is conditional on member **a** having failed, i.e. $Z_a \leq 0$. Thus, by following a deterministic load path the event $E_{b|a}$ is physically undefined unless $Z_a \leq 0$. The shaded region in Figure 6 illustrates the intersection between the two events E_a and $E_{b|a}$; it represents the probability for the failure sequence $a \rightarrow b$.

However, the shaded region may still contain events which are physically inadmissible, for instance, in reality member **b** may fail first. Thus the probability that both members **a** and **b** fail is evaluated as:

$$P \{ (E_a \cap E_{b|a}) \cup (E_b \cap E_{a|b}) \} \qquad (17)$$

By using the member replacement/modification method the failure surfaces for the individual events E are defined explicitly as continuous mathematical functions, and the probability for the events can be evaluated individually using FOSM (or SORM) procedures. That is, the failure surfaces are approximated by linear or quadratic functions defined at a design point.

Of course, a design point may be on a section of the failure surface that is physically undefined (Turner and Baker, (1987)). However, since such a point is only used as a mathematical tool to define an approximate surface, it does not usually cause concern that the point is not physically valid because it is eliminated when the system reliability is evaluated.

Unfortunately, a problem occurs when the safety margin is not defined explicitly, for instance, when a nonlinear analysis program is used; since calculations cannot be undertaken in inadmissible regions. Hence, design points on physically undefined failure surfaces cannot normally be evaluated.

Failure Tree Enumeration Using VDM

In principle, it is straightforward to incorporate the VDM program into a standard reliability analysis algorithm, and at Imperial College it has been used with the advanced FOSM program RELY. However, for general applicability two modifications must be made.

First, as already mentioned, the failure sequence must be controlled. For a deterministic set of values for the basic variables the failure sequence is unique. Thus, a standard nonlinear program will automatically produce the complete failure sequence for the structure under the given set of material properties, geometry, loads and load path. A small change in the value of some of the basic variables may produce a different sequence.

A reliability procedure for failure tree enumeration requires the margin of safety for a chosen member following a given sequence of failures; this needs to be evaluated repeatedly for different sets of values. Thus the VDM program has been modified to re-distribute the stresses for some/all the overstressed members in a prescribed sequence.

Secondly, the problem of physically inadmissable design points must be tackled. It is incorrect to ignore them because, even for valid design points, convergence problems will occur with conventional FOSM/SORM routines if the search for the closest point strays into a physically inadmissible region. Also, some of the events with inadmissible design points may have a significant effect on the system reliability. Indeed, it is not impossible for all the significant events to have inadmissable design points.

One solution, which has been shown to be feasible, is to "trick" the VDM program to change the state of member(s). Without altering the global loading on the structure, a set of virtual distortions can be introduced which have the effect of changing the state of a member from, say, an elastic state to a fictitious plastic state, i.e. the stresses are re-distributed such that the stress state for the "failed" member is forced into the post-elastic branch of the member behaviour. Figure 7 shows the typical effect - the new state is often forced onto an extended branch of the behaviour. This is further demonstrated in the next section with the analysis of an example. The results can be shown to correspond to those calculated using a conventional member replacement/modification method.

However, from an engineering view point "tricking" VDM is not a particularly satisfying solution, since who wants to analyse the impossible?

An alternative approach, which is under investigation, is to directly estimate the joint probability of a failure sequence by searching for a global design point. The method would be based on optimization techniques under multiple constraints. Such a procedure, if it can be developed, would need to deal with implicitly defined constraints (failure boundaries) which may not be defined continuously throughout the variable space.

A different solution is to use Monte Carlo methods or simulation. However, although recently proposed sampling techniques have been shown to be efficient with simple examples, they are largely untested with large multi-mode systems.

The problem is still under investigation and any suggestions would be welcome.

Reliability Analysis of an Offshore Jacket

The same structure as used in the first part of the paper (Figure 1) has been partially analysed probabilistically.

Load multiplier data:

$E[\lambda_V] = 1.0$ CoV = 0.05
$E[\lambda_A] = 1.0$ CoV = 0.267
$E[\lambda_B] = 1.0$ CoV = 0.267
$E[\lambda_C] = 1.0$ CoV = 0.267
$E[\lambda_D] = 1.0$ CoV = 0.267

The correlation coefficient between the horizontal loads on the legs A,B,C,D was assumed to be 0.5, and independent of the vertical loads, V. In this example the loading was applied proportionally.

Member data:

$E[F_y] = 190\ N/mm^2$ CoV = 0.055 $F_a = 0.895\ F_y$

where F_y = limit stress in tension F_a = limit stress in compression

$E[\gamma] = 1.0 \qquad CoV = 0.05$

where γ is a model uncertainty coefficient.

The yield stress and model uncertainty coefficient were assumed uncorrelated and independent between members.

All variables were normally distributed.

Failure function:

$g_i^- = \gamma_i + f_i/F_{ai}$ compression $\qquad g_i^+ = \gamma_i - f_i/F_{yi}$ tension

where f_i is the axial stress for member i.

Member post-failure behaviour:

For the analysis, three types of behaviour were used.

A. Compressive softening/tensile hardening as in Figure 2
B. Compressive ductile/tensile ductile
C. Compressive softening/tensile ductile

It must be mentioned that the three member behaviours were chosen to demonstrate the versatility of the method, and to show the effect on the reliability. In reality, a pin-ended member with deterministic geometry has a unique behaviour.

For simplicity, only four members are assumed to be potential failure elements. These are braces, 97, 98, 99 and 100.

Results

The structure was analysed using a version of the VDM program with the Imperial College advanced FOSM program RELY. VDM was modified to accept prescribed failure sequences, and to "trick" members into failing.

A typical failure sequence with an inadmissible design point is 97 → 98. Figure 8 shows the stress strain behaviour for both members evaluated at the design point for case A. Position 1 in the figure represents the elastic stress state after the full application of the load. Note the stress state for member 97 is below the limit state. Thus, ordinarily, member 97 would not fail. By "tricking" VDM, the state for 97 is forced to position 2. This involves strain reversal and is obviously physically impossible. The redistribution causes unloading in member 98.

The values of the basic variables at the design point are:

λ_V^*	λ_A^*	λ_B^*	λ_C^*	λ_D^*	γ_{97}^*	F_{y97}^*	γ_{98}^*	F_{y98}^*
.9981	1.6246	1.6965	1.5172	1.6366	.9798	185.2	.9611	180.8

These values can be checked using the member replacement/modification method. From the results of an elastic analysis for the intact structure the failure function for member 97 can be written as:

$$g_{97} = \gamma_{97}^* F_{y97}^* - k\,(-\,2.958\,\lambda_V^* + 23.51\,\lambda_A^* + 41.76\,\lambda_B^* + 2.973\,\lambda_C^* + 27.65\lambda_D^*) = 0$$

where k is a load factor, and at the design point k = 1.164.

By replacing 97 by its post-failure stiffness and re-analysing the structure, and by using the results of the intact analysis, the safety margin for member 98 given the failure of 97 can be written as:

$$g_{98|97} = 0.895\,\gamma_{98}^* F_{y98}^* + k\,(6.936\lambda_V^* - 28.41\lambda_A^* - 41.72\lambda_B^*$$

$$- 5.380\lambda_C^* - 30.11\lambda_D^*)$$

$$+ (1.0 - k)\,(8.343\lambda_V^* - 39.60\lambda_A^* - 61.58\lambda_B^*$$

$$-6.794\lambda_C^* - 43.26\lambda_D^*)$$

Hence at the design point $g_{98|97} \approx 0$

Part of the failure tree for the three member behaviour assumptions is shown in Figure 9. The reliability indices for case B are shown in brackets ( ) and for case C in braces { }.

The results show that there is very little difference between the reliabilities for case A (softening/hardening behaviour) and case B (ductile behaviour). The reliability increases steadily at higher levels of damage. However, there is considerable difference in the results for case C (softening/ductile behaviour). After two members fail there is very little reserve of strength left and the structure effectively fails.

All semi-slender members exhibit member "softening" behaviour in compresssion due to Euler buckling. Thus, reliability methods which assume perfectly ductile failure behaviour in compression may significantly overestimate the system reliability.

CONCLUSIONS

The Virtual Distortion Method of nonlinear structural analysis is a very efficient tool. It can analyse structures with generalised member behaviour, non-proportional load paths, cyclic loading, and is ideally suited to repetitive analysis such as optimal design and reliability analysis.

VDM has been modified for use with conventional first-order reliability methods and failure tree enumeration. However, improvements are needed to the system reliability procedure.

ACKNOWLEDGEMENTS

This paper represents part of the results of BRITE Project P1270, "Reliability Methods for Design and Operation of Offshore Structures". This project is funded through a 50% contribution from the Directorate General for Science, Research and Development of the Commission of the European Communities and another 50% contribution from the partners in the project and five industrial sponsors. The partners in this project are: TNO (NL), d'Appolonia (I), RINA (I), Snamprogetti (I), Aalborg University (DK), Imperial College (UK), Atkins ES (UK), Elf (F), Bureau Veritas (F), CTICM (F) and Ifremer (F). The industrial sponsors in this project are: Shell Research, The Hague (NL), Shell UK (UK), Maersk (DK), NUC (DK) and Ramboll and Hannermann (DK).

REFERENCES

1. Amdahl, J., Leira, B. and Wu, Yu-Lin. (1987). "On the application of a nonlinear finite element formulation in structural systems reliability". Proc. 1st IFIP WG 7.5 Conf., Aalborg, Denmark, Springer-Verlag, 1987.

2. Edwards, G.E., Heidweiller, A., Kerstens, J. and Vrouwenvelder, A. (1985). "Methodologies for limit state reliability analysis of offshore jacket platforms". Proc. 4th BOSS Conf., Delft, The Netherlands.

3. Holnicki-Szulc, J. and Gierlinski, J.T. (1988), "Structural Modifications simulated by Virtual Distortions". Int. J. Num. Methods Eng., to be published.

4. Melchers, R.E. and Tang, L.K. (1984). "Dominant failure modes in stochastic structural systems". Structural Safety, 2 (2).

5. Moses, F and Rashedi, M.R. (1983). "The application of system reliability to structural safety". Proc. 4th ICASP Conf., Florence, Italy.

6. Thoft-Christensen, P. and Murotsu, Y. (1986). Application of structural systems reliability theory. Springer-Verlag.

7. Turner, R.C. and Baker, M.J. (1987). "Structural system reliability analysis using multi-dimensional limit state criteria". Proc. 1st IFIP WG 7.5 Conf., Aalborg, Denmark, Springer-Verlag, 1987.

Table 1

Load Increment	Member Status			
1	all elastic			
2	all elastic			
3	98P			
4	98U			
5	all elastic			
6	98P	97P	99P	100P
7	98U	97U	99U	100U
8	all elastic			
9	98P	97P	99P	100P
	120P	118P ;	global collapse	

P – plastic

U – unloaded

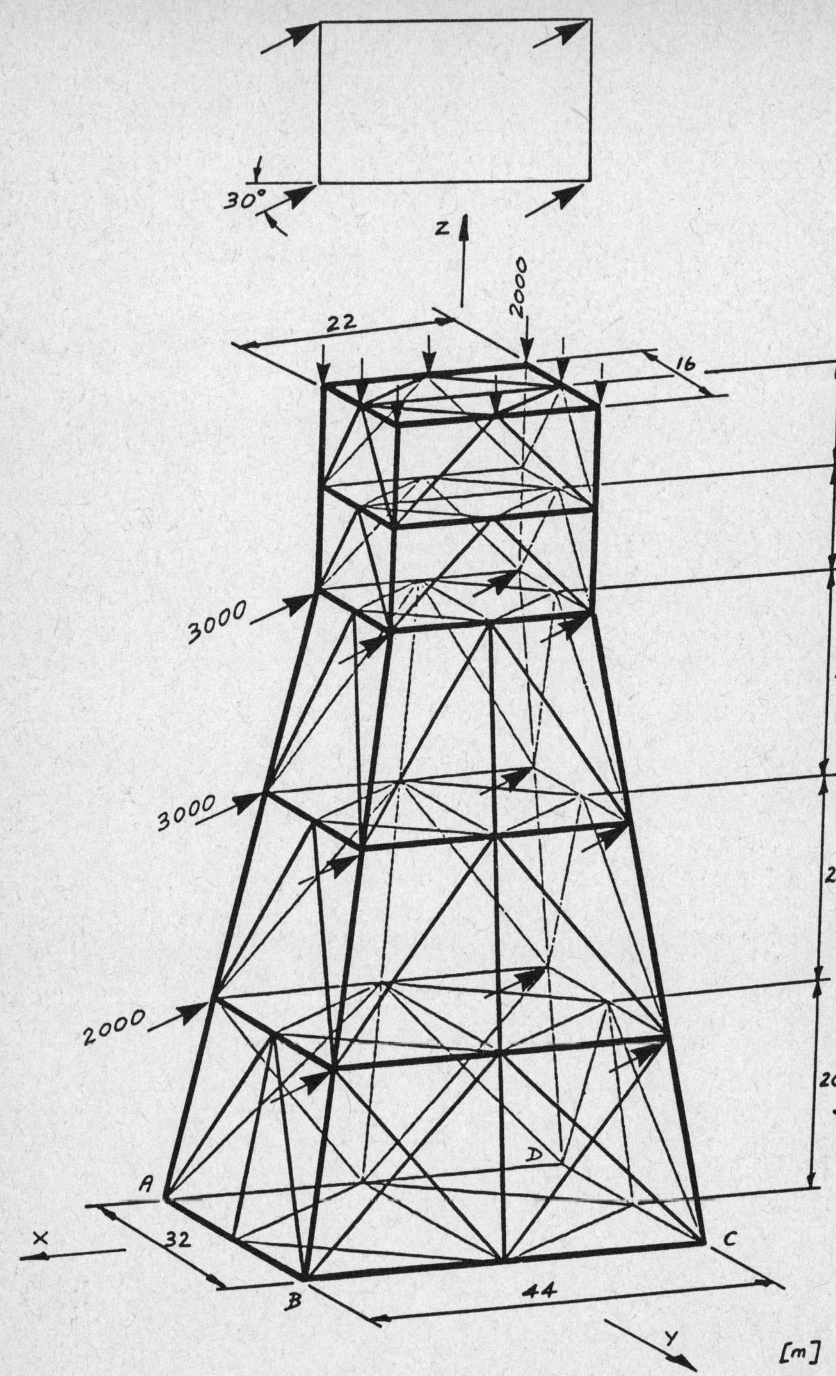

Figure 1: Offshore Jacket

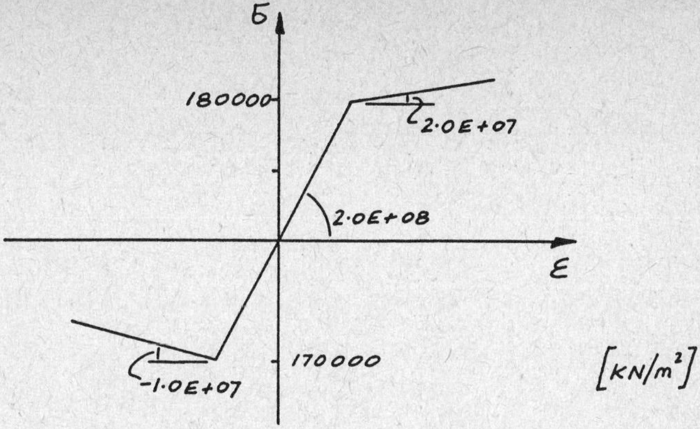

Figure 2: Non–linear Material Model for Members; Tension and Hardening, Compression and Softening

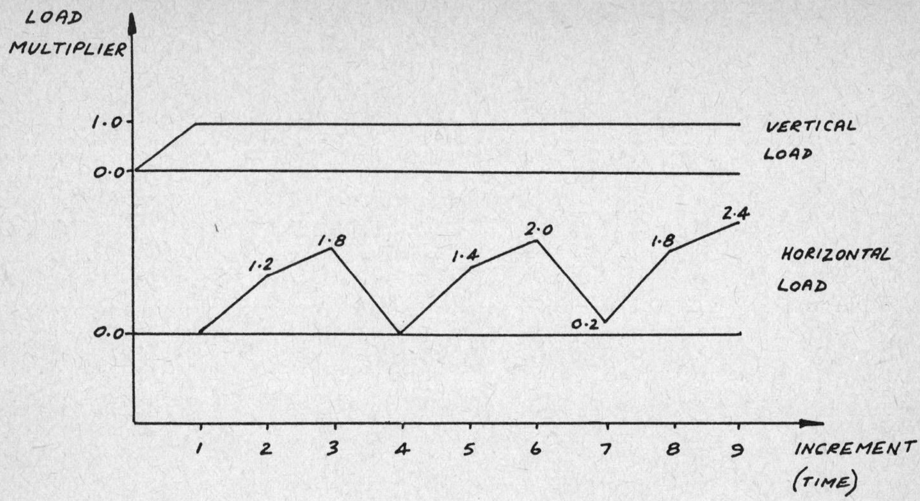

Figure 3: Load History: Non-proportional Loading

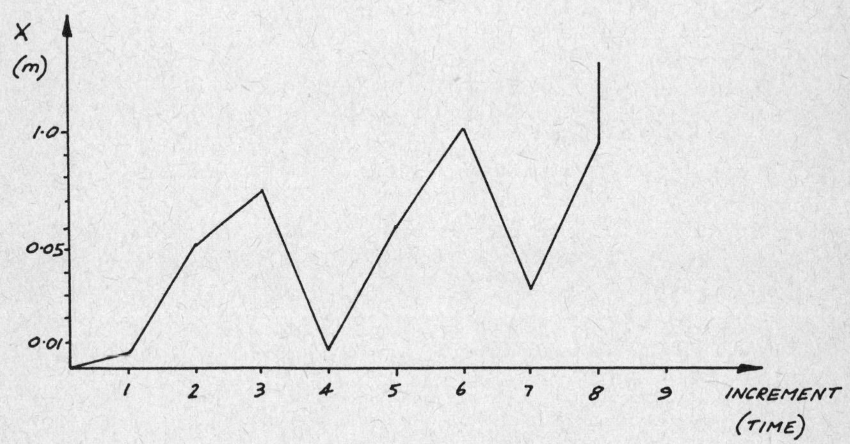

Figure 4: Displacement History of Top Deck

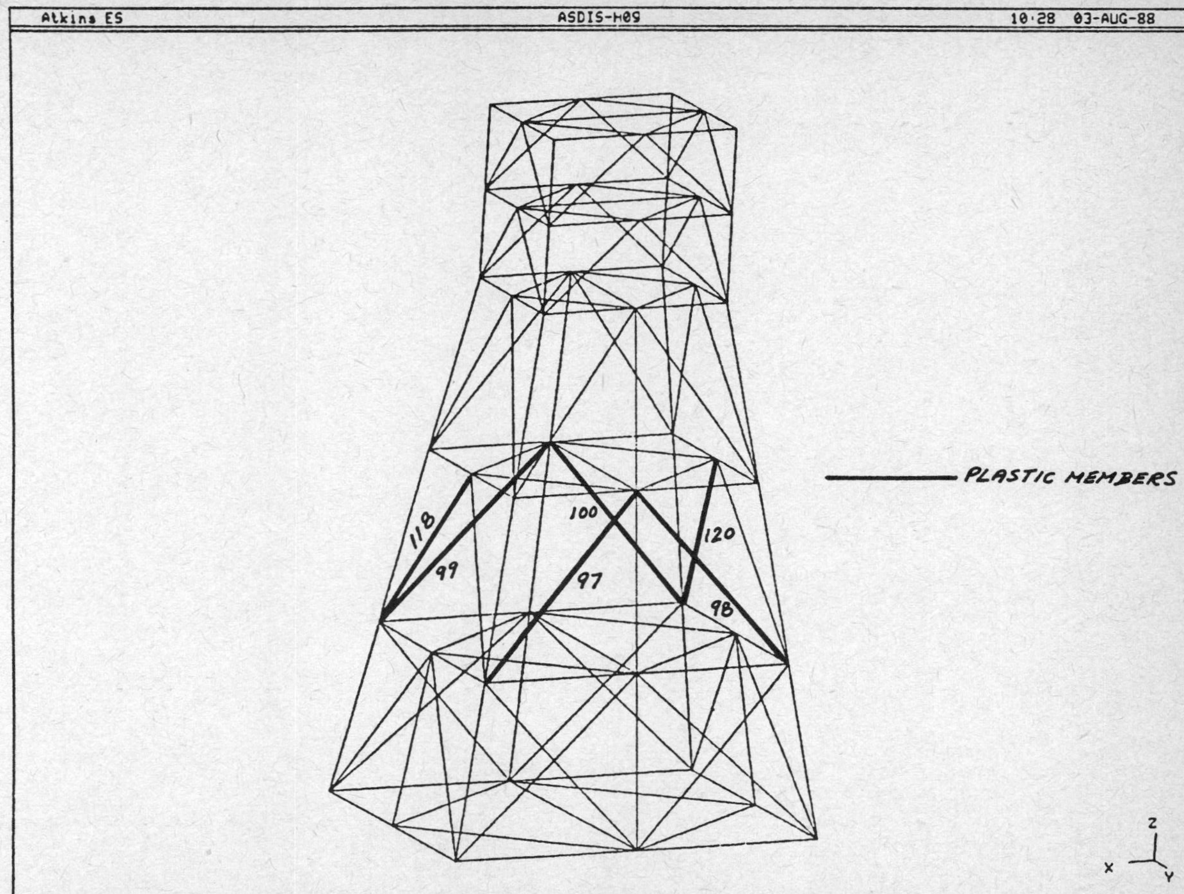

Figure 5: Position of Plastic Members in the Collapse Stage

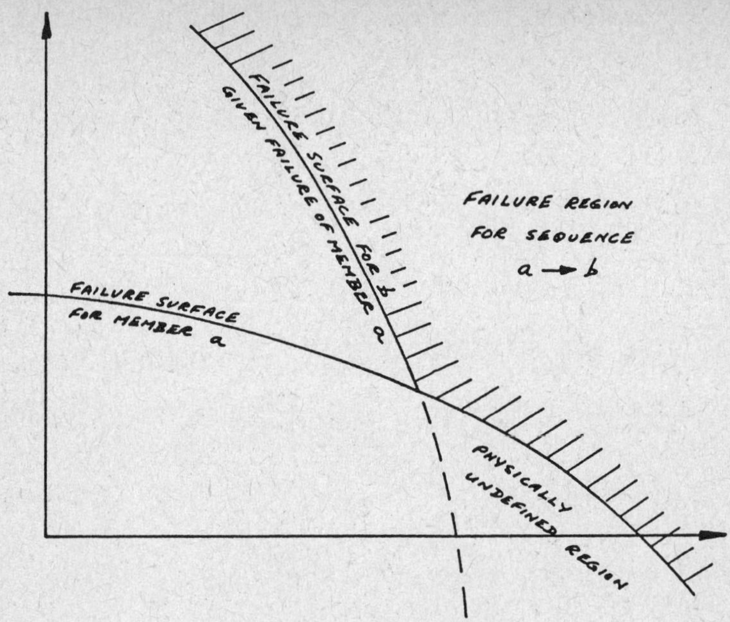

Figure 6: Failure Surfaces in 2-D Space

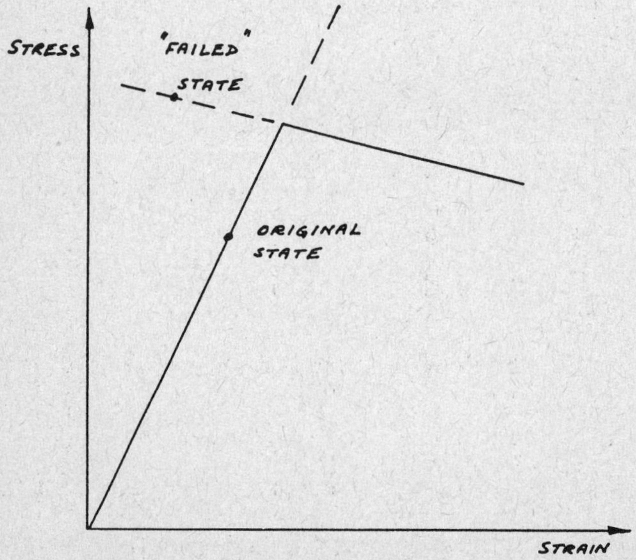

Figure 7: Stress-strain Diagram for "Tricked" Behaviour

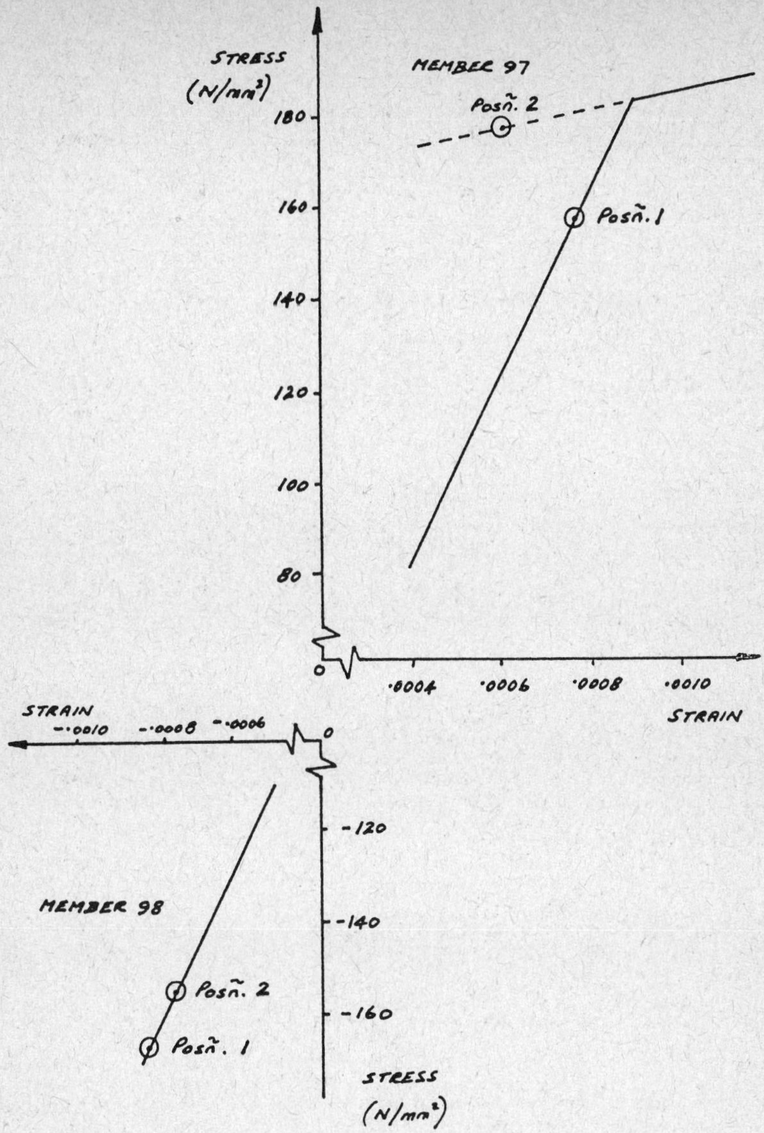

Figure 8: Stress-strain Diagrams for Members 97 and 98

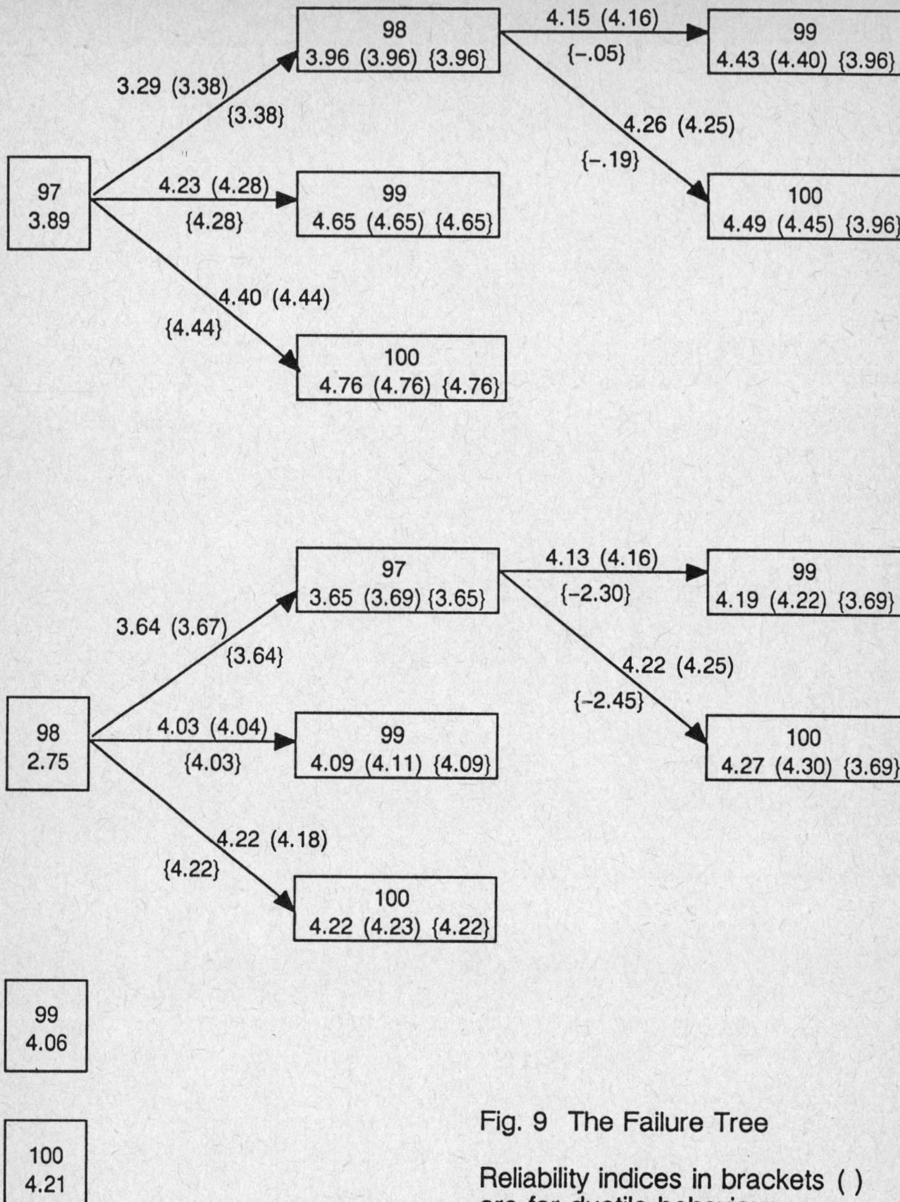

Fig. 9 The Failure Tree

Reliability indices in brackets ( )
are for ductile behaviour

Reliability indices in braces { }
are ductile tension behaviour and
softening compression behaviour

A PROBABILISTIC WAVE MODEL FOR RELIABILITY ANALYSIS

R. C. Turner* ** & M. J. Baker*
*Imperial College of Science, Technology and Medicine, London, UK
**W. S. Atkins Engineering Sciences, Epsom, UK

SYNOPSIS

A probabilistic wave model is suggested which is suitable for use in the reliability analysis of fixed offshore structures. The model is based on a joint distribution for wave height and period and is conditional on the sea state. The model is applied to the first order reliability analysis of an offshore structure, and some typical results are given.

INTRODUCTION

The probabilistic modelling of the hydrodynamic loading on offshore structures poses a difficult problem for the reliability analyst. Wave loading is a complex stochastic process, it varies in time and space, and is neither truly Gaussian or stationary. Yet, hydrodynamic loading is by far the most important stochastic variable to be treated in an offshore reliability analysis and has the largest influence on structural damage and collapse.

For much of the published research concerning the systems reliability analysis of offshore-type structures (e.g. Crohas *et al* (1984) and Edwards *et al* (1985)), the emphasis has been on the development of methods for systems analysis. In these earlier studies, the loading has been simplified, and has often been idealised as point loads acting at the structural nodes. Now that reliability procedures are beginning to be used for the assessment of real structures, much more care must be exercised over the modelling of the environmental loading.

Wave loads may be modelled either as a stochastic process, or by using a point-in-time distribution of the process – that is as a random variable. For extreme loading analysis, the latter is the most suitable approach, and for fixed structures the most extreme loading occurs with the highest wave.

The usual deterministic design approach is to assume a "design wave". This is a single large wave with a long return period, often 50 or 100 years. The method used to determine the parameters of the "design wave" is well established as design practice and forms the basis of the proposed probabilistic model.

Instead of a deterministic design wave, distributions for four variables are used to define a probabilistic model for the most extreme wave occuring within a specified exposure period, e.g. the design life of the structure.

BACKGROUND

Wave Data

A fundamental requirement for the structural analysis of an offshore platform is to obtain reliable wave data for the site. A variety of wave measuring devices are used, and historically, waves are generally recorded for a 1000 second interval on a 3 hourly cycle for at least a complete year.

Each wave record is examined to determine the mean water level for the sample, and is then analysed statistically to obtain at least:

(a) the significant wave height H_s. In this paper, H_s is defined as

$$H_s = 4\sqrt{m_o} \tag{1}$$

where $m_o = \int_0^\infty S_{\eta\eta}(\omega)\, d\omega$ and $S_{\eta\eta}(\omega)$ is the spectrum of water elevation η.

H_s corresponds very closely to the mean height of the highest third of the waves occurring in the sample, which was the original definition of H_s.

(b) the mean zero up-crossing period T_z. The zero up-crossing period is the period of time between one upward crossing of the mean water level and the next, and for a random Gaussian sea is given by

$$T_z = 2\pi\ (m_o/m_2)^{\frac{1}{2}} \tag{2}$$

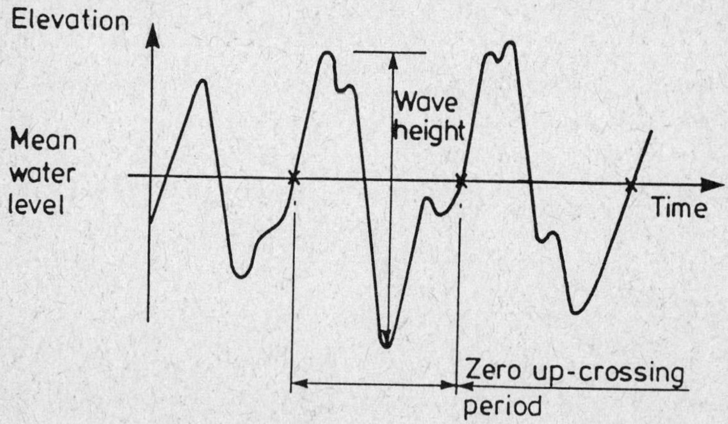

Figure 1: Sketch of part of a typical record

The data are usually presented as a histogram or scatter diagram of H_S against T_Z covering the period that measurements were taken. Figure 2 shows a typical scatter diagram for a southern North Sea site.

Deterministic Design Practice

As mentioned, shallow water jacket structures are generally designed using the "design wave" approach. The design wave is taken to be the wave having a specified return period, usually 50 or 100 years. It is assumed that this wave will produce the greatest load on the structure, and that it may approach the structure from any direction. The design wave is also assumed to take the form of a single regular wave for which the kinematics can be developed using an appropriate wave theory, and for which the wave loads can be obtained using Morison's equation.

The generally accepted method for evaluating the design wave involves firstly extrapolating the wave data to obtain the maximum sea state occurring in the return period. Cumulative probability diagrams are plotted for the significant wave height, H_S, and various distributions may be fitted to the data including Gumbel (Extreme Type I) and Weibull. The fitted distributions are extrapolated to predict the maximum significant wave height for the desired return period.

The next step is to obtain the maximum wave height that is most likely to occur within a period of maximum sea state. The sea state intensity is generally assumed to remain constant between intervals of wave sampling, and during this period it is assumed that the heights of all the individual waves are Rayleigh distributed. An estimate is made for the number of waves occurring during the interval, and the most probable maximum wave height is evaluated. An associated period is then chosen, often using some arbitrary wave steepness. In the future, revised UK Department of Energy Guidance Notes will also allow for the possibility that the maximum wave is not associated with the maximum H_S.

There are a number of limitations with this method, but it is well established as design practice, and, from the experienced gained with existing offshore installations, yields structures which are sufficiently strong to resist the extreme loads imposed. Unfortunately, the margin between the exceedance of the design load and the occurrence of structural failure is not known, and so the efficiency of individual structures cannot be examined by this method.

PROBABILISTIC WAVE MODEL

The wave model proposed for the probabilistic method is similar in approach to the deterministic method. The maximum wave loading on the platform is assumed to be caused by the most extreme wave, and the maximum wave is derived using the parameters for the most extreme sea state occurring within a specified exposure period - i.e. the design life of the structure.

The proposed wave model uses four variables each with its own probability distribution. Two of the variables - wave height, H, and wave period, T - are used to define the extreme wave for use within the analysis. The other two variables define the sea state, and are used as parameters for the distributions of H and T.

Extreme Significant Wave Height, H_s

As with the deterministic approach, the significant wave height data is extrapolated to predict the H_s with a specified return period of, say, 30 years. This leads to a point estimate of the mean value of the 30 year maximum distribution for H_s; and so for an actual 30 year exposure period the measured maximum significant wave height may be higher or lower than this predicted value. However, since the aim is to determine the probability of failure for the structure, we need the probability distribution for $(H_s)_E$ - the maximum significant wave height for the exposure period, E.

The distribution function for the maximum values in a large number of periods tends towards an asymptotic extreme-value distribution. A frequently used distribution is the Fisher-Tippett Type I extreme value or Gumbel distribution.

A particularly useful property of this distribution is that, within limits, the standard deviation, σ, is independent of the reference period, i.e. $(\sigma)_{30}$ for the 30 year maxima is the same as $(\sigma)_1$ for the annual maxima. This considerably aids the parameter estimation.

The Mean Zero Up-crossing Period, T_z

In order to define the sea state, both the significant wave height and the mean zero up-crossing period must be known. Ideally in a reliability analysis, a conditional distribution would be used to model T_z for given values of H_s. Unfortunately, a suitable distribution is not available for high values of H_s. This is an area requiring further research (Carter *et al* 1986).

Often, a constant linear relationship is assumed between H_s and T_z^2. However, it can be seen from the scatter diagram (Fig. 2) that this is not very satisfactory.

The characteristic wave steepness, s, is defined as:

$$s = \frac{2\pi H_s}{g T_z^2} \qquad (3)$$

It is proposed that the characteristic wave steepness is treated as a basic random variable in the wave model. The assumption that wave steepness is independent of H_s is obviously not true for low significant wave heights (i.e. less than 2m), but it is a reasonable assumption for the higher sea states.

The mean zero crossing period can then be evaluated for given values of H_s and s.

The Joint Distribution of Wave Heights and Periods

A number of researchers have investigated the joint distribution of wave periods and amplitudes, and a number of models have been suggested, including, Cavanié *et al* (1976), Lindgren (1972), and Longuet-Higgins (1983).

Srokosz and Challenor (1987) is the only published work to compare these three models. Lindgren's model uses a numerical distribution and is expensive computationally. For narrow band seas, both the Cavanié and the Longuet-Higgins distributions are adequate. Longuet-Higgins distribution is better for higher normalised waves, and has been used here.

In common with many other theories, Longuet-Higgins' distribution assumes that the sea state intensity is statistically stationary over short periods, i.e. the distribution parameters are invariant to translations in time. The sea surface is assumed to be formed from an infinite sum of independent sinusoidal wave trains, i.e. the sea surface elevation is regarded as a Gaussian stochastic process. Furthermore, the sea state is assumed to be relatively narrow banded, i.e. the energy of the wave spectrum is concentrated in a narrow range of frequency.

Longuet-Higgins, in his 1983 and 1975 papers derives a joint distribution for wave amplitude and period. The bivariate density function will simply be stated here as:

$$f(\theta,\tau) = \frac{2\,L(\nu)}{\nu\sqrt{\pi}}\,\frac{\theta^2}{\tau^2}\,\exp\left[-\theta^2\left(1+(1-1/\tau)^2/\nu^2\right)\right] \qquad (4)$$

where θ and τ are dimensionless parameters for wave amplitudes, ρ, and period, T, given by:

$$\theta = \rho/(2m_o)^{\frac{1}{2}}$$

$$\tau = Tm_1/2\pi m_o$$

with $m_n = \int_o^\infty \omega^n\, S_{\eta\eta}(\omega)\,d\omega \qquad n = 0,1,\ldots.\infty$

and where ν is the spectral narrowness parameter defined as

$$\nu = (m_o m_2/m_1^2 - 1)^{\frac{1}{2}}$$

and $L(\nu)$ is a normalisation factor, which allows for the fact that only positive values of τ are considered, and is given by:

$$1/L(\nu) = \tfrac{1}{2}\left[\,1 + (\,1+\nu^2)^{-\frac{1}{2}}\,\right] \qquad (5)$$

Longuet-Higgins' distribution is derived from wave amplitudes. However, in practice it is generally assumed that the wave height H is twice the wave amplitude ρ. This assumption is exact for narrow band spectra ($\nu = 0$), and has been shown to be acceptable for ocean waves. Srokosz and Challenor (1987) demonstrate clearly that it is important how wave height is defined. Wave height H is defined here as the maximum vertical distance between adjacent up-crossings at the mean level of the process.

Using these assumptions, it is clear that

$$\frac{\rho}{\rho_{1/3}} = \frac{\theta}{\theta_{1/3}} \approx \frac{H}{H_{1/3}} = \frac{H}{H_S} \qquad (6)$$

where $\rho_{1/3}$ is the mean of the amplitudes of the third highest waves, etc.

$$\text{Hence } \theta \approx \theta_{1/3}\, H/H_S \qquad (7)$$

Similarly, by manipulation of the above equations and by noting that the average number of up-crosssings of level zero per unit time is given by

$$E[N'_{+}(0)] = \frac{1}{2\pi}(m_2/m_o)^{\frac{1}{2}} \qquad (8)$$

it can be shown that

$$\tau = \frac{1}{(1 + \nu^2)^{\frac{1}{2}}} \; \frac{T}{T_Z} \qquad (9)$$

The theory is valid for low values of spectral narrowness, ν; Longuet-Higgins suggests $\nu << 0.6$. Values of ν range from 0.234 for an example given by Longuet-Higgins (1976), to 0.35 - 0.45 for the JONSWAP spectrum, to 0.425 for the Pierson-Moskowitz spectrum. Thus, for a typical value of $\nu = 0.4$ it can be shown that:

$$L(\nu) = 1.037$$

$$\theta = 1.414\, H/H_S$$

$$\tau = 0.928\, T/T_Z$$

The Longuet-Higgins distribution can then be considered as a conditional joint distribution in terms of wave height and period for a given sea state; i.e. values of significant wave height and mean zero crossing period. However, the bivariate distribution cannot be used in a first order reliability analysis directly and must be re-arranged into marginal and conditional distributions for the individual variables. For clarity, the dimensionless parameters θ and τ will be retained.

The distribution for wave height

The marginal distribution of the normalised wave amplitude θ is found by integrating the bivariate distribution over the range $0 \leqslant \tau \leqslant \infty$.

$$f(\theta) = \frac{2\ L(\nu)}{\nu\ \sqrt{\pi}} \theta^2\ e^{-\theta^2} \int_0^\infty \frac{1}{\tau^2} \exp\left[-\theta^2\ (1-1/\tau)^2/\nu^2 \right] d\tau \qquad (10)$$

By means of the substitution $\theta(1-1/\tau)/\nu = t/\sqrt{2}$

$$f(\theta) = 2\ L(\nu)\ \theta\ e^{-\theta^2} \Phi \left\{ \frac{\theta\sqrt{2}}{\nu} \right\} \qquad (11)$$

where Φ { } is the standard normal distribution function.

The maximum wave height in a stationary interval

Fig. 3 shows the distribution of normalised wave amplitudes over a sampling interval for a value of ν = 0.4. The number of waves, N, occurring in the interval can be estimated from T_Z. i.e. in a 3 hour interval

$$E[N] = 3 \times 3600/T_Z$$

For reliability analysis, the probability distribution for the most extreme wave amplitude is needed. This can be obtained by using order statistics

$$F\ (\theta_{max}\ |\ N\ ) = \left[F\ (\theta)\ \right]^N \qquad (12)$$

$$f\ (\theta_{max}\ |\ N\ ) = \frac{d}{d\theta} \left[F\ (\theta)\ \right]^N = N \left[F(\theta) \right]^{N-1}\ f(\theta) \qquad (13)$$

where $F(\theta) = \int_0^\theta f(\alpha)\ d\alpha$

$$= \int_0^\theta 2L(\nu)\ \alpha e^{-\alpha^2}\ \Phi \left\{ \frac{\alpha\sqrt{2}}{\nu} \right\} d\alpha$$

$$= 1 - L(\nu) \left[e^{-\theta^2}\ \Phi \left\{ \frac{\theta\sqrt{2}}{\nu} \right\} + \frac{1}{(1+\nu^2)^{\frac{1}{2}}} \left[1 - \Phi \left\{ \frac{\theta\sqrt{2}}{\nu} (1+\nu^2)^{\frac{1}{2}} \right\} \right] \right] \qquad (14)$$

For values of $\theta > 1.0$ $\qquad \Phi \left\{ \frac{\theta\sqrt{2}}{\nu} \right\} \approx 1.0$

Hence, $F(\theta) \approx 1 - L(\nu)\ e^{-\theta^2}$

For large values of N $$(1-x)^N \approx e^{-Nx}$$

Hence, $$F(\theta_{max} \mid N) \approx \exp\left[-N\, L(\nu)\, e^{-\theta^2} \right] \quad (15)$$

$$f(\theta_{max} \mid N) \approx 2\, N\, \theta\, L(\nu) \quad \exp\left[-\theta^2 - N\, L(\nu)\, e^{-\theta^2} \right] \quad (16)$$

Figure 4 is a plot of probability density functions for different values of N.

The Distribution for Wave Period

The conditional distribution for wave period can be obtained, dependent on a given value of θ.

$$f(\tau \mid \theta) = \frac{f(\theta,\tau)}{f(\theta)} \quad (17)$$

$$F(\tau \mid \theta) = \frac{1}{f(\theta)} \int_0^\tau f(\theta,\tau)\, d\tau \quad (18)$$

The numerator can be evaluated by using the same substitution as before.

Hence, $$\int_0^\tau f(\theta,\tau)\, d\tau = 2\, L(\nu)\, \theta\, e^{-\theta^2} \quad \Phi\left\{ \frac{\theta\sqrt{2}}{\nu} \left[1 - \frac{1}{\tau} \right] \right\} \quad (19)$$

Thus, $$F(\tau \mid \theta) = \frac{1}{\Phi\{\theta\sqrt{2}/\nu\}} \quad \Phi\left\{ \frac{\theta\sqrt{2}}{\nu} \left[1 - \frac{1}{\tau} \right] \right\} \quad (20)$$

Fig. 5 shows the conditional distribution of τ for various different values of θ.

THE USE OF THE WAVE MODEL IN A FIRST ORDER RELIABILITY ANALYSIS

One application of the wave model is to evaluate the reliability of individual components of fixed offshore structures using first order reliability methods.

The safety margin Z for a component i is

$$Z_i = g(\overline{X}) \quad (21)$$

where g is the failure function,
$\overline{X}$ is the set of basic variables.

and where
$Z_i \leqslant 0$ corresponds to failure
$Z_i > 0$ is a safe state.

The set of basic variables $\bar{X}$ may include:

- resistance parameters for member i
- post-failure behaviour parameters for any failed members
- deck loading variables
- environmental parameters, e.g. drag and inertia coefficients, marine growth, current velocity, wave height and wave period, etc.

The two sea state parameters H_s and s^{-1} are not functionally related to the safety margin. However, they do influence the probability of failure, and are introduced later as variables conditioning the parameters of the distributions for H and T.

Typical failure functions include criteria for ultimate strength of members and joints. They are usually expressed in terms of member forces/moments or member stresses. With deterministic design, and with the approach adopted here, the member stresses are implicit functions of the basic variables.

From a Taylor series expansion of the safety margin about an estimate of the design point, $\bar{X}$*,

$$g\,(\bar{X}^* + \delta\bar{X}) = g\,(\bar{X}^*) + \sum_{i=1}^{n} (a_i\ \delta X_i)$$

where $a_i = \dfrac{\partial g}{\partial X_i}\ \Big|\ \bar{X} = \bar{X}^*$

Hence, the reliability index β can be shown to be

$$\beta = \frac{g\,(\bar{X}^*) + \sum_{i=1}^{n} a_i\ (\mu_i' - X_i^*)}{\left\{ \sum_{i=1}^{n} (a_i\ \sigma_i')^2 \right\}^{1/2}} \qquad (22)$$

where μ_i' and σ_i' are the mean and standard deviation of the equivalent normal distribution.

The derivatives a_i of the safety margin are evaluated numerically by incrementing each of the basic variables in turn and re-evaluating the safety margin.

However, the distributions for the wave height and period are conditional on the two sea state parameters H_s and s^{-1}.

Consider the terms in (22) involving the variable H

i.e. $\dfrac{\partial g}{\partial H}\ (\mu_H' - H^*)$ and $\left[\dfrac{\partial g}{\partial H} \cdot \sigma_H' \right]^2$

Now, since $H = f(H_s)$, approximate expansions for μ'_H and σ'_H are:

$$\mu'_H = \mu'_{H|Hs} + \frac{\partial H}{\partial H_s}(\mu'_{Hs} - H_s{*}) \tag{23}$$

$$[\sigma'_H]^2 = [\sigma'_{H|Hs}]^2 + [\frac{\partial H}{\partial H_s} \cdot \sigma'_{Hs}]^2 \tag{24}$$

Hence it is necessary to evaluate $\partial H/\partial H_s$, i.e. the variation of H with respect to H_s evaluated at the design point.

For a small increase δH_s in the significant wave height H_s, the maximum predicted wave height H will increase by an amount δH. Formally it is necessary to satisfy the following condition:

$$P\{H \leqslant H^* \mid H_s\} = P\{H \leqslant (H+\delta H)^* \mid H_s + \delta H_s\}$$

where $P\{H \leqslant H^* \mid H_s\}$ is the probability that the wave height is less than H* for a given significant wave height, H_s.

By considering the form of the conditional distribution $F\{H_{max} | H_s, N\}$, it is clear that the above condition is met when

$$\frac{H}{H_s} = \frac{H + \delta H}{H_s + \delta H_s} \quad \text{i.e.} \quad \frac{\delta H}{\delta H_s} = \frac{H}{H_s} \tag{25}$$

The assumption is exact if the number of waves in the interval are equal for sea states H_s and $H_s + \delta H_s$. However, since the number of waves in a 3-hour period is generally above 1000, any error will be small.

Consider now the terms in (22) involving the variable T

$$\text{i.e.} \quad \frac{\partial g}{\partial T}(\mu'_T - T^*) \quad \text{and} \quad [\frac{\partial g}{\partial T} \cdot \sigma'_T]^2 \tag{26}$$

Again, since $T = f(T_Z) = f'(s^{-1})$

$$\mu'_T = \mu'_{T|s^{-1}} + \frac{\partial T}{\partial s^{-1}}[\mu'_{s^{-1}} - (s^{-1})^*] \tag{27}$$

$$[\sigma'_T]^2 = [\sigma'_{T|s^{-1}}]^2 + [\frac{\partial T}{\partial s^{-1}} \sigma'_{s^{-1}}]^2 \tag{28}$$

Now $$\frac{\partial T}{\partial s^{-1}} = \frac{\partial T}{\partial T_Z} \cdot \frac{\partial T_Z}{\partial s^{-1}} \tag{29}$$

Since $S = \dfrac{2\pi H_s}{g T_z^2}$ (30)

Thus $\dfrac{\partial T_z}{\partial s^{-1}} \Big|_{H_s = H_s^*, T_z = T_z^*} = \dfrac{\pi H_s^*}{g T_z^*}$ (31)

$\dfrac{\partial T}{\partial T_z}$ is evaluated under the condition

$P(T \leqslant T^* | H_z^*, H^*, T_z^*) = P(T \leqslant (T+\delta T)^* | H_s^*, H^*, (T_z+\delta T_z)^*)$

Hence it can be shown that

$$\frac{\partial T}{\partial T_z} \Big|_{H_s^*, H^*} = \frac{T^*}{T_z^*} \qquad (32)$$

EXAMPLE

The aim in developing the model described above, was to produce a procedure which would be closely compatible with existing design practice and which could be used with current industry standard computer programs for wave loading and structural analysis. Furthermore, it was intended that the model should be used in conjunction with computer programs for the first order reliability analysis, both at the member and system level. To illustrate these capabilities, the following example is included.

The structure shown in Figure 6 respresents a typical southern North Sea gas platform. Table 1 shows the basic variables and parameters. Table 2 shows the results for the reliability analysis of a brace member - the failure function was based on the API Code. A constant phase angle of -6° was assumed, and Stokes Vth wave theory was used. The reliability index was evaluated as 2.22.

It should be stated that the parameters used in the analysis were for demonstration purposes only. However, the results do show that the most sensitive variables are H_s and H, followed by the other hydrodynamic variables.

CONCLUSIONS

The proposed wave model is intended to be used with the reliability analysis of fixed offshore structures with negligible dynamic response. It is based on general practice in which it is assumed that the random behaviour of the sea can be split into a sequence of stationary intervals lasting several hours. It is assumed further, that each interval is reasonably narrow banded and Gaussian and that the sea surface can be described by the Longuet-Higgins bivariate distribution. By using extreme value distributions and order statistics the probability of occurrence of the most extreme maximum waves can then be estimated. It should be noted, however, that the model does not include the contribution of sea states other than the highest to the probability of occurence of the highest individual wave. In this respect the model is unconservative.

The accuracy of the predictions from the model are highly dependent on the accuracy of the wave data, and on the number of years of recordings. However, one advantage of the model is that it can be used directly with level 2 reliability methods to indicate how sensitive the reliability is to each variable.

The model has been used in a level 2 reliability analysis program to determine the probability of failure of members in a number of structures. From this limited experience the following observations have been made:

- From all the variables that were included in the analyses, the reliability was usually most sensitive to significant wave height, followed by the individual wave height.

- The probability of failure of main structural members is generally low, and involves very extreme events.

- Occasional problems were experienced with breaking waves. These could not be analysed using standard wave theory (Stokes Vth, Deans 9th etc.) and had to be excluded from the analyses.

This last point is a serious one, and can affect the suitability of the model as a more general analysis tool. It occurred because the extreme events being analysed were stretching the limit of validity of the theory. The bivariate distribution can be truncated to exclude breaking waves, but it is difficult to examine the consequences of this.

However, in deeper water, the proposed model is suitable for use in a reliability analysis to investigate the probability of failure of members. Unfortunately, more work is required to determine the accuracy of the predicted values, particularly for extreme events. The model can be improved in a number of areas, some of which are discussed in Turner and Baker (1987).

ACKNOWLEDGEMENTS

The work described in this paper was carried out as part of the programme of the London Centre for Marine Technology on "System Reliability of Fixed Offshore Platforms", and was jointly funded by SERC (MTD Ltd), W.S. Atkins ES and Shell UK, Exploration and Production.

REFERENCES

Cavanié, A., Arhan, M. and Ezraty, R. (1976), "A statistical relationship between individual heights and periods of storm waves", 1st BOSS conference, Trondheim, Norway.

Carter, D.J.T., Challenor, P.G., Ewing, J.A., Pitt, E.G., Srokosz, M.A. and Tucker, M.J. (1986), Estimating wave climate parameters for engineering applications. Off. Tech. Rpt. OTH 86 228. HMSO.

Crohas, H., Ali-Asghar Tai, Hachemi-Safai, and Barnouin, B., (1984), "Reliability analysis of offshore structures under extreme environmental loading". Proc. 16th OTC, Paper 4826, Houston, Texas.

Edwards, G.E., Heidweiller, A., Kerstens, J. and Vrouwenvelder, A. (1985), "Methodologies for limit state reliability analysis of offshore jacket platforms". Proc. 4th BOSS conference, Delft, The Netherlands.

Houmb, O.G. and Overvik, T. (1976), "Parameterization of wave spectra and long term joint distribution of wave height and period". 1st BOSS conference, Trondheim, Norway.

Lindgren, G. (1972), "Wave-length and amplitude in Gaussian Noise". Adv. appl. prob. Vol 4.

Longuet-Higgins, M.S. (1975), "The statistical analysis of a random moving surface". Proc. of Royal Society, London, series A. Vol 230.

Longuet-Higgins, M.S. (1975), "On the joint distribution of the periods and amplitudes of sea waves". Journal of Geophys. Research, Vol 80, No. 18.

Longuet-Higgins, M.S. (1983), "On the joint distribution of wave periods and amplitudes in a random wave field". Proc. of Royal Society, London, series A, Vol 389.

Srokosz, M.A. and Challenor, P.G. (1987), "Joint distribution of wave height and period - a critical comparison". Ocean Eng. Vol 14, No. 4.

Thoft-Christensen, P. and Baker, M.J. (1982), Structural reliability theory and its applications. Springer-Verlag.

Turner, R.C. and Baker, M.J. (1987), "A wave model for the reliability analysis of offshore structures". Imperial College Report SRRG/3/1987.

Table 1: Basic variables and parameters

Variable	Distribution Type	Mean Value	Standard Deviation
H_s	Extreme Type I	8.1 m	0.8 m
s^{-1}	Shifted Log normal	19.95	6.15
Marine growth	Truncated normal	0.04 m	0.02 m
H	Longuet-Higgins Bivariate	9.415* m	-
T		10.09* s	-
C_d	Normal	0.68	0.25
Current velocity	Truncated normal	0.400 m/s	0.40 m/s
Yield stress	Lognormal	380,000 kN/m²	38,000 kN/m²
Model uncertainty	Lognormal	1.019	0.2

* Parameters based on sea state design point values.

Table 2: Design point values and sensitivity coefficients

Variable	Design Value	Sensitivity Coefficient
H_s	9.415 m	0.6799
s^{-1}	16.90	-0.1670
Marine growth	0.0412 m	-0.0145
H	20.19 m	0.4475
T	10.90 s	-0.1403
C_d	0.9190	0.4311
Current velocity	0.4860 m/s	0.1398
Yield stress	377,855 kN/m²	-0.0431
Model uncertainty	1.132	0.2871

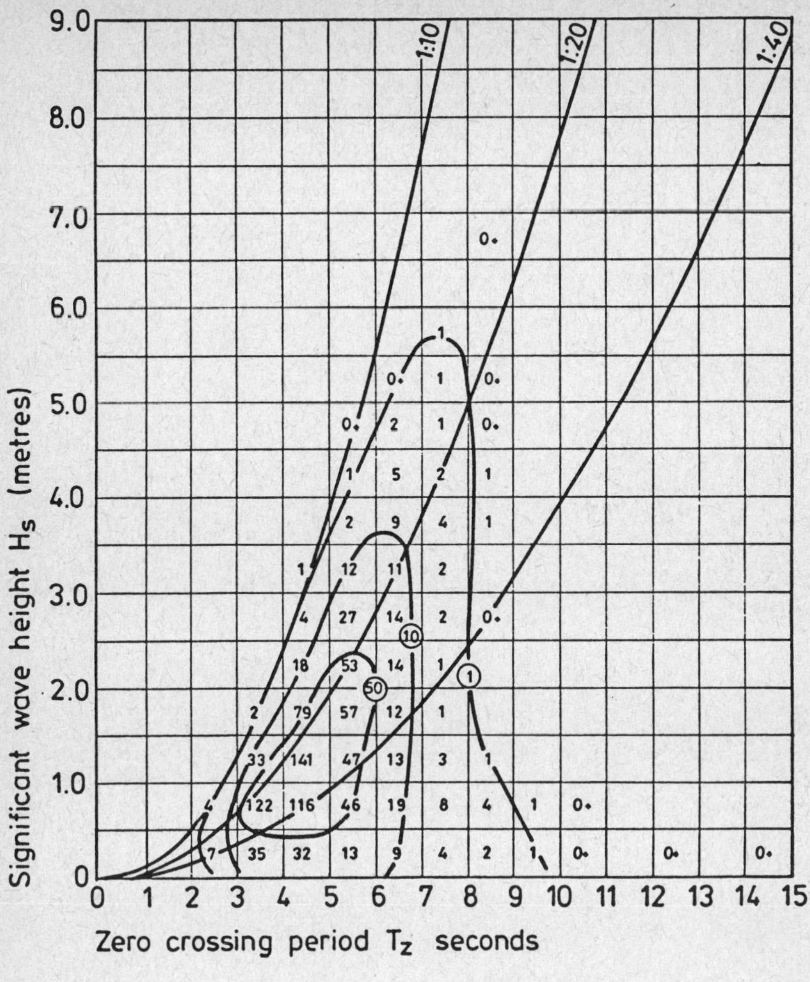

Figure 2: Typical wave scatter diagram (units indicated in parts per 1000)

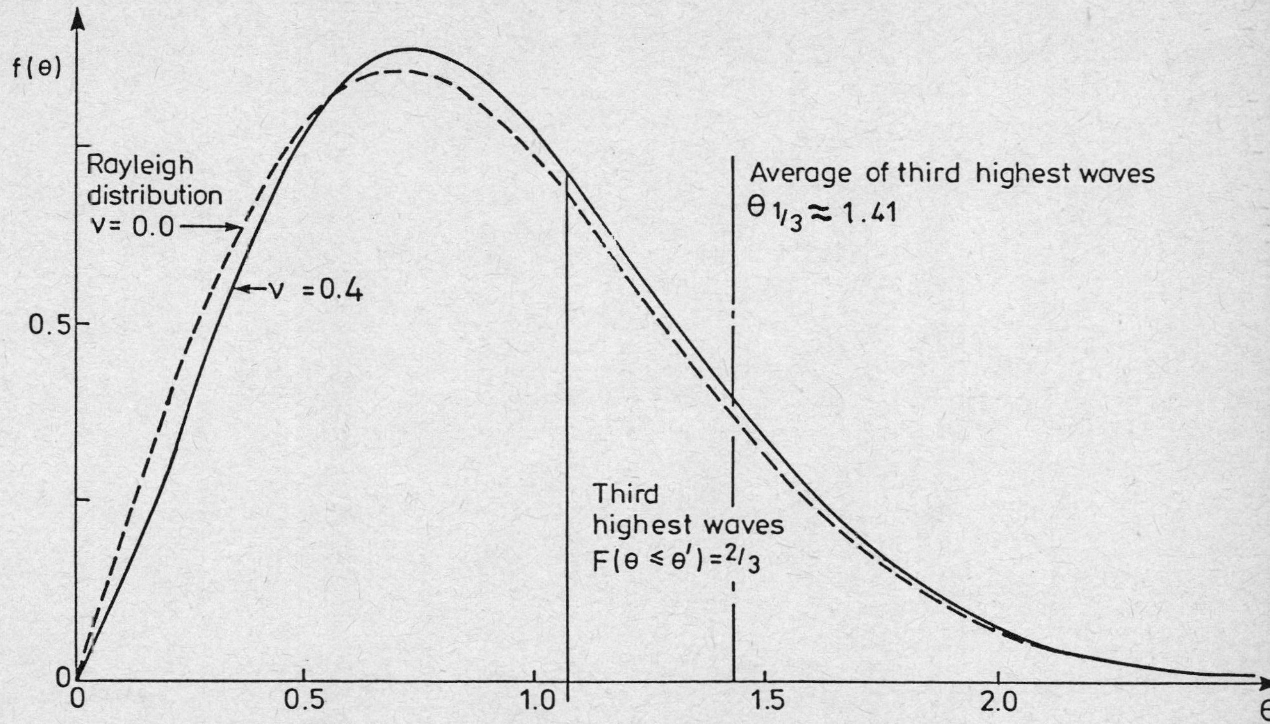

Figure 3: The marginal distribution for θ, with $\nu = 0.4$

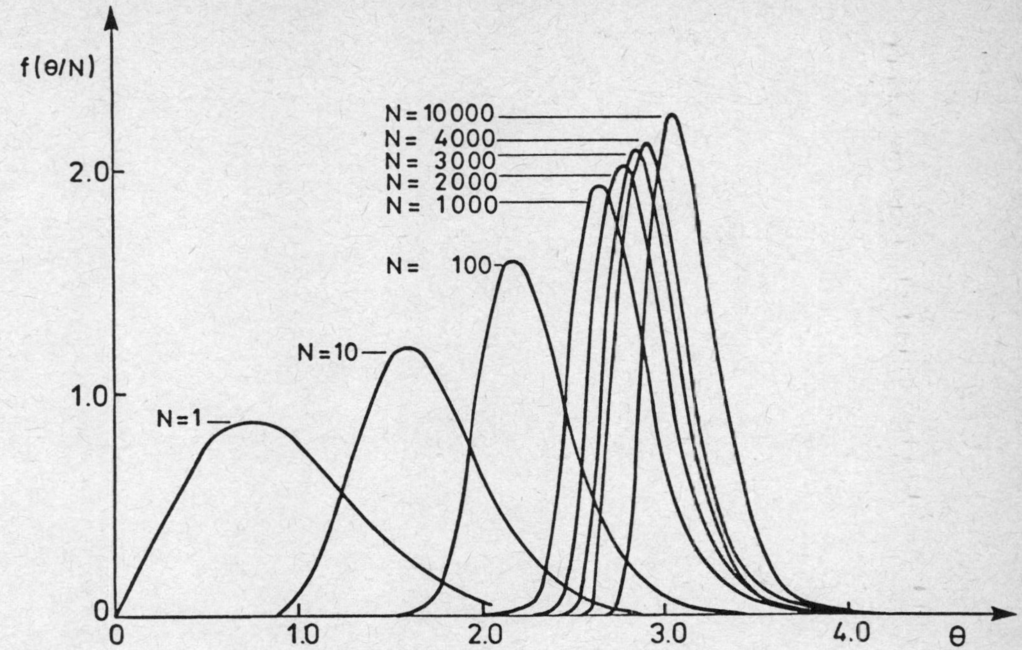

Figure 4: Density functions of extreme values of normalised wave height θ, for various values of N

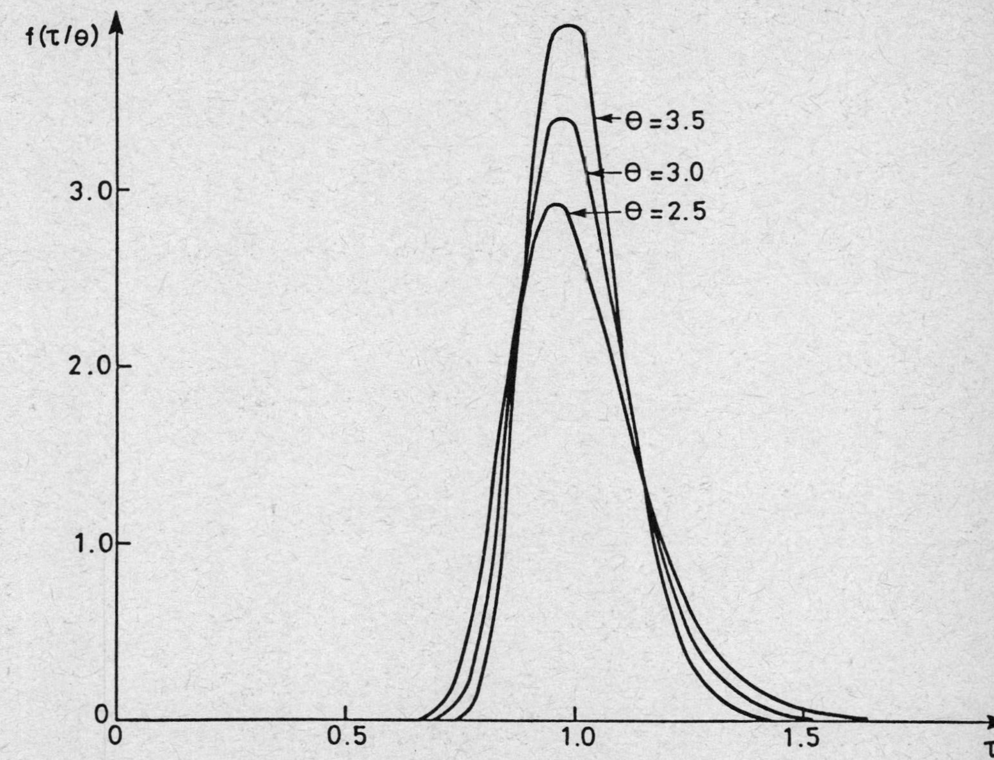

Figure 5: Density functions of wave period τ, for various values of θ

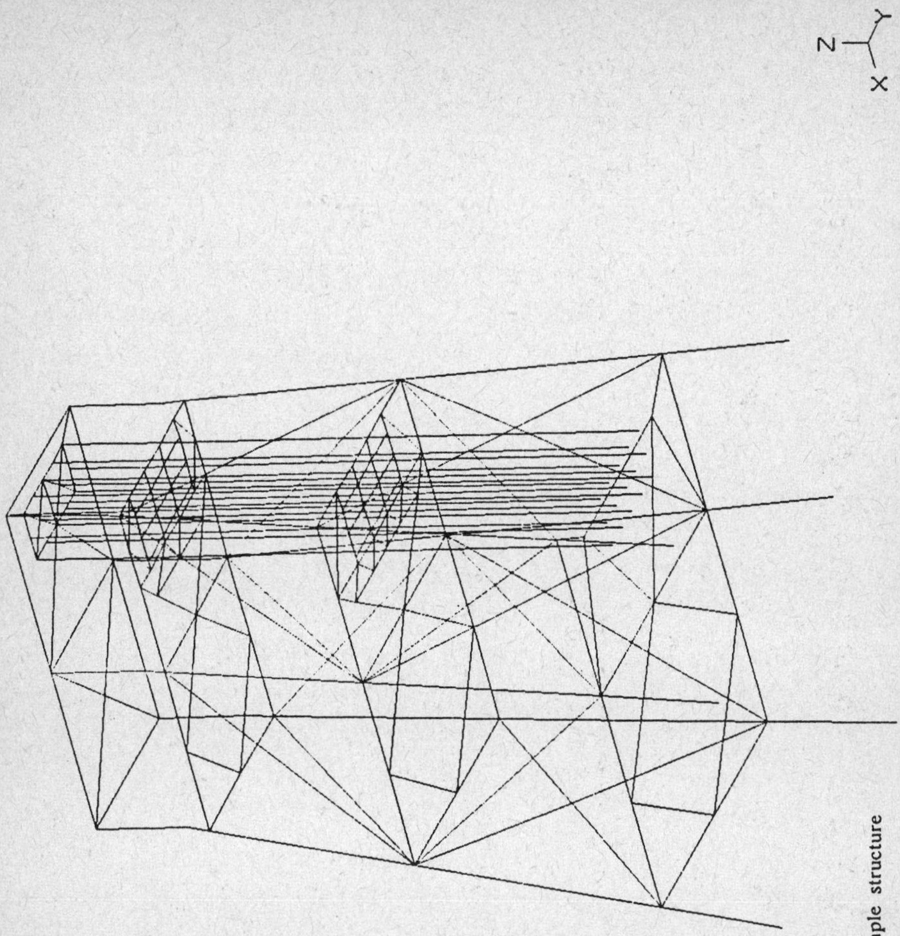

Figure 6: The example structure

SECOND ORDER RELIABILITY BY AN EXACT INTEGRAL

Lars Tvedt
A.S. Veritas Research, P. O. Box 300, N-1322 Høvik, Norway

ABSTRACT

The probability density function and the cumulative distribution function of a parabolic function of independent standard normal random variables are computed by inversion of the corresponding characteristic function. The method uses the saddle-point method in conjunction with the trapezoidal rule. The result is useful in second order reliability analysis.

1. Introduction

A fundamental result due to K.Breitung, [1], shows that the probability content of a curvature fitted parabolic approximation to the failure set, under weak restrictions on the curvatures, asymptotically yields the probability of the true failure set.

In [2], A.der Kiureghian suggests a parabolic approximation to the failure surface, using a point fitting method in orthogonal directions only. The method does not satisfy the asymptotic behaviour of Breitung's parabolic approximation but is justified in many cases by heuristic arguments.

In [3] some exact results and bounds for quadratic forms are presented.

None of the cited references, however, provide an exact result for the probability of the parabolic set.

The event of interest in second order reliability analysis is

$$Y \geq \beta+\sum_{i=1}^{n}\lambda_i X_i^2 \tag{1}$$

where Y and the X_i's are mutually independent standard normal random variables. How to obtain the parabolic approximation to the failure domain can be found [1] or in [4],p66. It should be noted that due to symmetry it is sufficient to consider $\beta \geq 0$.

In [5] a three term formula approximating the probability of this event was developed:

$$P_s \approx A_1+A_2+A_3 \tag{2}$$

where the three terms are

$$A_1 = \Phi(-\beta)\prod_{j=1}^{n}(1+2\beta\lambda_j)^{-½} \tag{3.a}$$

$$A_2 = [\beta\Phi(-\beta)-\phi(\beta)]\left\{\prod_{j=1}^{n}(1+2\beta\lambda_j)^{-1/2}-\prod_{j=1}^{n}(1+2(\beta+1)\lambda_j)^{-1/2}\right\} \tag{3.b}$$

$$A_3 = (\beta+1)[\beta\Phi(-\beta)-\phi(\beta)]\left\{\prod_{j=1}^{n}(1+2\beta\lambda_j)^{-1/2}-\mathrm{Re}\left\{\prod_{j=1}^{n}(1+2(\beta+i)\lambda_j)^{-1/2}\right\}\right\} \tag{3.c}$$

This result is also published in [4],p67. The first term, (3.a), in the formula is the asymptotic result by Breitung, [1].

The formula above is good for most practical cases in structural reliability analysis. However, if a curvature is close to the curvature of a circle with center at the origin and radius β, the formula does not work. This is because there are singularities in the formula for $\lambda_j=-1/(2\beta)$ and $\lambda_j=-1/(2(\beta+1))$. Also, if β is small, the asymptotic result by Breitung tends to the first order result $\Phi(-\beta)$ and the terms (3.b) and (3.c) are not always providing a good correction.

We are thus led to study the numerical evaluation of the probability, to obtain, in a numerical sense, exact results. In this paper the probability of the parabolic set is computed by inversion of the corresponding Laplace transform. The method yields 4 and more significant digits in the result for quadratures of 5 to 20 points in cases typical for reliability analysis. The method works well also in more general cases and is well suited for implementation on a computer.

In connection with Bayesian updating of reliabilities from findings during inspections, [6], there is an interest for computing the density of the distribution of the parabolic form with β as the distributed variable and λ fixed. The problem of computing this density is therefore addressed as well.

2. Formulation of the problem and general results.

The probability to be computed is for the event:

$$Y-\sum_{i=1}^{n}\lambda_i X_i^2 \le c \tag{4}$$

where the random variables Y and X_i are independent and standard normally distributed.

In Appendix A the characteristic function for the random variable

$$C = Y-\sum_{i=1}^{n}\lambda_i X_i^2 \tag{5}$$

is shown as

$$G_C(u)=\exp(u^2/2)\prod_{i=1}^{n}(1+2u\lambda_i)^{-1/2} \tag{6}$$

The probability density function of $Y-\sum_{i=1}^{n}\lambda_i X_i^2$ at c is then:

$$f_C(c,\lambda) = \mathrm{Re}\{\frac{1}{2\pi i}\int_{-i\infty}^{i\infty} \exp(-cu)G_C(u)du\} \tag{7}$$

and the cumulative distribution function is:

$$F_C(c,\lambda) = 1-\int_c^{\infty} f_C(t,\lambda)dt = 1-\mathrm{Re}\{\frac{1}{2\pi i}\int_{-i\infty}^{i\infty} \frac{\exp(-cu)G_C(u)}{u}du\} \tag{8}$$

For the integration in (8) to be valid the path of integration for u is indented to the right at $u=0$.

The relation of our problem, (1), to the distribution of C is, after some manipulation:

$$f_C(-\beta,-\lambda)=\phi(\beta)\,\mathrm{Re}\{-i(\frac{2}{\pi})^{1/2}\int_0^{i\infty} exp((u+\beta)^2/2)\prod_{i=1}^{n}(1-2\lambda_i u)^{-1/2}du\} \tag{9}$$

for the density and

$$F_C(-\beta,-\lambda)=\phi(\beta)\,\mathrm{Re}\{i(\frac{2}{\pi})^{1/2}\int_0^{i\infty} \frac{exp((u+\beta)^2/2)}{u}\prod_{i=1}^{n}(1-2\lambda_i u)^{-1/2}du\} \tag{10}$$

for the probability. The symmetry of the real part of the integrands about the real axis has been used. In (10) the path of integration for u is indented to the left at $u=0$.

3. Saddle point integration

S. O. Rice, [7], demonstrates the saddle point method, using the the trapezoidal rule, to be useful in connection with the inversion integrals of the indefinite quadratic form $\sum_{j=1}^{n}\lambda_j(X_j-w_j)^2$. We shall use the method of [7] in a modified form introduced by C. W. Helstrom, [8].

The motivation for using the saddle point method is that the paths of integration of (9) and (10) can be shifted parallel to the imaginary axis as long as a pole is not passed. The idea is to find a path most likely to yield an efficient quadrature. The point searched for is the one which minimises the integrand on the real axis. In the present case, the poles are on the real axis and the search interval for the saddle-point of the integrand of (9) is restricted by the poles

$$u_l = \max_{\{\lambda_i<0\}} \frac{1}{2\lambda_i} \tag{11}$$

and

$$u_u = \min_{\{\lambda_i>0\}} \frac{1}{2\lambda_i} \tag{12}$$

The search interval for the saddle-point of the integrand of (10) is restricted by (11) and the origin:

$$u_u = 0 \tag{13}$$

The saddle-point, u_s, is found by solving:

$$\frac{d\psi(u)}{du} = 0 \tag{14}$$

where $\psi(u)$ in the case of (9) is:

$$\psi(u) = \frac{1}{2}(u+\beta)^2 - \frac{1}{2}\sum_{j=1}^{n}\log(1-2\lambda_j u) \tag{15}$$

and in the case of (10) is:

$$\psi(u) = \frac{1}{2}(u+\beta)^2 - \frac{1}{2}\sum_{j=1}^{n}\log(1-2\lambda_j u) - \log u \tag{16}$$

The form of the trapezoidal rule to be used is:

$$\int_0^{\infty} f(x)dx = hb\left(\frac{f(u_s)}{2} + \sum_{j=1}^{\infty} f(u_s + jbhi)\right) \tag{17}$$

Thus, the path of integration is parallel to the imaginary axis starting at the saddle-point.

The scaling factor, b, is an estimate of the first zero point for the real part of the integrand. The estimate is based on the curvatures of the integrand at u_s in the direction of the path of integration. The scaling factor is:

$$b = \left[\frac{2}{\dfrac{d^2\psi(u_s)}{du^2}}\right]^{\frac{1}{2}} \tag{18}$$

An upper bound on the error made in truncating the series of (17) at a finite number of terms is, in the case of (9):

$$|E(s)| = \left|\phi(\beta)\mathrm{Re}\left\{-i\left(\frac{2}{\pi}\right)^{1/2} \int_{u_s+is}^{u_s+i\infty} exp((u+\beta)^2/2) \prod_{j=1}^{n}(1-2\lambda_j u)^{-1/2} du\right\}\right| \tag{19}$$

$$\leq \left(\frac{2}{\pi}\right)^{1/2} exp(u_s^2/2+\beta u_s)\Phi(-s)\prod_{j=1}^{n}((1-2\lambda_j u_s)^2 + 4\lambda_j^2 s^2)^{-1/4}$$

Thus

$$|E(s)| \leq \frac{\Phi(-s)}{\phi(s)}\,|f(u_s+is)| \tag{20}$$

A similar derivation in the case of (10) also leads to (20). The bounds are due to the monotonous decrease of the absolute value of the integrands as functions of s. If $s<1$, the ratio $\Phi(-s)/\phi(s)$ is replaced by the value 2. If $s \geq 1$, the ratio is replaced by the asymptotic result $1/s$ which is also an upper bound.

A further numerical efficiency is noteworthy. If, in the case of (10), the scaling factor b is less than 0.5, the saddle-point of the complementary problem is considered. The problem yielding the largest value of b is selected for numerical evaluation, since the oscillations of the exponential function are smallest in this case. In evaluating the complement, (10) is used, however with a shift in sign of β and of λ since

$$F_C(\beta,\lambda) = 1 - F_C(-\beta,-\lambda) \tag{21}$$

It is important to note that the search interval for the saddle point is the same for both problems. Also, the value 0.5 is judicial, but reflects that above this value the trapezoidal rule performs well. Since the performance in general deteriorates as b tends to zero, it is worth wile to check if the complementary problem performs better for b less than 0.5.

4. Numerical examples

As examples are chosen a number of cases where the asymptotic result does not work. The parabolic form is described in terms of β and n_λ quadratic terms. In the tables below the step length is h=0.5 in all cases. If the scaling factor b is less than 0.5 the complementary problem is solved. This is the case only in row 1 and in row 2 of Table 1. The quadrature has been truncated at a point where the relative error in the integral is at most 10^{-6}. n_p is the number of points in the quadrature and δ is the relative precision in the result as compared to the result using twice as many points in the quadrature. The performance of the quadrature is satisfactory in all cases. More correct figures in the results are obtained by a modest increase in the computational effort.

Table 1: Probabilities.

β	n_λ	λ	u_s	b	P_f	n_p	δ
0	5	-.500	-1.592	1.064	0.93658	10	2.2E-8
1	5	-.500	-1.099	0.914	0.79034	11	1.5E-6
2	5	-.250	-0.886	0.683	0.26378	13	3.1E-7
3	5	-.167	-1.685	0.844	0.31166E-1	11	1.4E-7
4	5	-.125	-2.600	0.909	0.12675E-2	11	2.2E-7
5	5	-.100	-3.553	0.938	0.18211E-4	11	3.1E-7

Table 2: Densities.							
β	n_λ	λ	u_s	b	p_f	n_p	δ
0	5	-.500	1.158	1.408	0.93848E-1	9	1.4E-7
1	5	-.500	0.581	1.000	0.19620	10	1.9E-7
2	5	-.250	-0.419	1.000	0.24765	10	2.0E-7
3	5	-.167	-1.416	1.000	0.56621E-1	10	1.9E-7
4	5	-.125	-2.419	1.000	0.34725E-2	10	1.8E-7
5	5	-.100	-3.419	1.000	0.67390E-4	10	1.9E-7

In Table 1 and in Table 2, for $\beta \geq 1$, the n_λ equal quadratic terms have curvatures corresponding to those of a sphere with radius β centered at the origin. The factors $(1-2\lambda_j u_s - 2i\lambda_j s)$ in (9) and in (10) suggest that the curvatures of the sphere represent a worst case in the performance of the numerical procedure when applied to reliability problems. The reason is that, for the same integration path, a greater value of λ_j yields a greater constant term $(1-2\lambda_j u_s)$. (If shifted to the saddle point, the performance is expected to be even better). Also, in reliability problems, $\lambda_j \geq -1/(2\beta)$ always.

Table 3: Probabilities.									
β	$n_{\lambda,1}$	λ_1	$n_{\lambda,2}$	λ_2	u_s	b	P_f	n_p	δ
0	5	-.1	5	.1	-0.910	0.908	0.50000	11	6.6E-6
1	10	-.2	10	.2	-0.805	0.649	0.26279	12	1.7E-7
2	5	-.1	5	.1	-2.105	1.126	0.33987E-1	9	1.2E-7
2	10	-.2	10	.2	-1.020	0.656	0.10450	12	1.9E-7
5	5	-.1	5	.1	-3.674	0.889	0.44391E-5	11	5.2E-7

Table 4: Densities.									
β	$n_{\lambda,1}$	λ_1	$n_{\lambda,2}$	λ_2	u_s	b	p_f	n_p	δ
0	5	-.1	5	.1	0.000	1.290	0.36558	9	1.0E-7
1	10	-.2	10	.2	-0.379	0.858	0.20555	10	1.1E-7
2	5	-.1	5	.1	-1.634	1.254	0.68243E-1	9	7.8E-8
2	10	-.2	10	.2	-0.728	0.807	0.11073	11	5.7E-8
5	5	-.1	5	.1	-3.558	0.946	0.17012E-4	11	4.2E-7

The results presented in Table 3 and in Table 4, with mixed signs on the curvatures, further demonstrates the performance of the method. In all the cases included above, 7-8 correct figures in the result are obtained by reducing the truncation limit to 10^{-10} and using approximately $2n_p$ points in the quadrature.

5. Summary and conclusion

The paper has presented a study of the use of the saddle-point method in connection with the inversion integrals of the Laplace transforms of the probability function and of the density function of a central parabolic form. The purpose has been to provide a numerical method that can be made automatic on a computer. The included results demonstrates that the suggested method works well and that a number of significant digits in the estimates, sufficient for reliability analysis, is obtained through a minor computational effort.

6.References

[1] Breitung, K., "Asymptotic Approximation for Multinormal Integrals," *Journal of Engineering Mechanics,* ASCE, Vol. 110, No. 3, Mar., 1984, pp. 357-366.

[2] Der Kiureghian, A., Lin H.Z. and Hwang, S.J., "Second Order Reliability Approximations," *Journal of Engineering Mechanics,* ASCE, Vol. 113, No. 8, Aug., 1987, pp. 1208-1225

[3] Fiessler, B., Neumann H.-J., and Rackwitz R., "Quadratic Limit States in Structural Reliability", *Journal of the Engineering Mechanics Division,* ASCE, Vol 105, 1979, pp. 661-676.

[4] Madsen, H.O., Krenk,S. and Lind,N.C, *Methods of Structural Safety* Prentice-Hall, Inc., Englewood Cliffs, NJ 07632, 1986

[5] Tvedt, L., "Two Second-Order Approximations to the Failure Probability," Veritas Report RDIV/20-004-83, Det norske Veritas, Oslo, Norway, 1983.

[6] Madsen, H.O., "Model Updating in Reliability Theory", Proc. ICASP-5, Vol. 1, Vancouver, 1987, pp. 564-577

[7] Rice, S.O., "Distribution of Quadratic Forms in Normal Random Variables- Evaluation by Numerical Integration," *SIAM, J. SCI. STAT. COMPUT.,* Vol. 1, No. 4, December 1980, pp. 438-448

[8] Helstrom, C.W., "Comment: Distribution of Quadratic Forms in Normal Random Variables - Evaluation by Numerical Integration," *SIAM, J. SCI. STAT. COMPUT.,* Vol. 4, No. 2, June 1983, pp. 353-356

[9] Johnson N. L. and Kotz S., *Distributions in Statistics - Continuous Univariate Distributions, 2* John Wiley, New York, 1970, Chapter 29.

Appendix A: Derivation of the characteristic function

The random variable C is defined as

$$C = Y - \sum_{j=1}^{n} \lambda_j X_j^2 = Y + V \tag{A1}$$

where Y and the $X_j, j=1,2,\ldots,n$ are mutually independent standard normal random variables.

The characteristic function of Y is

$$G_Y(u) = E[\exp(uy)] = \exp(u^2/2) \tag{A2}$$

and the characteristic function of V is according to Johnson and Kotz, [9], p152,

$$G_V(u) = \prod_{j=1}^{n}(1+2\lambda_j u)^{-1/2} \tag{A3}$$

The characteristic function of C is therefore

$$G_C(u) = \exp(u^2/2)\prod_{j=1}^{n}(1+2\lambda_j u)^{-1/2} \tag{A4}$$

STRUCTURAL RELIABILITY ASSESSMENT BY MINIMAL REPRESENTATIONS IN A MECHANISM APPROACH

Adrian Vulpe & Alexandru Carausu
Polytechnic Institute of Iasi, R-6600 Iasi-6, Romania

1. INTRODUCTION

A great variety of methods and techniques for structural reliability assessment has been developed in the literature of the last decade. They essentially depend on the approach adopted, that is, on the formulation of structural system failure, but also on the way the structural failure event is represented. Such representations can be naturally expressed in terms of set-theoretic operations,that are isomorphic to appropriate graph-theoretic models. The latter ones bring not only the advantage of synoptic / economic descriptions of system behaviour at failure, but also the opportunity to use some of the existing and rather powerful techniques implemented within the field of graph theory. Moreover, it seems possible to obtain even more synthetic representations in terms of extended structures such as the directed labeled hypergraphs and the labeled dags (see, e.g., Vulpe & Cărăuşu [1,2]).

As regards the approach to structural failure formulation, it has been argued by P.Thoft-Christensen (et al.) [3,4,5,6] that the most realistic and the least time consuming one is - in the case of elasto - plastic structures with ductile elements - the mechanismapproach. More precisely, the failure of the structure is defined as the formation of a mechanism. It has been shown that the powerful β-unzipping method developed by the authors can be used to identify the most significant plastic mechanisms in an efficient way [5,6]. The unzipping is based on reliability indices associated with fundamental mechanisms and linear combinations of fundamental mechanisms.

In this paper we are taking into account three ways to obtain and use minimal representations of the structural system failure event for the reliability assessment in a mechanism approach.

(i) Firstly, the system failure representation at the mechanism level is by itself minimal (when the β - unzipping method is used) since only the most significant, that is, the most probable failure mechanisms are considered. A SSPS (series system of parallel subsystems) representation [7] of the system failure event can thus be obtained, but with fundamental failure mechanisms instead of element / structural component failures. The appropriate graph-theoretic model corresponding to this kind of representations has been found to be the one in terms of labeled quasi-dags (Section 2).

(ii) The system failure modelling in a mechanism approach [4,5,6] involves a <u>failure tree</u> which is constructed level by level; a formal definition of failure trees that is compatible with this kind of F-trees was proposed in our paper [1]. However, such a representation may be rather highly redundant. It can be minimized either by applying an adapted CUTALG procedure developed at T.U.M. [8], or by turning the SSPS into a PSSS representation [7], further subjected to minimization by use of <u>disjoint</u> cut sets (Section 3).

(iii) After a min-cut representation has been derived for the system failure event F_S, the most probable compound failure mechanisms can be identified by sorting out the FMs whose failure probabilities fall below a given / prescribed level p_o (Section 4).

2. SSPS / PSSS REPRESENTATIONS OF STRUCTURAL FAILURE AT THE MECHANISM LEVEL

Let us consider an elasto-plastic frame structure S with N potential failure elements (e.g., plastic hinges) and M possible fundamental failure mechanisms (FMs). In what follows, the index (subsript) $i \in \{1,2,\ldots,M\}$ will correspond to a FM, while $j \in \{1,2,\ldots,N\}$ will identify a potential failure element. f_{ij} will denote the failure of element j within the FM i. As in [7], we also assume that event f_{ij} can be analytically described in terms of a failure indicator function (or state function) $g_{ij}(\underline{u})$ by

$$f_{ij} \text{ occurs iff } g_{ij}(\underline{u}) \leq 0 \tag{2.1}$$

where $\underline{u}$ is the set of possible values of the random vector $\underline{U} = (U_1,U_2,\ldots, U_n)$; its random components can represent both the structural properties and the applied loads.

Similarly, a state function g_i may be associated to each FM F_i. In [5,6] such a function is the <u>safety margin</u>

$$M_i = \sum_{j=1}^{N} |a_{ij}| R_j - \sum_{k=1}^{K} b_{ik} P_k \tag{2.2}$$

where R_j is the yield strength of F-element j, P_k is load number k, and a_{ij}, b_{ik} are the influence coefficients.

The distributional properties of the random parameters R_j and P_k can be described by their expected values and their coefficients of variation.

Let us denote by F_S the event of structural system failure and by P_f its probability, that is, the failure probability of the system. Then

$$P_f = P(F_S) = P(\bigcup_{i=1}^{M} \bigcap_{j=1}^{N_i} f_{ij}) = P(\bigcup_{i=1}^{M} \bigcap_{j=1}^{N_i} \{g_{ij}(\underline{U}) \leq 0\}) \qquad (2.3)$$

which follows from the well-known decomposition of F_S in terms of events f_{ij} and from analytical characterization (2.1). Let us note that eq.(2.3) differs from eq.(2.2) of [7] in that it is of dual form (union of intersections). Moreover, eq.(2.3) should be written (more rigorously) as

$$P_f = P(\bigcup_{i=1}^{M} \bigcap_{j \in I_i} f_{ij}) = \dots \qquad (2.4)$$

where I_i = the set of indices (subscripts) of the failure elements whose intersection defines the failure mechanism i ; obviously,

$$\bigcup_{i=1}^{M} I_i = \{1, 2, \dots, N\} \qquad (2.5)$$

and we may naturally assume that

$$i \neq k \Longrightarrow I_i \neq I_k \quad , \qquad (2.6)$$

i.e., two failure mechanisms differ iff the sets of failure elements they consist of are different.

Therefore, the natural set-theoretic (logical) decomposition of the system failure event F_S is given by

$$F_S = \bigcup_{i=1} \bigcap_{j \in I_i} f_{ij} \,. \qquad (2.7)$$

A decomposition like (2.7) can be readily represented by a <u>failure tree</u> of the form

$$T = (FE, U) \qquad (2.8)$$

where FE is the set of nodes / failure elements, that is,

$$FE = \{f_{ij} : i \in \{1, \dots, M\} \text{ and } j \in I_i\} \qquad (2.9)$$

and U is the set of arcs such that

$$u = (f_{ij}, f_{i\ell}) \in U \quad \text{iff} \quad j, \ell \in I_i \qquad (2.10)$$

(and they are consecutive in this set of subscripts, when the order of failing elements within every F-mechanism i is relevant).

It is easy to remark that a representation of F_S like (2.7) with (2.8) is redundant, since identical failure elements may occur in different FM's ; in other words, there may exist $i' \neq i''$ such that $f_{i'j} = f_{i''\ell}$ (with $j \neq \ell$). This may happen when $I_{i'} \cap I_{i''} \neq \emptyset$. Therefore, a first way to minimize the set-theoretic/graph-theoretic representation of F_S is to take each of the N failure elements only once. However, since it is necessary to distinguish between different FM's sharing common (subsets of) failure elements, it appears that we should employ a specific mathematical structure to represent F_S, namely the one of labeled directed acyclic graphs (ℓ-dag). We have also proposed its use in our recent paper [2]. The dags are involved in the theory of knowledge representation (within the context of artificial intelligence research) and they are formally studied in, e.g., the thesis [9]. However, the kind of representation structure that we need should be slightly more general than the one of a -dag, as it should accept loops ; let us call it "ℓ-qdag" (labeled quasi-dag).

The ℓ-qdag representing F_S (and isomorphic to the associated failure tree (2.8), see [5,6] for examples) is a 4-tuple of the form

$$D = (X, U, I, \ell) \tag{2.11}$$

where : X is the set of nodes (in one-to-one correspondence with the set FE of failure elements / with the set FM of fundamental mechanisms) ; $U \subset \mathcal{P}(X \times X)$ is the set of arcs ; $I = \{\pm 1, \pm 2, \ldots, \pm M'\}$ is the set of indices (labels) of possible failure mechanisms, such that the sign "+" and "-" are optional but exclusive ; and $\ell : U \longrightarrow I$ is the labeling monomorphism. Note that the ℓ-qdags accept multiple arcs as well as loops, provided that the labels on different arcs are distinct, and that labels on loops are signless. That is, if arcs $u_k = (x_i, x_j)$ and $u_m = (x_i, x_j)$ are both in U, then $k \neq m \Longrightarrow \ell(u_k) \neq \ell(u_m)$. And if $u = (x_i, x_i)$ is a loop, then $\ell(u) = j$.

As we have already pointed out, the representation of the event F_S in terms of ℓ-qdags is – if not yet minimal – however more economic than the ones by failure / fault trees because :

1° Every node, corresponding to one of the FM's, is taken only once, and

2° there are not introduced additional nodes for the "AND"/"OR"-gates as in the models based on fault trees [8].

The compound failure mechanisms will correspond to complete labeled paths in the ℓ-qdag D that models event F . Let us denote by path(D) the set of all paths in D which satisfy the following conditions :

(i) $u_j, u_k \subset p \in \text{path}(D) \Longrightarrow \ell(u_j) = \pm\ell(u_k)$; (2.12)

(ii) for $p = (x_1,\ldots,x_{i'},\ldots, x_{i''},\ldots,x_m) \in \text{path}(D)$,
$i' < i'' \Longrightarrow x_{i'} \neq x_{i''}$ (D is cycle-free) ; (2.13)

(iii) $p_1, p_2 \in \text{path}(D)$ and $p_1 \neq p_2 \Longrightarrow \ell(p_1) \neq \ell(p_2)$; (2.14)

(iv) if $p = (x, x) \in \text{path}(D)$ then $\ell(p)$ is signless . (2.15)

The way how a ℓ-qdag is associated to a subset of compound failure mechanisms is illustrated by

Example 2.1 Let S be an asymmetric two-bay two-story frame structure as the one represented in Figure 2.1. A set of ten fundamental mechanisms for this system has been considered in [6] (see Figure 52, page 59). By use of the β-unzipping method, a number of 12 most significant (compound) failure mechanisms have been identified in [6]. We list them below ; the fundamental mechanisms involved in the linear expansions of these mechanisms are indicated simply by "j" instead of "x_j".

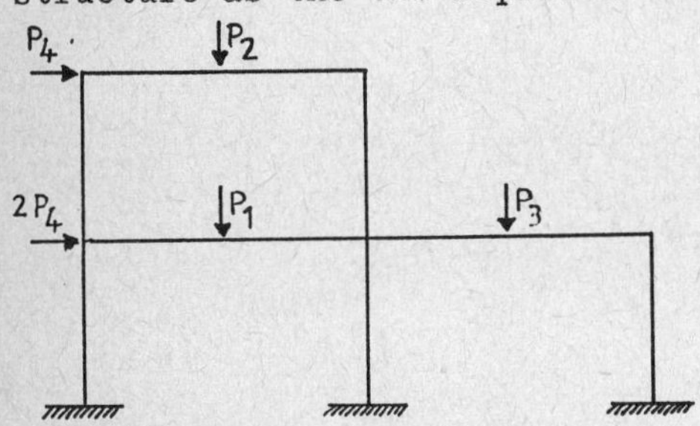

Figure 2.1

$$
\begin{aligned}
F_1 &= 1\\
F_2 &= 1 + 6 + 2\\
F_3 &= 1 + 6 + 2 + 5\\
F_4 &= 1 + 6 + 2 + 5 + 7 + 3 + 8\\
F_5 &= 1 + 6 + 2 + 5 + 7 + 3 - 8\\
F_6 &= 1 + 6 + 2 + 5 + 9 + 4 - 10\\
F_7 &= 1 + 6 + 2 + 5 + 7 + 3 + 8 + 9 + 4 - 10\\
F_8 &= 1 + 6 + 2 + 5 + 7 + 3 + 9 + 4 - 8 - 10\\
F_9 &= 2\\
F_{10} &= 3\\
F_{11} &= 3 + 7 - 8\\
F_{12} &= 4 - 10 \quad .
\end{aligned}
\tag{2.16}
$$

Therefore, the main part of the system failure event F_S may be represented by the union

$$F_S = (\bigcup_{i=1}^{12} F_i) \bigcup \ldots \qquad (2.17)$$

where less significant failure mechanisms are not explicitly taken. The ℓ-qdag corresponding to decomposition (2.17) with (2.16) is given below. In order to avoid overloading of the representation, several arcs between the same two nodes are replaced by a single arc with several (different) labels on it. The failure mechanisms F_1, F_9 and F_{10} which reduce to single FM's correspond to the three loops of the ℓ-qdag in Figure 2.2.

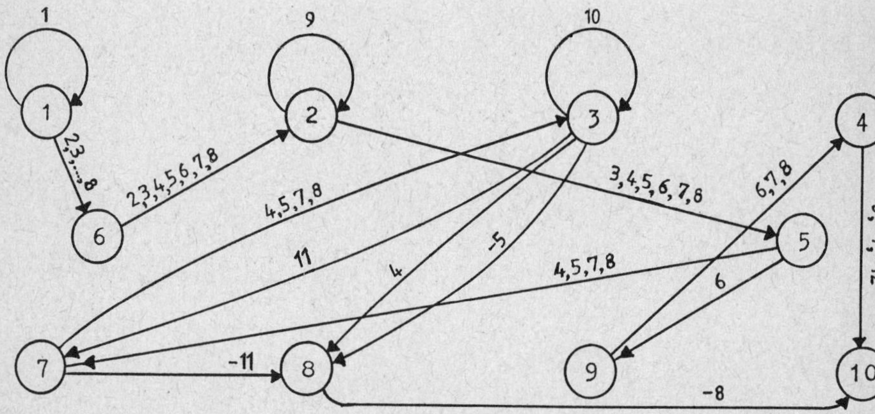

Figure 2.2 The ℓ - qdag representing F_S for the structure S of Fig.2.1

Remarks. When a representation as the one in Figure 2.2 is built up, a label j on an arc $u = (x_{i'}, x_{i''}) \subset p$ means that the FMs $x_{i'}$ and $x_{i''}$ are involved as $x_{i'} + x_{i''}$ (or simply i' + i" as denoted in the list (2.16)) ; the sign "-" before j corresponds to i' - i" in the structure of F_j that is represented by path p. The labels on the loops are always signless.

A ℓ-qdag can be analytically described by means of a $(M \times M)$-matrix $A = [a_{ij}]$ such that

$$a_{ij} = \ell(x_i, x_j) \subset 2^I \quad \text{and} \qquad (2.18)$$

(i) if (x_i, x_j) is not an arc in the ℓ-qdag D(F), then $a_{ij} = \emptyset$ (the corresponding entry is left empty in A) ;

(ii) if $p = (x_i, x_i)$ is a loop in D(F), then $a_{ii} = \ell(p)$;

(iii) if $(x_i, x_j) \subset p_1, \ldots, p_m \in \text{path}(D)$, then $a_{ij} = \{\pm 1, \ldots, \pm m\} \subseteq I$.

Thus, the matrix A associated to D(F) of Figure 2.2 will be

$$A = \begin{bmatrix} a_{11} & & & & & a_{16} & & & & \\ & a_{22} & & & a_{25} & & & & & \\ & & a_{33} & & & & a_{37} & a_{38} & & \\ & & & & & & & a_{48} & & a_{4\,10} \\ & & & & & & a_{57} & & a_{59} & \\ & a_{62} & & & & & & & & \\ & & a_{73} & & & & & a_{78} & & \\ & & & & & & & & & a_{8\,10} \\ & & & a_{94} & & & & & & \\ & & & & & & & & & \end{bmatrix} \qquad (2.19)$$

where

$$\begin{aligned} a_{11} &= \{1\} & a_{4\,10} &= \{-6,-7,-12\} \\ a_{16} &= \{2,3,4,5,6,7,8\} & a_{57} &= \{4,5,7,8\} \\ a_{22} &= \{9\} & a_{59} &= \{6\} \\ a_{25} &= \{3,4,5,6,7,8\} & a_{62} &= \{2,3,4,5,6,7,8\} \\ a_{33} &= \{10\} & a_{73} &= \{4,5,7,8\} \\ a_{37} &= \{11\} & a_{78} &= \{-11\} \\ a_{38} &= \{4,-5\} & a_{8\,10} &= \{-8\} \\ a_{48} &= \{-8\} & a_{94} &= \{6,7,8\} \end{aligned} \qquad (2.20)$$

The tenth row in the matrix A consists of empty entries only.

It can be readily seen that, conversely, a matrix A of the form (2.19) with (2.20) determines a unique ℓ - qdag (2.11) D(F) (as the one in Figure 2.2), which makes possible to reconstruct a partial failure tree as the one of Figures 57 & 58 of [6] (but without the β - values specified).

3. MINIMAL REPRESENTATIONS OF THE STRUCTURAL FAILURE EVENT BY SSPS / PSSS MODELS

It is obvious that a representation of F_S given by eq.(2.7) or – correspondingly – by a ℓ - qdag (2.11) / by a matrix of the form of eq.(2.19) is actually a SSPS model of this event. In the terminology of [8] it is a cut-set representation, since every term in the union (2.17) is an intersection of FMs.

Let us therefore assume that F_S is decomposed as in eq.(2.7). Two possibilities are now at hand for deriving minimal representations and both of them are considered in the next two subsections.

3.1 Expanding F_S into disjoint unions

A method to obtain expansions of F_S into disjoint unions (of minimal cut vectors) is presented in [7], provided that F_S is available as an intersection of unions, that is, it can be modeled as a PSSS ; more precisely,

$$F_S = \bigcap_{i=1}^{M} \bigcup_{j \in J_i} f_{ij} \quad . \tag{3.1}$$

A representation of the form (3.1) is obviously dual to one as given by eq.(2.7). They can be turned one into another by using the subroutine YCUT of CUTALG presented in [8]. We therefore may assume that F_S is decomposed as in (3.1) (if not, a procedure like YCUT should be first applied). With these representation preliminaries and specific distributional assumptions on the state functions g_{ij} (and their random arguments), minimal cut-vectors $\mathbf{k}$ and $P(F_S) = P_f$ could be derived by the techniques described in [7,8]. However, we have to see how an idealization of the form (2.7) can be actually turned into one like (3.1).

If F_S is represented as in (2.7), that is as a SSPS, it should first be turned into a cut-set fault tree ($c\,s_f$, according to [8]). Without going into details, the necessary transformations are included in the procedures :

I. <u>CSR ⟶ CST</u>

I.1 Every term $F_i = \bigcup_{j \in I_i} f_{ij}$ (3.2)

of (2.7) is asigned the $\triangle_*$ - gate node having as its successors the elements in $\{f_{ij} : j \in I_i\}$ as a support subset of $\{1,2,\ldots, M'\}$;

I.2 An $\triangle_+$ -gate node is introduced having as its M successor nodes the $\triangle_*$ - nodes defined at Step I.1.

M' in I.1 equals, e.g., the number of fundamental failure mechanisms.

II. <u>CST $\xrightarrow{\text{YCUT}}$ TST</u>

The cut-set fault tree obtained by I is turned into the corresponding tie-set fault tree by applying the YCUT procedure.

III. <u>TST ⟶ PSSS</u>

Informally, the $\triangle_+$ / $\triangle_*$ -gate nodes are erased and a PSSS representation of the form (3.1) is readily obtained.

IV. PSSS ⟶ DISJUN

The procedure of [7] is applied to the output of III to obtain

$$F_S = \bigcup_{\mathbb{k}} \{cut_{\mathbb{k}}\}, \tag{3.3}$$

where $cut_{\mathbb{k}}$ are the disjoint cut sets (cut vectors), together with the system failure probability

$$P_f = P(F_S) = \sum_{\mathbb{k}} P(cut_{\mathbb{k}}) . \tag{3.4}$$

The problem to compute the last member of eq.(3.4) is not too simple, because of the high number of possible vectors $\mathbb{k}$, but it becomes more tractable under the (restricting) assumptions accepted in [7].

3.2 Direct Derivation of Minimal SSPS Representations

To apply the sequence of four transformations I,...,IV presented in Subsection 3.1 seems to be rather much time consuming. It would be however justified by the advantage that the difficulty to evaluate P_f (by use of, e.g., Ditlevsen bounds) is transferred to automatic procedures that can render disjoint union representations of the form (3.3), for which exact evaluations like (3.4) could be applied.

An alternative way would consist in reducing to minimum the use of intermediate fault-tree models. As we have already seen, a natural set-theoretic representation of F_S within a mechanism approach is a two-level one, of the form (2.7). We have shown (see Section 2)how it can be turned into a ℓ- qdag as in (2.11), and then into a matrix A of the form (2.19) with (2.20).

Once available such representations, it remains to search for a way to obtain minimal ones (but of the same nature), that is,minimal cut-set representations of F_S. Every term occurring in such a decomposition should not be a subset of another term. This task can be accomplished by using an adapted version of CUTALG presented in [8] which we denote by MCUTALG (modified CUTALG). The input parameters are :

KN = number of gates = M+1 ;

KB = number of fundamental mechanisms = N ;

MEB = 0 ;

0 = +, 1,..., M is the (only) "OR-gate" with its successors

being the gates No. 1,2,..., M ;

I = *, J1, J2,..., JNI is the "AND-gate" with its successors being the fundamental mechanisms

$f_{ij_1}, f_{ij_2}, \ldots, f_{ij_{N_i}}$.

More condensed, the input file would read

M+1, N, 0

0 = +, 1, 2,..., M

1 = *, J1,..., JN1

⋮

M = *, J1,..., JNM .

Further the algorithm follows similarly as CUTALG - YCUT (see [8] for details), giving the min-cut representation of F_S at the mechanism level. Now, the m c s fault-tree can be readily turned into a representation of the form (2.7) which is isomorphic to a ℓ- qdag (2.11). An extended ℓ- qdag representation is described in Section 4 ; it is constructed by a procedure that renders the minimal/ most significant (probable) cut-set decompositions of F_S.

4. IDENTIFICATION OF PROBABILISTICALLY DOMINANT FAILURE MECHANISMS

As we have seen in Section 2, a cut-set representation (of the form (2.7)) can be formulated at the mechanism level, that is, the cuts in this decomposition are intersections of FMs giving compound failure mechanisms. The algorithm presented in Section 3 make possible to obtain representations of this kind that are minimal from the logical / set-theoretic point of view. However, their size may be rather large and – on another hand – the selection of the terms (i.e. mechanisms) in such representations does not take into account their probabilities of occurrence. That is why it is necessary to propose another, extended way to represent F_S which could be used to identify the probabilistically dominant failure mechanisms.

Let F_S be represented as a disjoint union of cut-sets of the form

$$F_S = \bigcup_{i=1}^{M} \bigcap_{j \in I_i^*} f_{ij} \quad , \tag{4.1}$$

that is, $i' \neq i''$ implies

$$\Big(\bigcap_{j\in I^*_{i'}} f_{ij}\Big)\cap\Big(\bigcap_{j\in I^*_{i''}} f_{ij}\Big) = \emptyset \quad . \tag{4.2}$$

A decomposition of the form (4.1) with (4.2) can be obtained by use of algorithms presented in Section 3.

Let us further assume that failure probabilities can be evaluated by usual techniques (see [3,4,..., 8]) for the intersections, that is, cut-sets

$$F_i = \bigcap_{j\in I^*_i} f_{ij} \longmapsto P(F_i) \underset{\text{not}}{=} P_i \tag{4.3}$$

on the basis of available distributional (second order) information on the random parameters occuring in eqs.(2.2) and (2.3) and their correlation structure.

In fact, since the cut terms in union decomposition (4.1) are disjoint, it may be equivalently written as

$$F_S = \bigcup_{i=1}^{M}\bigcap_{k\in K^*_i} f_k \tag{4.4}$$

where

$$i' \neq i'' \Longrightarrow K^*_{i'} \cap K^*_{i''} = \emptyset \quad \text{and} \tag{4.5}$$

$$\bigcup_{i=1}^{M}\{f_k : k\in K^*_i\} = FM = \text{the set of fundamental mechanisms.} \tag{4.6}$$

The above assumption is equivalent to the possibility of evaluating probabilities

$$P_i = P\Big(\bigcap_{k-K^*_i} f_k\Big) = P(f_{k_1})P(f_{k_2}|f_{k_1})\ldots \tag{4.7}$$

as functions of the FM probabilities $P(f_k)$ and of the correlation matrix $[\varrho]$.

A probabilistic characterization of F_S at the mechanism level can now be built up, in terms of "augmented" or rather probabilistic ℓ - qdags, by the following procedure :

Algorithm P-QDAG

Input data : FM, partition (4.4), P(FM) and $[\varrho]$, P_f-level p_o.

Let

$$D_P(F) = (X, U, \varkappa, \pi, p_o, M^*) \underset{\text{not}}{=} D \tag{4.8}$$

where the components of D are defined / constructed as follows :

X (the set of the nodes) is in one-to-one correspondence with

FM (the set of fundamental mechanisms, see eq.(4.6)) ;

U is the set of arcs including possible loops (x_k, x_k) iff f_k is a failure FM ; the other arcs (which are not loops) are formally constructed by Steps 2, 3, 4 that follow ;

for $\text{card}X = \text{card}FM = M_o$, is the partitioning function

$$\varkappa : \{1, \ldots, M_o\} \longrightarrow \mathbb{N} \tag{4.9}$$

such that $\varkappa(k) = i$ iff f_k is a cut term in $F_i = \bigcap_{k \in K_i^*} f_k$; available from input data (partition (4.4)) ;

p_o is a (prescribed) selecting level for failure probability.

All the remaining three components of D, that is $U \subset \mathcal{P}(X \times X)$, $\pi : \{F_i : i \in \{1, \ldots, M\}\} \longrightarrow [0,1]$, and M^* (the set of selected, i.e. probabilistically dominant failure mechanisms) are constructed by the following steps :

Step 1

Set SM = the set of <u>single</u> failure FMs, that is, $x_k \in SM$ if f_k is a <u>failure</u> FM ; check each f_k with x_k in SM by

$$P(f_k) \geqq p_o \quad ; \tag{4.10}$$

if YES store $u = (x_k, x_k)$ in $M^* \subset U$ with $\pi(u) = P(f_k)$, ELSE check x_{k+1} $(k+1 \leqq M_o)$; for $k+1 > M_o$ GO TO

Step 2

Set $\text{CMIND} = \{1, \ldots, M_o\} \setminus \{k : f_k$ is a <u>failure</u> FM$\}$ and (4.11)

$$m = \varkappa(\min\{k : k \in \text{CMIND}\}) \text{ if } \text{CMIND} \neq \emptyset , \tag{4.12}$$

and GO TO <u>Step 3</u> ; if CMIND = $\emptyset$ GO TO <u>Step 5</u> .

Step 3

Find $\text{path}(m) = (x_{m_1}, x_{m_2}, \ldots, x_{m_n}) \in \text{path}(D)$ (see Section 2 and (4.4)) and take all the arcs $(x_{m_j}, x_{m_{j+1}})$ to U ; evaluate $P_m = P(F_m)$ by (4.7), set $\pi(F_m) = P_m$ and check whether $P_m \geqq p_o$; if YES take F_m to M^*, ELSE let M^* unchanged and GO TO

Step 4

Set $\text{CMIND} = \text{CMIND} \setminus \{m_1, m_2, \ldots, m_n\}$ and check whether $\text{CMIND} = \emptyset$; if YES GO TO <u>Step 5</u>, ELSE GO TO (4.12) of <u>Step 2</u>.

Step 5

Print the matrix

'LIST OF PROBABILISTICALLY DOMINANT FAILURE MECHANISMS'

$$\left[m \;\middle|\; \text{path}(m) = F_m \in M^* \;\middle|\; \pi(F_m) = P_m \right]_{M_1 \text{ rows}} \tag{4.13}$$

where $M_1 = \text{card}\, M^* \leqq M$ is the number of selected (probabilistically dominant) single or compound failure mechanisms ;

compute

$$P_f^* = P(F_S^*) = \sum_{F_m \in M^*} \pi(F_m) = \sum_{m=1}^{M_1} P_m \quad . \tag{4.14}$$

STOP

Remarks. 1° CMIND represents the set of indices of fundamental mechanisms which are involved in compound failure mechanisms (see eq. (4.4)). It is reduced at each cycle of the procedure by eq. of Step 4 which may also be written as $\text{CMIND} = \text{CMIND} \setminus \varkappa_{-1}(m)$ and assures the convergence of the algorithm, as $\text{card}\, FM = M_0$ is finite ; according to (4.11), card CMIND may be even less than M_0 = the number of fundamental mechanisms.

2° The probabilities P_m which appear in (4.14) can be evaluated by eq.(4.7) using, e.g., the methods of [3,4,5,6] or the FORM/SORM methods of [8] (reformulated at the mechanism level).

3° Obviously, the evaluation P_f^* of eq.(4.14) gives a lower bound for the "exact" $P(F_S)$, but it is expected to be rather close to the actual value of P_f for a sufficiently low selecting level p_0.

4° Algorithm P-QDAG differs from the β- unzipping method as concerns the way the significant failure mechanisms are selected ; on another hand, the use of Ditlevsen - type bounds on P_f is avoided because F_S is previously expanded as a disjoint union of failure mechanisms.

Application

Coming back to Example 2.1 with the list (2.16) of the most significant (single/compound) failure mechanisms, the corresponding failure probabilities are given in Table 4.1. F_S^* has been expanded into a disjoint union of failure mechanisms of the form

$$F_S^* = \bigcup_{m=1}^{3} F_m \tag{4.15}$$

where $F_1 = (1,6,2,5)$, $F_2 = (7,3,8)$, $F_3 = (9,4,10)$. (4.16)

Table 4.1 Failure mechanisms with their failure probabilities

F_i	F_1	F_2	F_3	F_4	F_5	F_6
$P(F_i)$	0.0281	0.0233	0.0146	0.0183	0.0301	0.0150

	F_7	F_8	F_9	F_{10}	F_{11}	F_{12}
	0.0174	0.0262	0.0188	0.0150	0.0239	0.0233

Taking now the selecting level $p_o = 0.0170$, the probabilistically dominant failure mechanisms are found to be

$$F_1, F_2, F_4, F_5, F_7, F_8, F_9, F_{11}, F_{12} \quad . \tag{4.17}$$

The minimal disjoint cut-set decomposition of M^* (which consists just of the mechanisms listed in (4.17)) has been found to coincide with the expansion (4.15) - (4.16). The corresponding (lower bound) evaluation of $P(F_S) = P_f$ is given by

$$P_f^* = P(F_S^*) = 0.0146 + 0.0239 + 0.000021 = 0.0385 \quad . \tag{4.18}$$

The P-qdag representation of F_S^* (associated to (4.15) with (4.16)) reduces — in fact — to a partial ℓ- dag of the ℓ- qdag illustrated in Figure 2.2 ; more precisely, it is a labeled graph consisting of three connected components (see Figure 4.1).

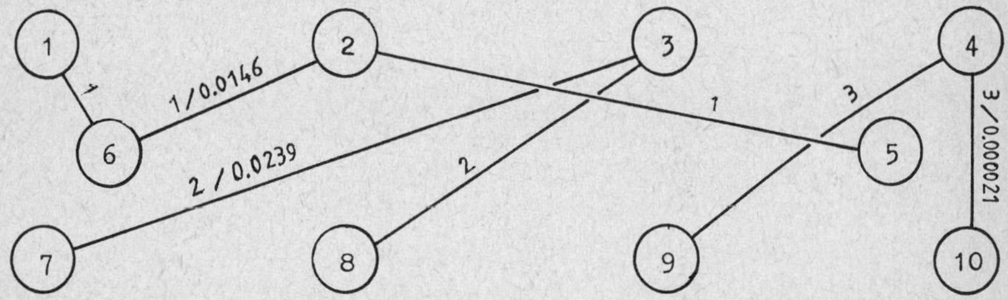

Figure 4.1 Labeled graph representation of the disjoint union decomposition of F_S^* (see (4.15) - (4.16))

5. CONCLUDING REMARKS

Certain mathematical structures emerging from the theory of graphs (labeled dags and labeled quasi-dags) have been proposed as rather well suited models for minimal representations of structural failure in a mechanism approach. A couple of techniques handling set-theoretic representations of F_S have been considered, and an algorithm that builds up the P-qdag associated to the structural failure event has been presented. It also renders a lower bound evaluation of system failure probability. Some of these representations have been illustrated on the case of an elasto-plastic frame structure.

REFERENCES

[1] Vulpe, A. & Cărăuşu, A.: On Some Graph-Theoretic Concepts and Techniques Applicable in the Reliability Analysis of Structural Systems. Lecture Notes in Engineering 33, Reliability and Optimization of Structural Systems (Proc. 1st IFIP WG 7.5 Working Conference, Aalborg, May 6-8, 1987), P.Thoft-Christensen (Ed.), Springer Verlag, Berlin, 1987, 399-416.

[2] Vulpe, A. & Cărăuşu, A.: Fragility Models in Seismic Probabilistic Risk Assessment of Systems with Multiple Failure Modes. 9th World Conference on Earthquake Engineering (submitted to), Tokyo - Kyoto, August 2-9, 1988.

[3] Thoft-Christensen, P. & Sørensen, J.D.: Reliability Analysis of Elasto-Plastic Structures. Structural Reliability Theory, Paper No.2, Instituttet for Bygningsteknik, Aalborg, IFB/A 8303 December 1983.

[4] Thoft-Christensen, P.: Reliability of Structural Systems. Structural Reliability Theory, Paper No.5, Instituttet for Bygningsteknik, Aalborg, R8403, June 1984.

[5] Thoft-Christensen, P. & Sørensen, J.D.: Optimization and Reliability of Structural Systems. Structural Reliability Theory, Paper No.6, Instituttet for Bygningsteknik, Aalborg, R8404, July 1984.

[6] Thoft-Christensen, P.: Reliability Analysis of Structural Systems by the β-unzipping Method. Structural Reliability Theory, Paper No.3, Instituttet for Bygningsteknik, Aalborg, R8401, March 1984.

[7] Egeland, T. & Tvedt, L.: Parallel Systems of Series Subsystems. Lecture Notes in Engineering 33, Reliability and Optimization of Structural Systems (Proc. 1st IFIP WG 7.5 Working Conference, Aalborg, May 6-8, 1987), P.Thoft-Christensen (Ed.), Springer Verlag, Berlin, 1987, 109-117.

[8] * * * User's Manual for Structural Reliability Programs CUTALG - FORM - SORM - SYSREL, SFB 96, L.K.I., Technische Univ. München, Heft 74, 1985.

[9] Warin, B.: Rationalité et reconnaissabilité dans les graphes. Thèse, Laboratoire d'Informatique Fondamentale, Université de Lille, 1986.

A RANDOM FIELD MODEL FOR THE ESTIMATION OF EARTHQUAKE HAZARD DUE TO ACTIVE FAULTS

M. Semih Yücemen
Department of Statistics
Middle East Technical University
Ankara, 06531 Turkey

ABSTRACT

A probabilistic method is described for estimating the earthquake hazard associated with potentially active faults. The methodology used in this study is based on a random field model developed for the assessment of seismic hazard along linearly extending seismic sources, like faults with well defined geometry.

In the probabilistic formulation, only a single fault is considered. An earthquake is assumed to occur when the total shear force developed over a portion of the fault plane of specified length and depth exceeds the counteracting total static friction force. The shear force is modeled as a space-time random field and the static friction force is taken as a spatial random field. Equations to evaluate the earthquake occurrence probability for a fixed and spatially varying shock origin are derived.

1. INTRODUCTION

The assessment of seismic hazard at a given site has received a great deal of attention and numerous probabilistic models have been developed for this purpose over the years. Most of these, however, studied the earthquake occurrence problem in the time domain, ignoring the spatial dimension. In the more recent studies (e.g. [2] and [8]) attempts were made to model the spatial characteristics of earthquakes.

In this study, the occurrence of earthquakes is treated as a space-time random field and the spatial and temporal correlations are taken into consideration. The application of the theory of random fields to the modeling of earthquake occurrences is relatively new, although the methodology of random fields provides the most appropriate tools for the development of such models. As a matter of fact, the theory of random functions is widely applied in biological research concerning the distribution of plants or animals over a given area and in the theory of turbulence more than two decades (e.g. [3] and [6]).

The aim of this paper is to model the occurrence of earthquakes in the space-time domain by making use of the theory of random functions. Equations for computing the earthquake occurrence probability for a fixed and spatially varying shock origin are derived based on the level crossing concepts of random functions.

2. A RANDOM FIELD MODEL FOR EARTHQUAKE OCCURRENCES

In the probabilistic formulation, we shall consider a single fault having a total length of L as shown in Figure 1. Faulting is assumed to occur on a vertical plane down to a depth of d, which remains constant along the total length of the fault. The assumption of a vertical fault plane (i.e., the angle of dip $\theta=90^0$) is for the sake of having a simple geometry, otherwise the model is applicable for any θ value. The value of d can be taken to equal twice the focal depth, implying a symmetric propagation of rupture around the shock origin (focus), in the vertical direction.

For the earthquake occurrence mechanism, it will be assumed that an earthquake originates when the local integral of the tangential stress random field exceeds the local integral of the static friction stress random field over a fault segment of specified length ℓ and having a depth of d.

Let $\tau_s(\underset{\sim}{u},t)$ be the shear stress developed at any point on the potential rupture plane. Speaking within the context of random functions, $\tau_s(\underset{\sim}{u},t)$ is a space-time random field with $\underset{\sim}{u}$ being the three-dimensional space vector $(\underset{\sim}{u}':\{x,y,z\})$, and t represents time. Similarly, $\tau_r(\underset{\sim}{u},t)$ will denote the counteracting static friction stress at any point on the potential rupture plane.

In the most general case, $\tau_s(\underset{\sim}{u},t)$ and $\tau_r(\underset{\sim}{u},t)$ are both four-dimensional random fields. However, by making some simplifying assumptions the dimension of this most general case will be reduced. Here, we shall idealize the fault as a line source with two spatial coordinates. The x-axis will extend along the axis of the fault and the z-axis along the edge of the fault plane as shown in Figure 1. The extent of the fault plane along the z-axis is assumed to be constant along the total length of the fault.

It has been indicated that the variability of τ_r immeadiately before shocks of similar magnitude is expected to be relatively small and thus, τ_r can be considered as constant in time [5]. Likewise, we shall assume τ_r to be constant in time, but we shall take into consideration the possible variability in space.

In view of the assumptions stated above, at any point in time and at any point on the fault plane, the state of the fault is fully described by the demand (shearing stress) and capacity (static friction stress) random functions, $\tau_s(x,z,t)$ and $\tau_r(x,z)$, respectively. At any point, (x',z'), on the fault plane and at any time, t', a **point rupture** will occur if $\tau_s(x',z',t')$ exceeds $\tau_r(x',z')$. However, earthquake generating mechanism depends on rupture over a finite area of the fault plane. This **areal rupture** mechanism requires that the total shear force developed over an area A exceeds the total static friction force developed over the same area of the fault plane. Accordingly, the local integrals of the original random functions $\tau_s(x,z,t)$ and $\tau_r(x,z)$ become of concern, which will be evaluated below.

Consider a potential rupture segment of length ℓ centered at x_0 and lying between $x_1=x_0-\ell/2$ and $x_2=x_0+\ell/2$. The extent of the rupture plane in the z direction is assumed to be equal to d (Figure 1). The ruptured portion of the fault plane will be rectangular in shape with dimensions ℓ by d. Here, x_0 can be interpreted as the projected location of the shock origin on the x-axis and the rupture extension is assumed to be symmetric about x_0 in the x direction. The total static friction force acting on this ruptured portion of the fault plane is written as follows:

$$R(x_0) = \int_0^d \int_{x_1}^{x_2} \tau_r(x,z)\, dx\, dz \qquad (1)$$

Here, x_0 in the argument of R shows the central location of the ruptured portion of the fault plane along the axis of the fault.

The mean and standard deviation of this spatial integral process are computed to be [4]:

$$\bar{R} = A\bar{\tau}_r = \ell d\bar{\tau}_r \qquad (2)$$

$$\tilde{R} = A\tilde{\tau}_r \Gamma_{r_2}(\ell,d) \qquad (3)$$

where, a **bar** and a **tilde** over a random variable denote, respectively, the mean and standard deviation of that random variable; $\Gamma_{r_2}(\ell,d)$ is the two-dimensional (areal) standard deviation reduction factor to be applied to the **point** standard deviation, $\tilde{\tau}_r$, due to the smoothing effect of the integration operation given by Equation 1.

It is assumed that the rupture length ℓ is fixed to the most critical value and d is constant and equals twice the average focal depth computed from the past seismic activity observed along the fault. It is also assumed that the static friction stress is probabilistically homogeneous over the fault plane. Furthermore, if we assume the correlation structure to be separable, then the two-dimensional standard deviation reduction factor can be expressed as the product of the respective one-dimensional standard deviation reduction factors as follows:

$$\Gamma_{r_2}(\ell,d) = \Gamma_{r_x}(\ell)\,\Gamma_{r_z}(d) \tag{4}$$

where, Γ_{r_x} and Γ_{r_z} are standard deviation reduction factors in the x and z directions, respectively. In the case of isotropic correlation structure, the standard deviation reduction factors in different directions will be the same, i.e., $\Gamma_{r_x}=\Gamma_{r_z}=\Gamma_{r_1}$ so that

$$\Gamma_{r_2}(\ell,d) = \Gamma_{r_1}(\ell)\,\Gamma_{r_1}(d) \tag{5}$$

In a similar way, the total shear force developed over the total potential rupture area on the fault plane (centered at x_0 and extending to a depth of d) is expressed as follows:

$$S(x_0,t) = \int_0^d \int_{x_1}^{x_2} \tau_s(x,z,t)\,dx\,dz = S(t) \tag{6}$$

In the above equation, S(t) denotes the two-dimensional (spatial) integral of the random field $\tau_s(x,z,t)$, at a given time t.

According to [5], the random field $\tau_s(x,z,t)$ may exhibit non-homogeneity in space and time. In the following, however, $\tau_s(x,z,t)$ is assumed to be a weakly homogeneous isotropic random field, with mean denoted by $\bar{\tau}_s$ and covariance function

$$B_{\tau_s}(x,z,t;\,x',z',t') = B_{\tau_s}(\eta,\xi) \tag{7}$$

where, $\eta=[(x-x')^2 + (z-z')^2]^{1/2}$ is the distance between two points on the fault plane and $\xi=t-t'$ is the time lag. The point variance of $\tau_s(x,z,t)$ is denoted by $\tilde{\tau}_s^2$ and is given by

$$\tilde{\tau}_s^2 = B_{\tau_s}(0,0) \tag{8}$$

Note that for homogeneous random fields the statistics of S(t) does not depend on x_0. In view of the above assumptions, the mean and standard deviation of S(t) are as follows:

$$\bar{S} = A\,\bar{\tau}_s = \ell d\,\bar{\tau}_s \tag{9}$$

$$\tilde{S} = A\,\tilde{\tau}_s\, r_{s_2}(\ell,d) \tag{10}$$

For the case of separable and isotropic correlation structure,

$$r_{s_2}(\ell,d) = r_{s_1}(\ell)\, r_{s_1}(d) \tag{11}$$

where, $r_{s_2}(\ell,d)$ and $r_{s_1}(.)$, denote, respectively, the two-dimensional and one-dimensional standard deviation reduction factors associated with the smoothing effect of the (spatial) integration operation carried out on the shear stress random field (see Equation 6).

It has been stated in [4] that a derived random field obtained by locally aggregating (integrating) a correlated random field converges to normality, provided that the size of the domain over which the original random field is aggregated, is sufficiently large compared to the correlation distances of the original random field. Hence, R and S are assumed to be Gaussian, based on the generalized central limit theorem, since both quantities result essentially from the summation (integration) of the point stresses (assumed to be weakly correlated) developed over the fault plane lying between x_1 and x_2 and extending for a depth of d.

3. EARTHQUAKE OCCURRENCE PROBABILITIES

An earthquake will occur when a certain portion of the fault ruptures over the fault plane. In our study this ruptured area is idealized to a rectangular shape extending over a length of ℓ along the x-axis and to a constant depth of d. Here, it is assumed that an earthquake will occur when the shear force developed over this ruptured area exceeds the counteracting static friction force. Below, we shall first consider the probability of earthquake occurrences at a specified location along the fault. This corresponds to the case where the shock origin along the x-axis is fixed. Later, the earthquake occurrence probability due to the shock origin taking place anywhere along the total length of the fault will be evaluated.

The x-coordinate becomes the only spatial coordinate to be considered in the subsequent discussions, since the location of the ruptured portion of the fault plane is to be referenced with respect to x_o, which is the projection of the shock origin on to the axis of the fault.

3.1 Probability of Earthquake Occurrences-Fixed Shock Origin

A portion of the fault plane centered at location x_0 will rupture causing an earthquake, whenever **S(t)>R** over the ruptured portion of the fault plane during the time interval (0,T). The depth of the ruptured portion will be taken, as explained in Section 2, constant and equal to d. Here, we are interested in the events where the random process S(t) upcrosses the level R. Although R is assumed to be constant in time, it is a random variable due to the variability of the static friction stress over the potential rupture area. The problem of upcrossing a random threshold can be solved by conditioning the mean upcrossing rate on the specific values of R. The mean number of upcrossings per unit time, conditional on the specified level R=r, is denoted by $\nu(r)$ and is computed from the following expression [8]:

$$\nu(r) = \frac{1}{2\pi} (\tilde{\dot{S}}/\tilde{S}) \exp[-(r-S)^2/2\tilde{S}^2] \tag{12}$$

where, $\dot{S}$ represents the derivative of S(t), assuming that S(t) is a differentiable random process in continuous time with continuous derivative.

The overall mean rate of upcrossings per unit time, ν is

$$\nu = \int_0^\infty \nu(r) \, f_R(r) \, dr \tag{13}$$

where, $f_R(r)$ is the probability density function of R which is shown to be approximately Gaussian with mean and variance as given by Equations 2 and 3, respectively.

The value of ν forms the basis for predicting the probability distribution of the extremes of the Gaussian process, S(t). For large earthquakes, which is of engineering interest, the threshold will be high. Then, using the fact that upcrossings of a high level by a smooth Gaussian process is asymptotically Poisson, we can express the probability of at least an earthquake with the shock origin located at x_0 occurring during the time interval (0,T) as [8]:

$$p_e(x_0) = 1 - \exp(-\nu T) \tag{14}$$

In deriving Equation 14, it is implicitly assumed that the process starts below the threshold with probability 1.0. The overall seismic hazard due to a single fault can be computed if the location of the shock origin along the axis of the fault is assumed to be random. This will be discussed in the next section.

3.2 Probability of Earthquake Occurrences-Random Shock Origin

In this case the location of the shock origin is not specified but could vary along the axis of the fault. The spatial randomness is restricted only to the x-coordinate, and the z-coordinate of the shock origin is implicitly assumed to be fixed and equal to d/2. Therefore, in the following the spatial variable z does not appear in the arguments of the random functions.

An earthquake will occur each time S(x,t) exceeds R(x) at any point of the region $D:\{x\varepsilon(\ell/2, L-\ell/2)$ and $t\varepsilon(0,T)\}$, where x denotes the projection of the location of the shock origin on to the x-axis and T is the total time period. Alternatively, we can consider the following margin between the two forces:

$$M(x,t) = S(x,t) - R(x) \tag{15}$$

Since, S(x,t) and R(x) are both assumed to be Gaussian, M(x,t) will be a Gaussian random field, with mean $\bar{M}=\bar{S}-\bar{R}$.

In order to simplify the computations we make the following transformation

$$V(x,t) = M(x,t) - \bar{M} \tag{16}$$

Accordingly, V(x,t) becomes a zero-mean Gaussian random field. We are now interested in the random number of excursions of the level $-\bar{M}=\bar{R}-\bar{S}$ by the random field V(x,t). For random fields, the concept of the number of upcrossings is lost, whereas the number of maxima above a given threshold is still valid [4]. For the region under consideration, D, the mean number of maxima above $\bar{R}-\bar{S}$ is

$$\bar{N} = [(L-\ell)T]\lambda \tag{17}$$

where, λ is the mean number of maxima above $\bar{R}-\bar{S}$ per unit distance along the fault line and per unit time for the random field V(x,t). The value of λ is to be evaluated from the following asymptotic formula given in [1] and [4]:

$$= |K|^{1/2} \frac{(\bar{R} - \bar{S})}{(2\pi)^{3/2} \tilde{V}^3} \exp[-(R-S)^2/2\tilde{V}^2] \tag{18}$$

In Equation 18, |K| is the determinant of the covariance matrix of the first partial derivatives of V(x,t). The covariance matrix [K], in this case is a 2x2 symmetric matrix and expressions to evaluate its elements are given in [7].

The earthquake occurrence probability can now be formulated as a first passage problem. The number of "upcrossings" at a high level is approximately equal to the number of maxima above that level [4]. This approximation will be valid especially in predicting the mean number of large magnitude earthquakes where the level to be crossed is significantly high.

The pattern of occurrence of local maxima and hence the occurrence of "upcrossings" above a high level tend to follow a planar Poisson process [8]. Accordingly, the probability of at least one earthquake occurrence associated with the fault during the time interval (0,T) is approximated by [8]:

$$p_E = 1 - \exp\left[-(L-\ell)\, T\lambda\right] \tag{19}$$

4. CONCLUDING COMMENTS

Earthquakes are caused by the accumulation of stress and strain to the point where rupturing occurs. In this study, it is assumed that an earthquake is generated whenever the total shear force developed over the ruptured area of the fault plane exceeds the total counteracting static friction force developed over the same area. These forces are modeled as weakly homogeneous Gaussian space-time random fields.

Equations to compute the earthquake occurrence probabilities are derived for the cases where the location of the shock origin is fixed and random. In these derivations the level crossing concepts of random functions and the related asymptotic equations are utilized.

The implementation of the proposed probabilistic model is hindered by the lack of data to assess the statistical parameters (such as mean, "point" standard deviation, standard deviation reduction factor, covariance function, e.t.c.) of the shear stress and static friction stress random fields. An alternative formulation based on the assumption that an earthquake of magnitude m occurs whenever the accumulated strain energy over the fault plane exceeds the energy level corresponding to m is currently under study. Such a formulation appears to be more suitable for implementation due to the availability of data on the energy random field through the magnitude-energy conversions.

REFERENCES

1. Hasofer, A.M., "The Mean Number of Maxima Above High Levels in Gaussian Random Fields", Jour. Appl. Probability, Vol. 13, pp. 377-379, 1976.

2. Kiremidjian, A.S. and Anagnos, T., "A Homogeneous Alternating Markov Model for Earthquake Occurrences", Proc. 7th World Conf. on Earthquake Eng., Vol. 1, pp. 355-362, 1980.

3. Panchev, S., Random Functions and Turbulence, Pergamon Press, New York, 1971.

4. Vanmarcke, E., Random Fields: Analysis and Synthesis, M.I.T. Press, Cambridge, Mass., 1981.

5. Veneziano, D. and Cornell, C.A., Earthquake Models with Spatial and Temporal Memory for Engineering Seismic Risk Analysis, Dept. Civil Eng., M.I.T., Cambridge, Mass., Rep. R74-18, 1974.

6. Whittle, P., "Topographic Correlation, Power-Law Covariance Functions and Diffusion", Biometrika, Vol. 49, pp. 305-314, 1962.

7. Yücemen, M.S., "Probabilistic Seismic Safety of Earth Slopes", Proc., Intern. Conf. an Soil Dynamics and Earthquake Eng., A.A. Balkema Publishers, pp.651-665, 1982.

8. Yüçemen, M.S., "Probabilistic Modeling of Earthquake Occurrences in the Space-Time Domain", Transactions, 8th Intern. Conf. on Structural Mechanics in Reactor Technology, Vol. K (a), pp. 19-24, 1985.

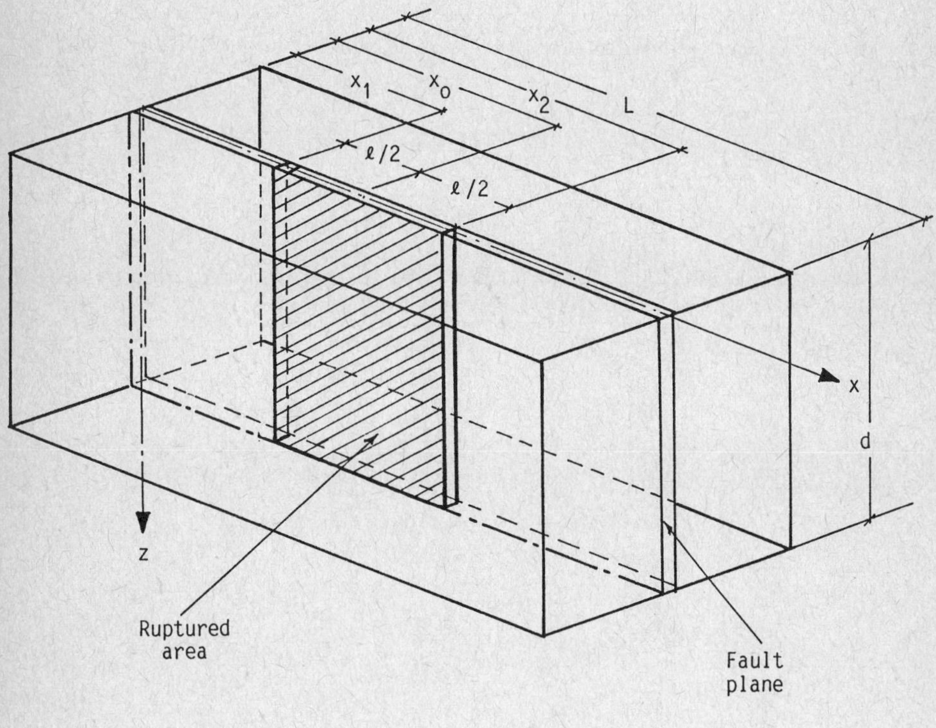

Figure 1: Side view into the earth's crust showing the fault plane and the geometry of the ruptured area

RELIABILITY ANALYSIS OF STOCHASTIC RIGID-IDEAL PLASTIC WALL BY FINITE ELEMENTS

Ove Ditlevsen, Torben Arnbjerg-Nielsen
Technical University of Denmark, Building 118, DK-2800 Lyngby, Denmark

Abstract The method of substitute polyhedral yield conditions is used iteratively in the method of invariant linear combinations of linearly associated lower bound safety margins. The purpose is to construct upper bound approximations to the reliability with respect to collapse of a finite element discretized rigid-ideal plastic wall. It is attempted to make a direct and fast identification of several of the weaker collapse mechanisms in the series system of all collapse mechanisms of the finite element system. This is to some extend achieved by use of the dominant safety margins in the invariant linear combinations. Among the linearly associated lower bound safety margins those are called dominant that have a common smallest simple reliability index (i.e. mean to standard deviation) made as large as possible by varying the element nodal forces over all statically admissible values.

For the homogeneous wall the yield condition is given on the stress level and it is of deterministic shape. The yield stress is specified to vary as a homogeneous gaussian random field over the plane of the wall. On the basis of these material properties an element yield condition in terms of the nodal forces is modelled by a suitable stress tensor averaging procedure over the finite element. Correspondingly the yield strength parameter for the element is obtained as an average of the random field over the element.

Introduction The object of study is a rigid-ideal plastic wall structure. The yield condition is of a given deterministic type but the yield strength parameter varies like a homogeneous gaussian random field over the plane of the wall. The load on the boundery of the wall has a given distributed configuration and a random intensity. The problem is to evaluate the reliability of the wall in terms of the reliability index, i.e. the standard normal fractile corresponding to the survival probability.

An exact solution to this problem is out of reach. One has to resort to a discretization of the wall into a finite number of triangular finite elements combined with a suitable modelling of the yield condition of each element in terms of the nodal forces. The yield parameters of the set of finite elements, one for each, then become a gaussian random vector of correlated elements generated from the given gaussian field. Yielding of an element means that the entire triangular element becomes plastic and deforms under preservation of the triangular shape in such a way that compatibility in the wall is preserved at the nodes and such that the associated flow rule (the normality condition) is obeyed. If an element is not yielding, it is in a state of infinite rigidity.

If the number of elements is not quite small there is a very large number of collapse mechanisms for this structural system. Generally it is impracticable to identify very many of these mechanisms. Also it is not of crucial interest to do so because only some few mechanisms usually dominate in their contribution to the probability of collapse by some mechanism.

In the reliability modelling sense the structural system behaves like a weakest link system (a series system) with the different collapse mechanisms as the links.

Most often only very few of these links contribute essentially to the total collapse probability. Therefore the problem is to identify these, or most of these, relatively weak collapse mechanisms without having to spend efforts on identifying too many less important (relatively strong) mechanisms.

A method of solving analogous problems for rigid-ideal plastic frame or truss structures with multidimensional yield conditions has been summarized in *Ditlevsen* (1988a). The detailed theory is given in *Ditlevsen* (1988b). This method will be applied to the present wall element system in an attempt to solve the stated wall reliability problem.

Identification of redundants For each of the r elements there are 2 nodal force components in each of the 3 nodal points. Among these 6 components there are 3 equations of equilibrium. This makes $3r$ equations. For each node there are 2 node equilibrium equations. With k nodal points this makes $2k$ equations. Thus $3r + 2k$ linearly independent equations are given among $6r$ unknown nodal force components and s support reaction components. This means that the degree of statical redundancy is $n = 3r + s - 2k$.

The n redundants are chosen among the $6r + s$ unknowns by use of Gauss elimination on the set of $3r + 2k$ equilibrium equations. During the Gauss elimination a successive pivoting procedure is applied such that the dominating terms of the equations are placed in the diagonal of the resulting equation system. The last n terms in the resulting equations contain the unknowns that constitute the vector of redundants $\boldsymbol{z}$. By solving these equations the vector of internal forces $\boldsymbol{Q}$ is obtained as a linear function of the vector of redundants $\boldsymbol{z}$ and the vector of external nodal forces $\boldsymbol{X}$. Among its components $\boldsymbol{Q}$ contains the redundants.

Element yield conditions In order to obtain a yield condition for the element in terms of the nodal forces and the yield strength of the material it is necessary to establish a relation between the nodal forces and some representative average stress tensor in the element. Since the element is infinitely rigid and the distributed stresses along the boundaries of the element, as it is sitting in the homogeneous wall, are replaced by the nodal forces in the discretized wall, there is no unique way to define such a representative stress tensor. The only guiding principle is that in the limit, as the element decreases in size, the representative stress tensor definition should approach the stress tensor of a statically admissible stress field in the homogeneous wall.

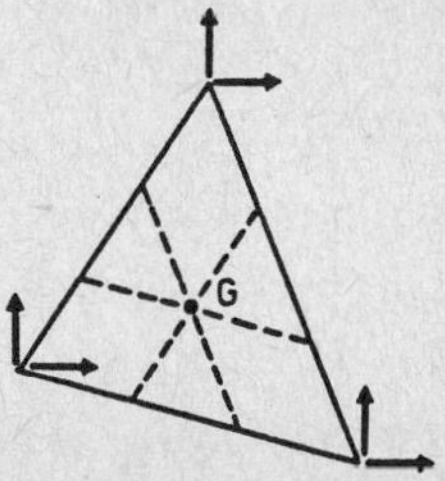

Fig. 1. Triangular element.

In this study it has been chosen to define the representative stress tensor by use of uniformly distributed normal and shear stresses along lines parallel to the sides of the triangle and intersecting in the geometrical centre G of the triangle, Fig. 1. These stresses are obtained as the average stress definede uniquely by projection equilibrium with the nodal forces. At the point G these stresses acting on the three intersecting cuts are generally not consistent with the requirement that the stress state at G should be defined by a stress tensor. This defect is corrected by changing the stresses at G on the three cuts by the smallest possible amount so that consistency is obtained.

The measure of total corrections is taken to be the sum of the squares of the corrections of the 6 stress components. By minimization of this sum the 3 components of the stress tensor $\sigma_{11}, \sigma_{22}, \sigma_{12}$ are obtained as linear functions of the 6 nodal forces. By use of the 3 equations of element equilibrium, 3 of the nodal forces can be eliminated from these linear functions.

Finally the stresses $\sigma_{11}, \sigma_{22}, \sigma_{12}$ expressed in terms of the nodal forces are substituted into the given yield condition leading to a yield condition

$$f_i(Q_i, Y_i) = 0, \qquad i = 1, ..., r \tag{1}$$

for each element. Y_i is a random variable that defines the yield strength of the ith element. (For some types of yield conditions there can be more than one strength parameter, in which case Y_i is the vector of these parameters).

Stochastic representation of the field vector $\boldsymbol{Y}$ of yield strengths The homogeneous gaussian random field that models the variation of the yield strength over the plane of the wall is completely defined by specifying a mean value μ, a standard deviation σ, and a correlation function $\rho(x_1 - x_2, y_1 - y_2)$ that defines the coefficient of correlation between the yield strengths at (x_1, y_1) and at (x_2, y_2).

As for the formulation of the element yield condition there is no unique way of finding the distribution of the vector $\boldsymbol{Y}$ of element yield strengths from the given gaussian field. Bearing in mind the average way of defining the stress tensor at the geometrical centre G it seems reasonable to define the element yield strength by averaging the gaussian yield strength field over the element. Thereby the vector $\boldsymbol{Y}$ becomes gaussian with the mean μ and a covariance matrix defined by

$$\mathrm{Cov}[Y_i, Y_j] = \frac{\sigma^2}{A_i A_j} \int_{E_i} dx_1 dy_1 \int_{E_j} \rho(x_1 - x_2, y_1 - y_2) dx_2 dy_2 \tag{2}$$

in which E_i, E_j denote the sets of points covered by elements No. i and j respectively and A_i, A_j denote the areas of these sets.

This completes the modelling of the continuous rigid-ideal plastic wall into a discretized structure of rigid-ideal plastic elements with correlated random yield strengths. The following paragraphs repeat the concepts and results summarized in *Ditlevsen* (1988a) but adapted to the present application.

Lower bound safety margins Assuming that the functions f_i in the r yield conditions (1) are defined such that they take positive values inside the yield surfaces (opposite the convention in plasticity theory), the functions $f_i[\boldsymbol{Q}(\boldsymbol{X},\boldsymbol{z}),Y_i], i = 1,...,r$, for each choice of the vector of redundants $\boldsymbol{z}$ define r random variables that are called the lower bound safety margins. According to the lower bound theorem of ideal plasticity theory (the "static theorem") a realization of the discretized structure is not in a state of collapse if and only if there is a choice of the vector of redundants $\boldsymbol{z}$ for which all the r lower bound safety margins take nonnegative values.

For each choice of $\boldsymbol{z}$ there is a unique probability of getting an outcome of the load and the yield strength variables for which all the lower bound safety margins are nonnegative. The largest value of this probability obtained by varying $\boldsymbol{z}$ over $\Re^n$ is obviously a lower bound on the probability of no collapse. Usually this lower bound is substantially smaller than the probability of no collapse (the gap is due to the parallel system effect).

Lower bound safety margins for polyhedral yield conditions In the particular case where the yield conditions for $Y_1,...,Y_r$ given are all polyhedral in $\boldsymbol{Q}$, the complete set of linear equations of the faces of the polyhedrons with normal vectors $\boldsymbol{\alpha}_{ij}, j = 1,...,n_i, i = 1,...,r$, can be used to split up the lower bound safety margins $f_i[\boldsymbol{Q},Y_i]$ into $n_1 + ... + n_r$ linear functions of $\boldsymbol{Q}$ that are all nonnegative if and only if all $f_i[\boldsymbol{Q},Y_i]$ are nonnegative. Thus these linear functions $M_i(\boldsymbol{Q}_i,Y_i,\boldsymbol{\alpha}_{ij})$ of $\boldsymbol{Q} = (\boldsymbol{Q}_1,...,\boldsymbol{Q}_r)$ can be considered as lower bound safety margins themselves.

Moreover, assume that the vectors $\boldsymbol{\alpha}_{ij}$ are independent of Y_i. Then the simple reliability indices

$$\beta_{ij}(\boldsymbol{z}) = \frac{\mathrm{E}[M_i(\boldsymbol{Q}_i(\boldsymbol{X},\boldsymbol{z}),Y_i,\boldsymbol{\alpha}_{ij})]}{\mathrm{D}[M_i(\boldsymbol{Q}_i(\mathrm{X},\boldsymbol{z}),Y_i,\boldsymbol{\alpha}_{ij})]} \tag{3}$$

($\mathrm{E}[\cdot]$ = expectation,$\mathrm{D}[\cdot]$ = standard deviation) are linear functions in the vector of redundants $\boldsymbol{z}$. This is because the terms that depend on $\boldsymbol{z}$ are nonrandom and separated from the terms that depend on $\boldsymbol{X}$ and Y_i. Only these last terms contribute to the standard deviation in the denominator of (3) (see the expression (4) for M_i below). If the functions $M_i(\boldsymbol{Q}_i,Y_i,\boldsymbol{\alpha}_{ij})$ are linear also in Y_i the simple reliability indices can be obtained by solely using the mean and the standard deviation for the intensity of $\boldsymbol{X}$ and the mean $\mathrm{E}[Y_i] = \mu$ and the variance of Y_i given by (2). Otherwise the information that $\boldsymbol{X}$ and Y_i are gaussian must be used to compute $\beta_{ij}(\boldsymbol{z})$.

A possibly weaker lower bound on the probability of no collapse than that mentioned at the end of the previous paragraph can be obtained by maximizing the smallest of the simple reliability indices $\beta_{ij}(\boldsymbol{z})$ as $\boldsymbol{z}$ varies over $\Re^n$.

For the corresponding $\boldsymbol{z}$ the probability of all the lower bound safety margins being nonnegative will at most be equal to the lower bound of the previous paragraph. The optimization procedure is seen to be dramatically simpler, since it is just a linear programming problem. There is another good reason for being interested in this suboptimization as it will become clear in the following.

Upper bound safety margins Let $\boldsymbol{\alpha}_i$ be an admissible strain rate vector for the ith element. This means that there is at least one point $\boldsymbol{P}_i$ on the yield surface (which now is not necessarily assumed to be polyhedral) at which the vector $\boldsymbol{\alpha}_i$ is orthogonal to the yield surface (the associated flow rule is obeyed). The scalar product $D(Y_i, \boldsymbol{\alpha}_i) =< \boldsymbol{P}_i, \boldsymbol{\alpha}_i >$ is the corresponding plastic dissipation. For Y_i given the dissipation is uniquely defined by $\boldsymbol{\alpha}_i$. The difference

$$M_i(\boldsymbol{Q}_i, Y_i, \boldsymbol{\alpha}_i) = D(Y_i, \boldsymbol{\alpha}_i) - < \boldsymbol{Q}_i, \boldsymbol{\alpha}_i > \tag{4}$$

is called a linearly associated lower bound safety margin to $f_i(\boldsymbol{Q}_i, Y_i)$.

The functions $M_i(\boldsymbol{Q}_i, Y_i, \boldsymbol{\alpha}_{ij})$ considered in the previous paragraph for polyhedral yield conditions are special cases of (4). At a vertex or edge point of the polyhedral yield condition any convex linear combination of the lower bound safety margins corresponding to the adjacent faces is a linearly associated lower bound safety margin with the strain rate vector given by a convex linear combination of the strain rate vectors corresponding to the adjacent faces and with the same coefficients, i.e.

$$M_i(\boldsymbol{Q}_i, Y_i, \textstyle\sum_{j\in J} a_j \boldsymbol{\alpha}_{ij}) = \textstyle\sum_{j\in J} a_j M_i(\boldsymbol{Q}_i, Y_i, \boldsymbol{\alpha}_{ij}) \tag{5}$$

with $\sum_{j\in J} a_j = 1$ and $a_j \geq 0$. J is the relevant set of indices.

A linear combination of linearly associated lower bound safety margins is said to be invariant, if it is constant for the vector of redundants $\boldsymbol{z}$ varying over $\Re^n$. In *Ditlevsen* (1988b) a proof is given that any invariant linear combination

$$\sum_{i=1}^{r} c_i M_i(\boldsymbol{Q}_i(\boldsymbol{X}, \boldsymbol{z}), Y_i, \boldsymbol{\alpha}_i) \tag{6}$$

of linearly associated lower bound safety margins with nonnegative coefficients $c_1, .., c_r$ has the property that the probability that (6) is positive is an upper bound on the probability of no collapse. Such a linear combination is called an upper bound safety margin.

It is observed that if (6) is invariant then the linear combination

$$\sum_{i=1}^{r} | c_i | M_i(\boldsymbol{Q}_i(\boldsymbol{X}, \boldsymbol{z}), Y_i, sign(c_i)\boldsymbol{\alpha}_i) \tag{7}$$

is invariant provided $sign(c_i)\boldsymbol{\alpha}_i$ is an admissible strain rate. Thus (7) is an upper bound safety margin if (6) is invariant.

All upper bound safety margins obtained by using the upper bound theorem of ideal plasticity theory (the "kinematic theorem") can be represented as invariant linear combinations (6) with nonnegative coefficients, *Ditlevsen* (1988b).

Construction of invariant linear combinations of linearly associated lower bound safety margins Choose n admissible strain rate vectors $\gamma_{i_1},...,\gamma_{i_n}$ for n different elements with numbers $i_1,...,i_n$ respectively. Make this choice such that the n scalar products $< \boldsymbol{Q}_{i_k}(\boldsymbol{X},\boldsymbol{z}),\gamma_{i_k} >, k = 1,...,n$, are linearly independent in $\boldsymbol{z}$. Then the n equations obtained from (4) by setting $i = i_k$ and $\alpha_i = \gamma_{i_k}, k = 1,..,n$, can be solved uniquely with respect to $\boldsymbol{z}$. The solution represents the components of $\boldsymbol{z}$ as inhomogeneous linear combinations of $M_{i_1}(\boldsymbol{Q}_{i_1},Y_{i_1},\gamma_{i_1}),...,M_{i_n}(\boldsymbol{Q}_{i_n},Y_{i_n},\gamma_{i_n})$ in which the coefficients depend linearly on $\gamma_{i_1},...,\gamma_{i_n}$ only, while the constant terms depend also on $\boldsymbol{X}$ and $Y_{i_1},...,Y_{i_n}$. Next substitute these linear combinations into the scalar product $< \boldsymbol{Q}_i(\boldsymbol{X},\boldsymbol{z}),\alpha_i >$ on the right hand side of (4). Hereby the scalar product becomes an inhomogeneous linear combination of $M_{i_1}(\boldsymbol{Q}_{i_1},Y_{i_1},\gamma_{i_1}),...,M_{i_n}(\boldsymbol{Q}_{i_n},Y_{i_n},\gamma_{i_n})$ in which the coefficients depend linearly on α_i. After moving the homogeneous part of this linear combination to the left hand side of (4) the left hand side becomes an invariant linear combination of the lower bound linearly associated safety margins $M_i(\boldsymbol{Q}_i,Y_i,\alpha_i), M_{i_1}(\boldsymbol{Q}_{i_1},Y_{i_1},\gamma_{i_1}),...,M_{i_n}(\boldsymbol{Q}_{i_n},Y_{i_n},\gamma_{i_n})$ with the first coefficient being 1.

The strain rate simplex of upper bound safety margins Since the coefficients in the obtained linear combination except the first are linear functions of the 3-dimensional strain rate vector α_i, all the coefficients will be nonnegative within the intersection of n half-spaces in $\Re^3$ with plane boundaries through the origin. If it is not empty this intersection is a simplex with vertex at the origin and defined by at most n strain rate vectors along the edges of the simplex. Each of the possible edge vectors are obtained by successively considering all the combinations of two coefficients out of the n coefficients and setting these two coefficients to zero. The corresponding two equations in three unknowns then determine a unit vector α which is an edge vector of the simplex if all the remaining $n - 2$ coefficients obtained by substituting α for α_i are nonnegative. If they are all nonpositive, $-\alpha$ is an edge vector of the simplex. If none of these conditions are satisfied the two considered coefficient equations cannot define an edge of the simplex. As mentioned, at most n of the $n(n-1)/2$ pairs of coefficient equations will define simplex edges.

The simplex edges are of particular interest because a search within the simplex for a strain rate vector that defines an upper bound safety margin with the smallest probability of being nonnegative can as an approximation be restricted to the set of edge vectors. This is an easy suboptimization procedure which in most cases hardly will give a much larger value of the upper bound on the probability of no collapse.

Search for upper bound safety margins of low reliability A convenient automatic way to choose $\gamma_{i_1},...,\gamma_{i_n}$ comes out of the solution of the linear programming problem of minimizing the smallest of the simple reliability indices $\beta_{ij}(z)$ provided the necessary linearity in z is present for all i, j. In fact, it can be shown that there is an optimal solution z for which the largest value of the smallest reliability index is shared by at least $n+1$ of the reliability indices $\beta_{ij}(z)$. Moreover, the corresponding subset of at least $n+1$ linearly associated lower bound safety margins $M_i(Q_i, Y_i, \alpha_{ij})$ (called dominating) has a coefficient matrix to z of the full rank n, *Ditlevsen and Bjerager* (1984). Thus there is at least one choice (and usually more than one) of n dominating lower bound safety margins that can be used to define $\gamma_{i_1}, \ldots, \gamma_{i_n}$. It can happen that these strain rate vectors correspond to less than n different elements. Simultaneously acting strain rate vectors for the same element necessarily correspond to adjacent sides of the polyhedral yield condition and combine as in (5). Since for the optimal z the dominating lower bound safety margins have the smallest reliability indices, it can be a reasonable strategy to use these for constructing upper bound safety margins of low reliability. Experience confirms the usefulness of this strategy, *Ditlevsen and Bjerager* (1984), *Ditlevsen* (1988b).

Method of substitute yield conditions For more general nonpolyhedral yield conditions a similar procedure for choosing $\gamma_{i_1}, \ldots, \gamma_{i_n}$ as in the previous paragraph leads to a complicated nonlinear programming problem. However, an upper bound analysis is preserved if the given yield conditions are replaced by suitably approximating substitute polyhedral yield conditions. If a set of strain rate vectors and nonnegative combination coefficients are obtained for the substitute yield conditions as described in the previous paragraphs, and these strain rate vectors are admissible for the given yield conditions, then the same strain rate vectors and combination coefficients can be applied for the given yield condition. This follows from (4) because the last term in (4) is common to the given and the substitute yield condition for the ith element. Thus the invariance property of the linear combination is independent of the yield conditions as long as the strain rates are admissible. It is only the dissipations that change.

In the following example all strain rate vectors are admissible. A first set of substitute yield conditions is chosen by replacing the given yield condition for each element by the polyhedral yield condition defined by the 6 linearly associated lower bound safety margins corresponding to the unit strain rate vectors in direction of the coordinate axes and their opposite vectors.

After a determination of a suitable set of strain rate simplex edge vectors corresponding to this first set of substitute polyhedral yield conditions, the given yield conditions are next replaced by substitute polyhedral yield conditions using the linearly associated lower bound safety margins corresponding to the identified strain rate simplex edge vectors and the strain rate vectors corresponding to the dominating lower bound safety margins from the first run. The linearly associated lower bound safety margins corresponding to certain other strain rate vectors are also included. These are the strain rate vectors obtained from combining dominating lower bound safety margins that belong to the same element and contribute to the same upper bound safety margin.

With these new substitute polyhedral yield conditions the computations start from scratch. However, after each run it is for the new optimal z checked whether some of the linearly associated lower bound safety margins from the previous run, but left out in the just terminated run, have simple reliability index values that are smaller than the obtained optimal value. If so, the substitute polyhedral yield condition is revised by including all sides corresponding to these linearly associated lower bound safety margins with too low simple reliability indices, and a new run is made without using any other results from the just terminated run. This is repeated until no change is observed. Next new substitute yield conditions are chosen as explained and the computations start all over. After some iterations of this procedure no new upper bound safety margins will be identified.

It can happen in any run that the strain rate simplex for an element is empty. In such a case a new simple try can be made by replacing one or more of the strain rate vectors corresponding to the dominating lower bound safety margins by the opposite strain rate vectors. As noted above, see (7), this replacement preserves the property of invariance.

This method of substitute yield conditions gives a set of upper bound safety margins. However, it does not guarantee that the resulting upper bound on the probability of no collapse of the element system is a narrow upper bound. The iteration method can fail to identify important upper bound safety margins because the search by its way of using dominant lower bound safety margins can have a tendency to become local in the strain rate vector space. Therefore another first choice of the substitute polyhedral yield conditions will most often lead to another upper bound on the probability of no collapse.

Finer element discretization For a consistent sequence of yield hinge discretizations of a given rigid-ideal plastic frame structure it is obvious that the probability of collapse does not decrease with the number of potential yield hinges. The sequence of discretizations is consistent if any finer discretization contains all the potential yield hinges of any courser discretization and if the yield conditions of these potential yield hinges are the same in both discretizations. The number of possibilities of having collapse is simply increased implying that the probability of collapse cannot decrease.

The situation is more complicated for the discretization into elements of the rigid-ideal plastic wall because any sequence of finer discretizations lacks consistency in the aforementioned sense. Each futher subdivision into finer elements implies a change of the degree of statical redundancy giving an increase in the number of geometrical constraints (compability conditions at the nodal points). Moreover, a single yield condition of one large element is replaced in a somewhat arbitrary way by as many yield conditions as the number of subelements. The increase in redundancy tends to strengthen the element system while, at the same time, the larger number of potential collapse possibilities tends to weaken the system. A general statement cannot be given about whether increase or decrease of the collapse probability will take place by subdivision at a given discretization level. An upper bound on the probability of no collapse of the element system can therefore turn out to be smaller than the probability of no collapse of the given continuous wall.

In any case, for the discretization method to be reasonable it should have the property that the collapse probability converges to a fixed value as the discretization becomes finer and finer. Of course, even if such convergence is assumed to take place it is not guaranteed that the upper bound on the probability of no collapse obtained by anything less than an exact method will show a corresponding tendency to convergence.

The present study has mainly been concerned with the method of substitute polyhedral yield conditions as a tool to construct invariant linear combinations of lower bound safety margins with nonnegative coefficients for a given element system. A more detailed study of the effect of the fineness of the discretization will be made in the future. It seems necessary to use extensive Monte Carlo simulations to check the goodness of the upper bound results.

Example Consider the wall sketched with element dicretizations in Fig. 2. The dicretization contains 24 elements, 21 nodal points, and 4 support points each with 2 reaction components. Thus the degree of redundancy is 38. The top edge of the homogeneous wall is uniformly loaded. The discretized wall is loaded at the boundary nodal points as if it was loaded indirectly through the actions of simply supported uniformly loaded beams with supports at the nodal points.

The uniform load intensity is a gaussian random variable with expectation 1 $[F]/[L]$ ($[F]$ = force unit,$[L]$ = length unit) and 50% coefficient of variation. The yield condition for the wall material is of the von Mises type (i.e. $\sigma_{11}^2 + \sigma_{22}^2 - \sigma_{11}\sigma_{22} + 3\sigma_{12}^2 = Y^2$) where the yield strength Y is a homogeneous and isotropic gaussian random field with mean $\mu = 3\ [F]/[L]$, and variance $\sigma^2 = 0.3\ ([F]/[L])^2$. The correlation function is $\rho(r) = \exp[-0.3r]$ with $r^2 = (x_2 - x_1)^2 + (y_2 - y_1)^2$ and r in the unit $[L]$.

The results of the 4 first iteration runs are shown in the following table. Figs. 3 and 4 show two essential collapse machanisms with indication of the movement of the upper wall edge.

For the substitute polyhedral yield conditions used in the fourth iteration, a Monte Carlo simulation with uniform bidirectional sampling (following the method in *Ditlevsen and Bjerager* (1989)) has been carried out in order to check the accuracy by which the reliability index has been obtained. With a generated sample size of 3500 (using 4.5 day of runtime on the same computer as specified in the table) the 90% confidence interval [2.54,2.66] and the 95% confidence interval [2.53,2.68] were estimated for the reliability index. The upper bound safety margin identification procedure failed to give the collapse mechanism symmetric to that shown in Fig. 4. If this is taken into account the reliability index decreases from 2.72 to 2.64 which is within the aforementioned 90% confidence interval.

iteration no.	1	2	3	4
# of l.b. margins	144	234	286	294
optimal β	0.7047	0.7047	0.6779	0.6779
# of u.b. margins	7	12	20	20
reliability index	3.00	2.87	2.72	2.72
runtime [sec.]	164	360	165	173

# of l.b. margins = number of lower bounds safety margins used in the definitions of the substitute polyhedral yield conditions.

optimal β = largest smallest value of the simple reliability indices shared by at least 39 of the lower bound safety margins.

# of u.b. margins = number of identified upper bound safety margins for which the simple reliability index is less than 7.5. Up to 3 nondominant lower bound safety margins have been allowed in the invariant linear combinations (use of (7)).

reliability index = standard normal fractile corresponding to the established upper bound on the probability of no collapse of the element system.

runtime [sec.] : The computations are made on a SUN 3/50 work station.

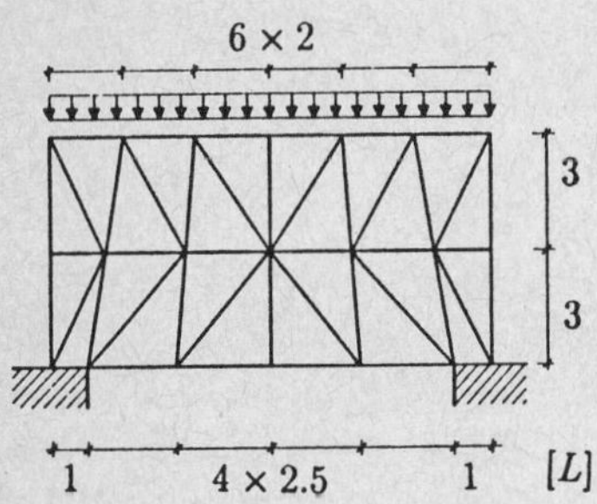

Fig. 2. Element system.

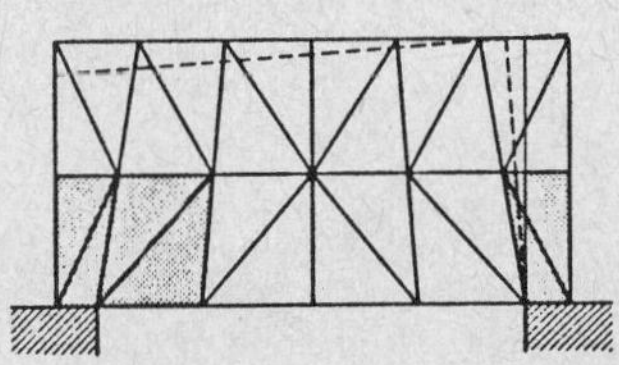

Fig. 3. $\beta = 3.00$.

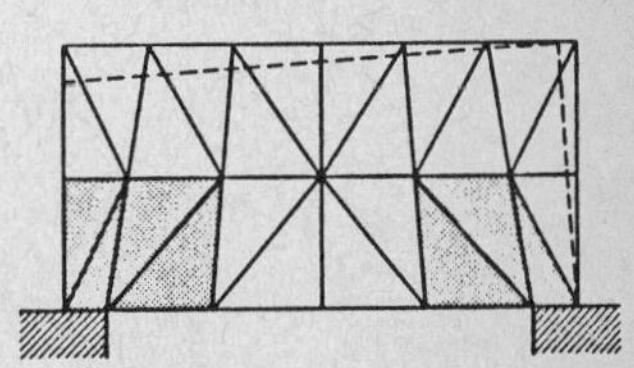

Fig. 4. $\beta = 2.76$.

References

Ditlevsen,O., and Bjerager,P.(1984), "Reliability of highly redundant plastic structures", Journal of Engineering Mechanics, ASCE, 110(5), 671-693.

Ditlevsen,O.(1988a), "Polyhedral substitute yield conditions in plastic reliability analysis", in Probabilistic Methods in Civil Engineering, (ed.: P.D.Spanos), ASCE, (proceedings of 5th ASCE EMD/GTD/STD Joint Specialty Conference on Probabilistic Mechanics and Structural and Geotechnical Safety, Blacksburg, Virginia, May, 1988), 82-85.

Ditlevsen,O.(1988b), " Probabilistic statics of discretized ideal plastic frames", Journal of Engineering Mechanics, ASCE, 114(12).

Ditlevsen,O., and Bjerager,P.(1989), "Plastic reliability analysis by directional simulation", Journal of Engineering Mechanics, ASCE.

LONG-TERM SYSTEM RELIABILITY OF OFFSHORE JACKET STRUCTURES

Carl Andreas Holm*, Peter Bjerager*, Henrik O. Madsen**

***A.S. Veritas Research, P. O. box 300, N-1322 Høvik, Norway**

****Danish Engineering Academy, Building 373, DK-2800 Lyngby, Denmark**

ABSTRACT

The system reliability of an offshore jacket structure modeled as a spatial truss structure is determined. Recent developments in system reliability of elastic ideal-plastic structures are employed together with a stochastic load model. The physical quantities, which can be modeled as stochastic variables comprise: compression and tension yield forces, extreme wave height and associated period, wind current and tidal current, nodal forces describing gravity, buoyancy and live load, drag and inertia coefficients in Morison's equation, marine growth thickness, and model uncertainty in the Morison load model.

1. INTRODUCTION

The paper describes a sea state based reliability analysis of an offshore jacket structure, modeled as a truss structure with quasi-static behavior and loaded by wave, wind, current, gravity, buoyancy and live load. A reliability analysis for first member yielding during a reference period is carried out based on a linear elastic analysis. A corresponding reliability analysis for global collapse is performed based on an ideal plastic analysis, where the static and kinematic theorems of ideal plasticity theory are applied in a probabilistic setting [11,7].

A long term reliability analysis with multiple environmental load processes can conveniently be carried out by an approach in which the processes (e.g., wave, wind and current) are assumed to be stationary within successive time intervals of constant duration [2]. These intervals are referred to as sea states within marine engineering. The long term analysis presented here is based on existing methods and a computer program [1] for systems reliability analysis of a jacket structure with respect to a loading described by deterministic water particle kinematics, i.e., for deterministic wave height, wave period, mean current velocity, etc. To extend the analysis to cover a full probabilistic load model, a coupling between the systems reliability program and an external probability computation module is established. In the present implementation some simplifying assumptions are made.

It is assumed that the most severe loading condition is for the maximum wave height within the reference period. It is further assumed that the maximum wave height occurs in the sea state with the largest significant wave height. The largest significant wave height together with the corresponding wave period in N sea states of duration D is determined from the joint distribution function. For the sea state defined by the maximum significant wave height and a corresponding peak period, the maximum wave height and a corresponding wave period is determined. The extreme wave model is combined with a random wind current, and a random tidal current. Alternatively, the wave model is combined with an extreme wind current and, a random tidal current. The reliability of the jacket for given maximum wave height and period, together with given tidal and wind current is computed by first order reliability methods [9]. By a nested application of these methods [12], the unconditional systems reliability with respect to the probabilistic environmental load model is computed.

The sensitivity of the structural system reliability with respect to the reference period is examined. In addition the conditional system reliability with respect to the "worst" wave position was compared with the system reliability with respect to all wave positions.

2. STRUCTURAL MODEL

The structure is modeled as a spatial truss structure. The procedures presented in the following are, however, directly extendable to a space frame model, and interaction between cross sectional forces can be accounted for. However for the same size structure the space frame model requires significantly more computational effort. The structural elements are assumed to exhibit an elastic ideal-plastic behavior. Failure of the structure is defined as plastic collapse, i.e. formation of a mechanism, and the lower and upper bound theorems of plasticity theory are applied.

The uncertainty in the structural resistance is modeled by $2m$ random variables describing the yield load in tension and compression, respectively for each of the m elements. The yield load in tension is denoted by N_i^+ and in compression by N_i^-, $i=1,2,....,m$. Each bar can fail in either tension or compression, so the structural system can be modeled by $2m$ component safety margins. Component failure is defined as yielding, and each component is either in a safe state or in a failure state depending on whether the bar is in the elastic or the plastic domain.

The structure is loaded by wave, current, buoyancy, wind, gravity and live loads. The loading is represented by a finite set of nodal forces $\mathrm{Q}=(Q_1,Q_2,....,Q_l)$, which induces normal forces in the bars. The normal forces in the bars are denoted by $\mathrm{N}=(N_1,N_2,\ldots,N_m)$. In the elastic analysis with respect to first yielding the normal forces are

$$\mathrm{N}=\mathrm{A_Q}\,\mathrm{Q} \tag{1}$$

where $\mathrm{A_Q}$ is the matrix of elastic influence coefficients. In the plastic analysis the normal forces may be expressed as

$$\mathrm{N}=\mathrm{B_Q}\,\mathrm{Q}+\mathrm{B_z}\,\mathrm{z} \tag{2}$$

where $\mathrm{z}=(z_1,z_2,....,z_n)\in \mathrm{R}^n$ is called the vector of redundants. Equation (2) expresses a choice of a statically determinate primary system, if each redundant corresponds to an axial force in a bar.

The geometry of the structure is assumed to be deterministic, and the complete set of random variables is Q, N^+ and N^-. Two different limit states of the structural system are considered, namely the limit state of initial yielding (first failure), and the limit state of failure by formation of a plastic mechanism. The two criteria are referred to as the *elastic* and the *plastic* limit state, respectively.

The system limit state function defining elastic failure of the system is given by (3)

$$g(\mathrm{N}^+,\mathrm{N}^-,\mathrm{Q})=\min_{i=1}^{2m}\,[g_i(N_i^+,N_i^-,N_i)] \tag{3}$$

where N_i is defined in (1), and $g_i(\cdot)$ is given by

$$g_i(N_i^+,N_i^-,N_i)=\begin{cases} N_i^-+N_i & \textit{compression}\,,\quad i=1,...,m \\ N_i^+-N_i & \textit{tension} \end{cases} \tag{4}$$

The system limit state function corresponds to a series system of $2m$ components.

The system limit state function defining plastic failure of the system is given by (5)

$$g(N^+,N^-,Q) = \max_{z \in R^n} [\min_{i=1}^{2m}[g_i(N_i^+,N_i^-,N_i(z))]] \tag{5}$$

where g_i is given by (4) with N_i defined by (2). The limit state function in (5) is based on the lower bound theorem of plasticity theory, and the max-operation expresses that an admissible equilibrium distribution of the internal forces (N_i) is sought for each set of values of the basic variables. The system limit state function for plastic collapse corresponds to a parallel system of series subsystems of $2m$ elements, with one parallel system for each choice of z.

An upper bound on the system reliability can be determined by the method of linear combination of lower bound safety margins, [4]. An upper bound safety margin is given as

$$L = \sum_{i=1}^{2m} \gamma_i \, g_i(N_i^+,N_i^-,N_i(z)) \tag{6}$$

If L is independent of z, and the coefficients γ_i are all non-negative, it can be shown that L is an upper bound safety margin corresponding to a plastic mechanism. Hence, the structural system is modeled as a series system of upper bound safety margins. In the linear combination method the upper bound is established by using the so-called dominant lower bound safety margins [4], as identified in a lower bound analysis.

3. LOAD MODEL

The gravity of the deck structure, and the live load, and the wind load on the deck structure are all modeled by random nodal forces. The current is divided into tidal current and wind current, and it is modeled as piece-wise constant functions of the elevation above sea bed. Random nodal forces from waves and currents are determined by Morison's equation. Water particle kinematics due to a single wave are determined from a specified wave theory, wave direction and wave crest position and a chosen extreme wave height, a corresponding wave period, tidal current and wind current. To include the uncertainty in wave and current forces the coefficients in Morison's equation and the thickness of the tube walls due to marine growth are modeled as spatial stochastic processes [7]. The uncertainty in the loads due to gravity and buoyancy of the jacket structure is neglected. The gravity and buoyancy forces are hence deterministic and they are added to the random nodal forces in the analysis.

4. RANDOM EXTREME SEA STATE MODEL

The most severe loading condition in a reference period is assumed to occur for the maximum wave height. The maximum wave height is further assumed to occur in the sea state with the largest significant wave height. The maximum wave height is then combined with wind and tidal current.

A conditional extreme sea state model is defined by a selected wave theory together with a wave direction, a wave crest position, an extreme wave height together with the corresponding wave period, a wind current and a tidal current. To reduce the amount of computations, the wave theory, wave crest position, and wave direction are selected so that the system reliability for the conditional sea state is minimized.

The sea state is described by the significant wave height H_S and the peak period T_P. The combined lognormal/Weibull distribution [6] is used for the significant wave height, i.e.

$$F_{H_S}(h_S) = \begin{cases} \Phi\left(\dfrac{\ln h_S - \lambda}{\kappa}\right) & ; h_S \leq \eta \\ \exp(-(h_S/\rho)^{\alpha}) & ; h_S > \eta \end{cases} \tag{7}$$

where κ, α, λ , and ρ are three parameters used to achieve good fit to the underlying data and continuity for the density function and the distribution function for $h_S = \eta$.

The distribution function for the largest significant wave height in the N sea states in the reference period T, is

$$F_{H_{S,max}}(h_S) = F_{H_s}(h_S)^N \tag{8}$$

when the sea states are assumed to be independent. The conditional distribution of the associated peak period, conditioned on the significant wave height, is taken as a lognormal distribution [6], i.e.,

$$F_{T_p}(t_P | H_{S,max}) = \Phi\left(\frac{\ln(t_P - \epsilon) - \mu}{\sigma} \right) \tag{9}$$

where $\mu = E(ln(T_P - \epsilon))$, $\sigma^2 = Var(ln(T_P - \epsilon))$, and ϵ is a possible lower limit for the spectral peak period due to wave breaking. When the Pierson-Moskowitz spectrum is chosen for the sea surface elevation process within a sea state, the mean zero-upcrossing period T_Z can be approximated as $T_Z \approx 0.7\, T_P$.

The distribution of the maximum wave height $H_{\max}$ within a stationary sea state characterized by $H_{S,max}$ and T_P is defined by an upcrossing problem [9]. Assuming a Gaussian sea surface elevation process, the distribution of $H_{\max}$ can be taken as

$$\begin{aligned} F_{H_{\max}}(h_{\max} | H_{S,max}, T_Z) = {} & (1 - \exp\{ -\frac{h^2_{\max}}{8\lambda_0} \}) \cdot \\ & \exp\{ -D \frac{1}{2\pi} \sqrt{\lambda_2/\lambda_0} \exp\{ -\frac{h^2_{\max}}{8\lambda_0} \} (1 - \exp\{ -\sqrt{2\pi}\, h_{\max} \frac{\sqrt{1 - \lambda^2_1/\lambda_0\lambda_2}}{\sqrt{\lambda_0}} \}) \} \end{aligned} \tag{10}$$

where λ_i is the i'th-spectral moment and D is the duration of one stationary sea state.

The conditional distribution for the associated maximum period $T_{\max}$ is modeled by integrating the Longuet-Higgins density function for jointly distributed $H_{\max}$ and $T_{\max}$, i.e.

$$\begin{aligned} F_{T_{\max} | H_{\max}}(t_{\max} | H_{S,max}, T_Z, H_{\max}) &= \frac{1}{f_{L-H}(h_{\max})} \int_0^{t_{\max}} f_{L-H}(v, h_{\max})\, dv \\ &= \frac{\Phi\left(\sqrt{2\pi}/2\, h_{\max} \frac{\sqrt{\lambda_1/\lambda_0}}{\sqrt{\lambda_2\lambda_0 - \lambda_1}} (1 - \frac{\lambda_0}{\lambda_1} \frac{2\pi}{t_{\max}}) \right)}{\Phi\left(\sqrt{2\pi}/2\, h_{\max} \frac{\sqrt{\lambda_1/\lambda_0}}{\sqrt{\lambda_2\lambda_0 - \lambda_1}} \right)} \end{aligned} \tag{11}$$

where f_{L-H} is the Longuet-Higgins joint distribution of maximum wave height and corresponding wave period [8].

From 50 m below mean sea level and up to the maximum surface elevation, the wind current velocity is modeled equal to C_W. Below 50 m it is assumed to be zero. C_W is described by a normal distribution where the mean value is a function of the maximum significant wave height, i.e.,

$$F_{C_W}(c_W) = \Phi\left(\frac{c_W - \mu_{C_W}(h_{S,max})}{\sigma_{C_W}} \right) \tag{12}$$

where μ_{C_W} and σ_{C_W} are the mean value, and standard deviation, respectively.

The maximum wind current within N sea states is defined by

$$F_{C_{W,max}} = (F_{C_W})^N \tag{13}$$

when the sea states are assumed to be independent.

Finally, the tidal current velocity is modeled as a constant C_T from the sea bed to the maximum wave height. C_T is described by a normal distribution.

$$F_{C_T}(c_T) = \Phi(\frac{c_T - \mu_{C_T}}{\sigma_{c_T}}) \tag{14}$$

where μ_{C_T} and σ_{C_T} are the mean value and standard deviation, respectively.

5. SYSTEM RELIABILITY FORMULATION

The probabilities that the structural system fails with respect to the elastic and the plastic limit state, respectively, during the reference period are considered. For deterministic water particle kinematics, the probability can be computed as described in [7]. In the following it is demonstrated how the long term failure probability can be conveniently computed based on the system reliability within a single sea state.

The failure probability for the structural system is given by

$$P_F = P[\, g(\mathrm{U}) \leqslant 0\,] = P[\, g(\mathrm{V}, \mathrm{W}) \leqslant 0\,] = \Phi(-\beta_T) \tag{15}$$

where $g(\cdot)$ is the limit state function, and W are the normalized Gaussian variables describing H_{max}, T_{max}, C_W and C_T , i.e. the basic random variables that describe the random sea state, V describes the random variables that are independent of the actual sea state, such as random yield forces, thickness due to marine growth, coefficients in Morison's equation and nodal loads not induced by wave or current. $\Phi(\cdot)$ is the standardized normal distribution function, and β_T is the reliability index for the reference period T.

Conditioning on the random sea state variables, the system reliability can be evaluated as a function of w,

$$P_F(\mathrm{w}) = P[g(\mathrm{V}, \mathrm{W}) \leqslant 0 \mid \mathrm{W} = \mathrm{w}] = \Phi(-\beta_S(\mathrm{w})) \tag{16}$$

where β_S is the reliability index for a given sea state. The unconditional failure probability is given by

$$P_F = \int_{all\,\mathrm{w}} P_F(\mathrm{w})\, f_{\mathrm{W}}(\mathrm{w})\, d\mathrm{w} \tag{17}$$

where $f_{\mathrm{W}}(\mathrm{w})$ is the density function for the sea state variables w.

As pointed out in [12], $P_F(\mathrm{w})$ can be written as

$$P_F(\mathrm{w}) = \Phi(u_0) = P[\, U_0 \leqslant u_0] = P[\, U_0 \leqslant \Phi^{-1}(P_F(\mathrm{w}))\,] \tag{18}$$

where U_0 is an auxiliary standard normal variable. Inserting (18) in (17) yields

$$\begin{aligned} P_F &= \int_{all\,\mathrm{w}} P[\, U_0 \leqslant \Phi^{-1}(P_F(\mathrm{w}))\,]\; f_{\mathrm{W}}(\mathrm{w})\, d\mathrm{w} \\ &= \int_{all\,\mathrm{w}} P[\, U_0 - \Phi^{-1}(P_F(\mathrm{W})) \leqslant 0 \mid \mathrm{W} = \mathrm{w}\,]\; f_{\mathrm{W}}(\mathrm{w})\, d\mathrm{w} \\ &= P[\, \mathrm{U}_0 - \Phi^{-1}(P_F(\mathrm{W})) \leqslant 0\,] \end{aligned} \tag{19}$$

Then the limit state function for a reference period is given by

$$g(\mathrm{v},\mathrm{w}) = u_0 + \Phi^{-1}(P_F(\mathrm{w})) = u_0 + \beta_S(\mathrm{w}) \tag{20}$$

The failure probability associated with the limit state function in (20) can be computed by a first-order reliability method, using the existing computer program RAPJAC (Reliability Analysis of Plastic JACket structures) to evaluate β_S [1].

The random variables describing the sea state is $H_{\max}, T_{\max}, C_T, C_W$ or, alternatively $C_{W,max}$, see example below. In the description above these variables were transformed into standardized normally distributed variables W. The relationships between the basic random variables described in Section 4 and the variables W are given by:

Maximum significant wave height:

$$\Phi(w_{H_{S,max}}) = (F_{H_S}(h_{S,max}))^N \tag{21}$$

Corresponding peak period:

$$\Phi(w_{T_P}) = F_{T_P | H_{S,max}}(t_P | H_{S,max}) \tag{22}$$

Maximum wave height in the sea state with the largest significant wave height:

$$\Phi(w_{H_{\max}}) = F_{H_{\max}}(h_{\max} | H_{S,max}, T_Z) \tag{23}$$

Wave period corresponding to the maximum wave height:

$$\Phi(w_{T_{\max}}) = F_{T_{\max} | H_{\max}}(t_{\max} | H_{S,max}, T_Z, H_{\max}) \tag{24}$$

Wind current:

$$\Phi(w_{C_W}) = F_{C_W}(c_W | H_{S,max}) \tag{25}$$

Maximum wind current:

$$\Phi(w_{C_{W,max}}) = (F_{C_W}(c_{W,max} | H_{S,max}))^N \tag{26}$$

Tidal current:

$$\Phi(w_{C_T}) = F_{C_T}(c_T) \tag{27}$$

For an outcome w of W the transformations above are performed, and thereby an outcome of the random sea state as defined by $h_{\max}$, $t_{\max}$, c_T and c_W or $c_{W,max}$ is generated.

6. NUMERICAL EXAMPLE

The structure analyzed is a spatial truss structure that consists of 48 bars and is 12 times redundant [3,5,7]. The structure and its geometry are given in Figure 1.

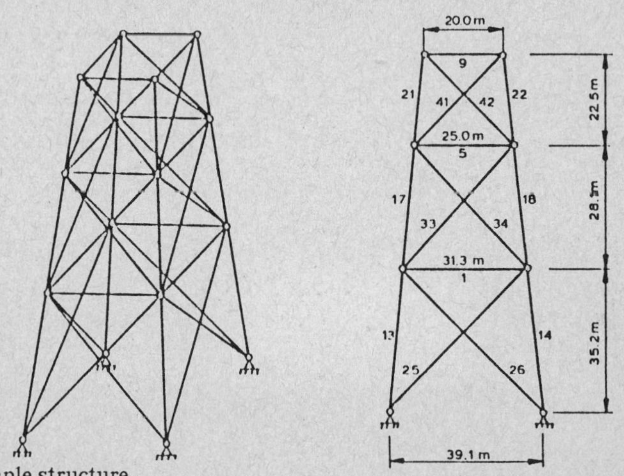

Figure 1. Example structure.

The dimensions of the bars are given in Table 1. The dimensions and the geometry of the structure are assumed to be deterministic.

Table 1. Dimensions and mean tension yield forces of bars.

The ratio between the diameter d and the wall thickness t is assumed to be $d/t = 60$ for all bars.

Bar No.	d $[m]$	Area $[m^2]$	$E[N_i^+]$ $[MN]$
1-4	2.0	0.210	67.20
5-8	1.5	0.116	37.07
9-12	1.0	0.053	16.80
13-24	2.5	0.324	103.78
25-32	1.5	0.116	37.07
33-48	1.2	0.074	23.73

The yield forces are assumed to be normally distributed with the following representation

$$E[N_i^-] = 0.75\,E[N_i^+]\,,\quad D[N_i^-] = 0.15\,E[N_i^-]\,,\quad D[N_i^+] = 0.10\,E[N_i^+]$$

All yield forces are assumed equi-correlated with a correlation coefficient 0.5.

Gravity and live loads from the deck structure are modeled by four vertical loads in each of the four top level nodes. Each force has a mean value 20 MN, a coefficient of variation 0.10, and the four forces are equi-correlated with a correlation coefficient 0.5. Gravity loads of the structure itself are referred to the nodes as single deterministic forces. These forces are calculated for a unit weight of the tubular members of $7.85 \cdot 10^3\, kg/m^3$. The unit weight of the interior of the members is $0.25 \cdot 10^3\, kg/m^3$. This includes gravity loads from internal stiffeners.

Wind load on the deck structure is modeled by a horizontal and a vertical force in each of the four top level nodes. These forces are proportional to a random variable with a mean of 1.0 MN and a coefficient of variation of 0.30. It is assumed that the wind acts in a direction of 30 degrees with one side of the truss structure, see Figure 2. Proportionality factors, relating the individual nodal wind forces to the total wind force, are given in Figure 2. Wind load on the jacket structure itself is neglected.

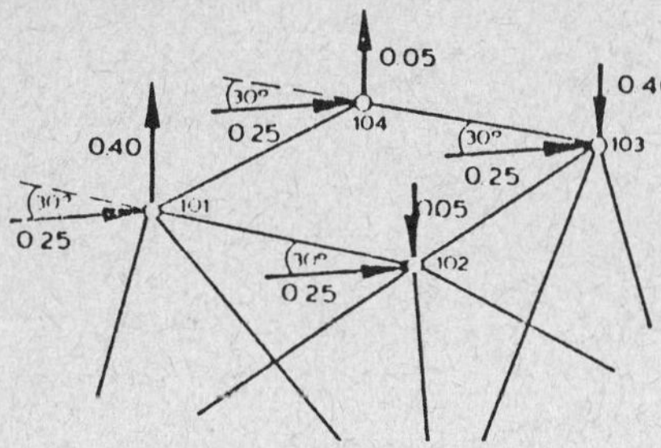

Figure 2. Wind load on the deck structure.

Buoyancy loads are assumed to be deterministic and are added to the random nodal forces.

The maximum wave height during N sea states of duration D and the corresponding wave period are determined from the distributions described in Section 4. The parameters needed to define F_{H_s} and $F_{T_p|H_{s\,max}}$ are selected equal to those in [6], based on a three years period of sampling from the Northern North Sea, see Table 2.

Table 2. Parameters defining the distributions for significant wave height and peak period.

Parameters defining F_{H_s} and $F_{T_P \mid H_s}$.						
η	λ	κ^2	ρ	α	ϵ	D
3.27 m	0.386	0.376	2.822	1.547	0.00	3 hours

Wave loads are calculated for a water depth of $70\,m$, using Stoke's 5th order wave theory, and with the wave crest approximately in the middle of the structure i.e. the phase angle is $\theta=20^\circ$.

The tidal current velocity C_T, and wind current C_W are modeled as independent. Mean values and standard deviations for C_W and C_T are taken as

$$E[C_W]=0.05H_{S,max}\ ,\quad D[C_W]=0.30E[C_W]\ ,\quad E[C_T]=0.30\,m/s\ ,\quad D[C_T]=0.20E[C_T]$$

The total current is modeled as a piece-wise constant function of the elevation above sea bed. The current velocities are given as functions of the elevation above sea bed, as given in Table 3.

Table 3. Current profile.

The mean values and standard deviations to the current velocities are functions of the elevation above sea bed. Between the given levels they are constant.		
Elevation z	$E[C_W]$	$E[C_T]$
00 m	0.0	0.3 m/s
50 m	0.05 $H_{S,max}$	0.3 m/s
120 m	0.05 $H_{S,max}$	0.3 m/s

The drag coefficient, the inertia coefficient and the marine growth thickness are modeled as Gaussian white noise processes. In Table 4 the second moment representation are shown, given as a function of the elevation above sea bed.

Table 4. Mean value and standard deviation of C_D, C_M and H.

Coefficients are a function of the elevation above sea bed. Between the given levels the quantities vary linearly, and above $70\,m$ they are constant. At a given position the three variables are assumed uncorrelated.						
Elevation z	$E[C_D]$	$D[C_D]$	$E[C_M]$	$D[C_M]$	$E[H]$	$D[H]$
00 m	1.0	0.4	2.0	0.3	0.00 m	0.000 m
30 m	1.0	0.4	2.0	0.3	0.01 m	0.002 m
65 m	1.0	0.4	2.0	0.3	0.05 m	0.010 m
70 m	1.0	0.4	2.0	0.3	0.10 m	0.030 m

6.1 Sensitivity analysis

In [7] a sensitivity study of the example structure is presented. Example results are presented here for the reliability against yielding in any point (elastic), and the reliability against plastic collapse (plastic) for selected sea states. Results are presented in terms of a conditional reliability index β_S, see (5.2).

The reliability is evaluated for three different wave periods. The data used in the analysis are given in Figure 3.

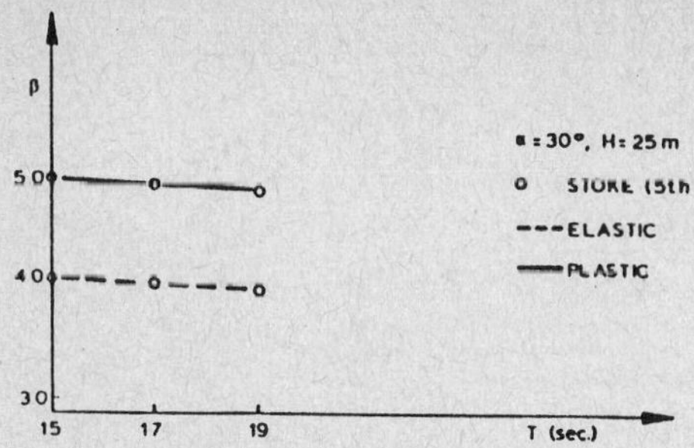

Figure 3. Conditional reliability index with varying wave period.

From the results it is seen that both the elastic and the plastic reliability for the structure are fairly insensitive for changes in the wave period.

The wave position is defined by the phase angle θ. For $\theta=0^0$ the wave crest is above the first support of the structure, and for $\theta=20^0$ it is approximately in the middle of the structure, Figure 4. The reliability for plastic collapse is evaluated for different phase angles, and the results are shown in Figure 4.

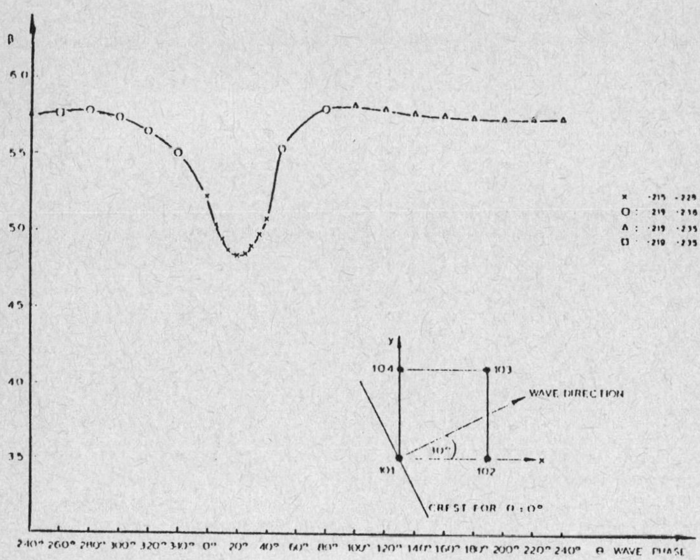

Figure 4. Conditional reliability index with varying wave position.

Different failure modes are dominant (or most likely) for different phase angles. From Figure 4 it is seen that the smallest reliability is for the wave crest approximately in the middle of the structure. The reliability with respect to all θ 's is evaluated using a series system method. The difference is negligible (4.773 vs. 4.771), because the failure modes are strongly correlated.

The reliability is also evaluated for four different wave directions. The data used are given in Figure 5.

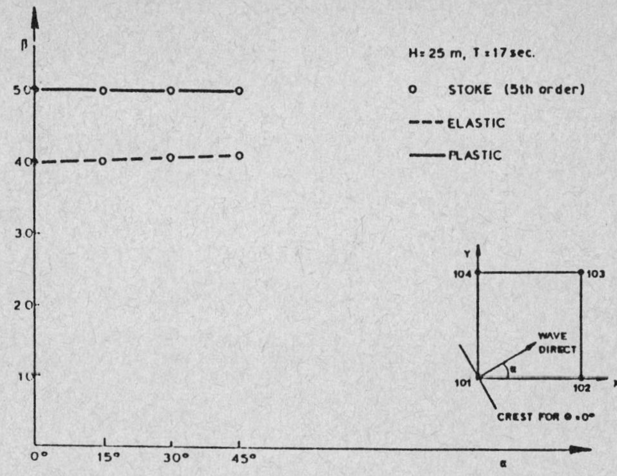

Figure 5. Conditional reliability index with varying wave directions.

From Figure 5 it is seen that the plastic reliability is almost unchanged and that the elastic reliability has a minor increase with the wave direction from 0^0 to 45^0.

6.2 Reliability under random Extreme Sea State Loading

The results above indicate that at least for this example structure, the uncertainty in wave period, wave position and wave direction are not essential. These quantities are therefore taken as deterministic in the long term analysis.

The system reliability for different numbers of sea states is evaluated. Example results are presented for the reliability against yielding in any point (elastic), and the reliability against plastic collapse (plastic) during a reference period. The results are presented in Figure 6 in terms of the long term reliability index β_T , see (5.1),

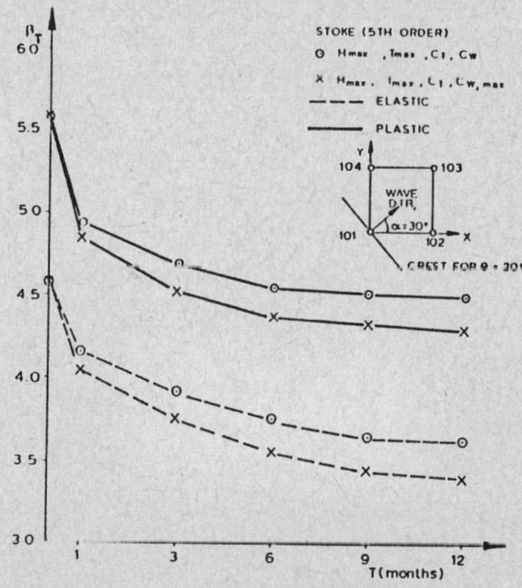

Figure 6. Reliability index with varying reference period.

The figure shows that the structural system reliability is sensitive to the duration of the reference period in particular for very short such periods. The difference between an arbitrary wind current and the maximum wind current is small, i.e. the example structure is not sensitive with respect to the wind current.

7. CONCLUSIONS

A jacket structure modeled as a spatial truss structure with elastic ideal-plastic force deformation characteristics of the bars is analyzed. It is assumed that the geometry is deterministic and that the random variables describing resistance and load parameters are normally distributed. The loads due to current and wave are described stochastically, based on a statistical description of the wave climate. The extreme wave load is modeled in a realistic way as a function of the number of sea states. The "maximum load" is assumed to appear for maximum wave height. The reliability with respect to initial yielding is evaluated, and an upper bound on the reliability with respect to plastic collapse is determined during a reference period.

Sensitivity analyze with respect to various input parameters have been carried out for the example structure. The wave position is defined by the phase angle. The analysis showed that the structure is sensitive with respect to the wave position. The worst case appears when the wave crest is near the middle of the structure. Different failure modes appear for different phase angles, and the conditional system reliability for a series system consisting of the failure modes corresponding to the different phase angles was evaluated. The conditional probability of failure with respect to all phase angles was found to be almost equal to the conditional probability that the system fails when the wave crest is in the middle of the structure. Therefore, for this example structure with the actual load, a good approximation will be obtained by considering only wave positions with the wave crest in the middle of the structure.

Changes in the wave direction and the wave period showed that for the actual structure the reliability is almost independent of these two parameters. The assumption that the most severe loading condition occur for the maximum wave height, is justified for this example structure since the dependence on the wave period is very small.

Two models for representation of the "maximum load" have been used. The maximum wave height was combined with an arbitrary current, and alternatively the maximum wave height was combined with an arbitrary tidal current and a maximum wind current. The results showed that for this example the current has only a minor effect.

The dependency of the structural system reliability on the reference period was analyzed. The results showed that the system reliability for six months is significantly smaller than the reliability for one month. The behavior of the structure is similar for initial failure and for global failure.

The present analysis represents the first study on combining a full structural system reliability program with a module defining a probabilistic environmental load model. The two program modules are separate and can therefore easily be modified. For example, another environmental model could be adopted. Also, any variable that enters the system reliability computation as a deterministic variable may be randomized. For example, a more complete description of the directionality of the load processes and of the combination of these could be made.

The outlined method consists of a full long term stochastic load description combined with a full stochastic system description. The long term analysis is based on the concept of a most severe loading condition as defined by the highest wave within the reference period. Of course, this may in fact not be the worst case, and the approximation may be (slightly) non-conservative. A rational long term reliability methodology for marine structures (based on extreme response in stead of extreme loading, say) is described in [2]. It would be preferable to use such a formulation on the structural system problem considered herein. However, the very high number of system parameters complicates such an analysis.

Finally it is noted that the ideal-plastic structural model adopted in this study represents an idealization of the actual structural behavior. Priority has been given to base the analysis on a mathematical and physical tractable model with consistency between probabilistic and mechanical computations. It would be preferable to be able to treat a more realistic mechanical behavior including geometric non-linearities as well as material non-linearities other than the ideal plastic model. However, research is needed before such analyses become feasible.

8. ACKNOWLEDGMENTS

Parts of the work presented in this paper have been financially supported by the joint research program "Reliability of Marine Structures" of A.S VERITAS Research, SAGA Petroleum a/s, STATOIL and CONOCO, Norway. These support is gratefully acknowledged. The opinions in the paper are those of the authors and they do not necessarily reflect the views of the sponsoring companies.

9. REFERENCES

[1] Bjerager, P. and Olesen, R., "RAPJAC User's Manual", A.S Veritas Research Report, No. 87-2014, Høvik, Norway, 1987.

[2] Bjerager, P., Løseth, R., Winterstein, S. R., and Cornell, C. A., "Reliability Method for Marine Structures under Multiple Environmental Load Processes", *Proceedings* of the International Conference on Behaviour of Offshore Structures, Vol. 3,Trondheim, June, 1988, pp.1239-1253.

[3] Bjerager, P., "Plastic Systems Reliability by LP and FORM", *Computers & Structures*, Vol. 31, pp.187-196, 1989.

[4] Ditlevsen, O. and Bjerager, P., "Reliability of Highly Redundant Plastic Structures", *Journal of Engineering Mechanics*, ASCE, Vol. 110, No. 5, 1984, pp. 671-693.

[5] Ditlevsen, O. and Bjerager, P., "Plastic Reliability Analysis by Directional Simulation", Accepted for *Journal of Engineering Mechanics*, ASCE, 1988.

[6] Haver, S., and Nyhus, K. A., "A Wave Climate Description for Long Term Response Calculations", *Proceedings* of 5th International Offshore Mechanics and Arctic Engineering Symposium, Vol. IV, ASME, pp. 27-34, 1986.

[7] Holm, C. A., Bjerager, P., Olesen, R., and Madsen, H. O., "System Reliability of Offshore Jacket Structures by Elastic and Plastic Analysis", *Proceedings* of the International Conference on Behaviour of Offshore Structures, Vol. 3,Trondheim, June, 1988, pp. 1255-1269.

[8] Longuet-Higgins, M. S., "On the Joint Distribution of Wave Periods and Amplitudes in a Random Wave Field", *Proceedings* of Royal Society, A389, pp. 241-258, 1983.

[9] Madsen, H.O., Krenk, S. and Lind, N.C., *Methods of Structural Safety*, Prentice-Hall Inc., New Jersey, 1986.

[10] Madsen, H.O., "PROBAN Theoretical Manual", A.S Veritas Research Report, No. 86-2036, Høvik, Norway, 1987.

[11] Malvern, L.E., *Introduction to the Mechanics of a Continuous Medium*, Prentice-Hall Inc., 1969.

[12] Wen, Y.-K., and Chen, H.-C., "On Fast Integration for Time Variant Structural Reliability", *Probabilistic Engineering Mechanics*, Vol. 2, No. 3, 1987, pp. 156-162.

INDEX OF AUTHORS

SUBJECT INDEX

Lecture Notes in Engineering

Edited by C.A. Brebbia and S.A. Orszag

Vol. 1: J. C. F. Telles,
The Boundary Element Method
Applied to Inelastic Problems
IX, 243 pages. 1983.

Vol. 2: Bernard Amadei,
Rock Anisotropy and
the Theory of Stress Measurements
XVIII, 479 pages. 1983.

Vol. 3: Computational Aspects of
Penetration Mechanics
Proceedings of the Army Research
Office Workshop on Computational
Aspects of Penetration Mechanics
held at the Ballistic Research Laboratory
at Aberdeen Proving Ground, Maryland,
27–29 April, 1982
Edited by J. Chandra and J.E. Flaherty
VII, 221 pages. 1983.

Vol. 4: W.S. Venturini
Boundary Element Method in Geomechanics
VIII, 246 pages. 1983.

Vol. 5: Madassar Manzoor
Heat Flow Through Extended
Surface Heat Exchangers
VII, 286 pages. 1984.

Vol. 6: Myron B. Allen III
Collocation Techniques for Modeling
Compositional Flows in Oil Reservoirs
VI, 210 pages. 1984.

Vol. 7: Derek B. Ingham,
Mark A. Kelmanson
Boundary Integral Equation
Analyses of Singular, Potential,
and Biharmonic Problems
IV, 173 pages. 1984.

Vol. 8: Linda M. Abriola
Multiphase Migration of Organic
Compounds in a Porous Medium
A Mathematical Model
VIII, 232 pages. 1984.

Vol. 9: Theodore V. Hromadka II
The Complex Variable Boundary
Element Method
XI, 243 pages. 1984.

Vol. 10: C. A. Brebbia, H. Tottenham,
G. B. Warburton, J. M. Wilson, R. R. Wilson
Vibrations of Engineering Structures
VI, 300 pages. 1985.

Vol. 11: M. B. Beck
Water Quality Management:
A Review of the Development and
Application of Mathematical Models
VIII, 108 pages. 1985.

Vol. 12: G. Walker, J. R. Senft
Free Piston Stirling Engines
XIV, 286 pages. 1985.

Vol. 13: Nonlinear Dynamics
of Transcritical Flows
Proceedings of a DFVLR International
Colloquium, Bonn, Germany, March 26, 1984
VI, 203 pages. 1985.

Vol. 14: A. A. Bakr
The Boundary Integral
Equation Method in Axisymmetric
Stress Analysis Problems
XI, 213 pages. 1986.

Vol. 15: I. Kinnmark
The Shallow Water Wave
Equation: Formulation,
Analysis and Application
XXIII, 187 pages, 1986.

Vol. 16: G. J. Creus
Viscoelasticity – Basic
Theory and Applications
to Concrete Structures
VII, 161 pages. 1986.

Vol. 17: S. M. Baxter
C. L. Morfey
Angular Distribution
Analysis in Acoustics
VII, 202 pages. 1986.

Vol. 18: N. C. Markatos,
D. G. Tatchell, M. Cross, N. Rhodes
Numerical Simulation of Fluid Flow
and Heat/Mass Tranfer Processes
VIII, 482 pages. 1986.

Vol. 19: Finite Rotations
in Structural Mechanics
Proceedings of the Euromech
Colloquium 197, Jablonna 1985
VII, 385 pages. 1986.

Vol. 20: S. M. Niku
Finite Element Analysis
of Hyperbolic Cooling Towers
VIII , 216 pages. 1986.

Lecture Notes in Engineering

Edited by C.A. Brebbia and S.A. Orszag

Vol. 21: B. F. Spencer, Jr.
Reliability of Randomly
Excited Hysteretic Structures
XIII, 138 pages. 1986.

Vol. 22: A. Gupta, R. P. Singh
Fatigue Behaviour
of Offshore Structures
XXI, 299 pages. 1986.

Vol. 23: P. Hagedorn, K. Kelkel,
J. Wallaschek
Vibrations and Impedances
of Rectangular Plates
with Free Boundaries
V, 152 pages. 1986.

Vol. 24: Supercomputers
and Fluid Dynamics
Proceedings of the First
Nobeyama Workshop
September 3-6, 1985
VIII, 200 pages. 1986.

Vol. 25: B. Hederson-Sellers
Modeling of Plume Rise
and Dispersion –
The University of Salford
Model: U. S. P. R.
VIII, 113 pages. 1987.

Vol. 26: Shell and Spatial Structures:
Computational Aspects
Proceeding of the International Symposium
July 1986, Leuven, Belgium
Edited by G. De Roeck, A. Samartin Quiroga,
M. Van Laethem and E. Backx
VII, 486 pages. 1987.

Vol. 27: Th. V. Hromadka, Ch.-Ch. Yen
G. F. Pinder
The Best Approximation Method
An Introduction
XIII, 168 pages. 1987.

Vol. 28: Refined Dynamical Theories
of Beams, Plates and Shells and
Their Applications
Proceedings of the Euromech-Colloquim 219
Edited by I. Elishakoff and H. Irretier
IX, 436 pages. 1987.

Vol. 29: G. Menges, N. Hövelmanns,
E. Baur (Eds.)
Expert Systems in Production Engineering
Proceedings of the International Workshop
Spa, Belgium, August 18-22, 1986
IV, 245 pages. 1987.

Vol. 30: R. Doležal
Simulation of Large State Variations
in Steam Power Plants
Dynamics of Large Scale Systems
X, 110 pages. 1987.

Vol. 31: Y. K. Lin, G. I. Schueller (Eds.)
Stochastic Structural Mechanics
U.S.-Austria Joint Seminar, May 4-5, 1987
Boca Raton, Florida, USA
XI, 507 pages. 1987.

Vol. 32: Y. K. Lin, R. Minai (Eds.)
Stochastic Approaches
in Earthquake Engineering
U.S.-Japan Joint Seminar, May 6-7, 1987
Boca Raton, Florida, USA
XI, 457 pages. 1987.

Vol. 33: P. Thoft-Christensen (Editor)
Reliability and Optimization
of Structural Systems
Proceedings of the First IFIP WG 7.5
Working Conference
Aalborg, Denmark, May 6-8, 1987
VIII, 458 pages. 1987.

Vol. 34: M. B. Allen III, G. A. Behie,
J. A. Trangenstein
Multiphase Flow in Porous Media
Mechanics, Mathematics, and Numerics
IV, 312 pages. 1988.

Vol. 35: W. Tang
A New Transformation
Approach in BEM
A Generalized Approach for
Transforming Domain Integrals
VI, 210 pages. 1988.

Vol. 36: R. H. Mendez, S. A. Orszag
Japanese Supercomputing
Architecture, Algorithms, and Applications
IV, 160 pages. 1988.

Vol. 37: J. N. Reddy, C. S. Krishnamoorthy,
K. N. Seetharamu (Eds.)
Finite Element Analysis
for Engineering Design
XIV, 869 pages. 1988.

Vol. 38: S. J. Dunnett, D. B. Ingham
The Mathematics of Blunt Body Sampling
VIII, 213 pages. 1988

Vol. 39: S.L. Koh, C.G. Speziale (Eds.)
Recent Advances in Engineering Science
A Symposium dedicated to A. Cemal Eringen
June 20–22, 1988, Berkeley, California
XVIII, 268 pages. 1989